# Student Solutions Manual

for Johnson and Mowry's

# Mathematics
## A Practical Odyssey

### Fifth Edition

# Student Solutions Manual

for Johnson and Mowry's

# Mathematics
## A Practical Odyssey

Fifth Edition

**Deann J. Christianson**
**Elaine M. Werner**

*University of the Pacific*

THOMSON

BROOKS/COLE

Australia • Canada • Mexico • Singapore • Spain • United Kingdom • United States

Printed in Canada
1  2  3  4  5  6  7   07  06  05  04  03

Printer: Webcom

ISBN: 0-534-40060-4

For more information about our products,
contact us at:
**Thomson Learning Academic Resource Center**
**1-800-423-0563**

**For permission to use material from this text,
contact us by:**
**Phone:** 1-800-730-2214
**Fax:** 1-800-731-2215
**Web:** http://www.thomsonrights.com

**Brooks/Cole—Thomson Learning**
10 Davis Drive
Belmont, CA 94002-3098
USA

**Asia**
Thomson Learning
5 Shenton Way #01-01
UIC Building
Singapore 068808

**Australia/New Zealand**
Thomson Learning
102 Dodds Street
Southbank, Victoria 3006
Australia

**Canada**
Nelson
1120 Birchmount Road
Toronto, Ontario M1K 5G4
Canada

**Europe/Middle East/South Africa**
Thomson Learning
High Holborn House
50/51 Bedford Row
London WC1R 4LR
United Kingdom

**Latin America**
Thomson Learning
Seneca, 53
Colonia Polanco
11560 Mexico D.F.
Mexico

**Spain/Portugal**
Paraninfo
Calle/Magallanes, 25
28015 Madrid, Spain

# PREFACE

This *Student Solutions Manual* has been prepared to accompany the textbook, *Mathematics: A Practical Odyssey*, Fifth Edition, by David B. Johnson and Thomas A. Mowry. Detailed solutions for every other odd problem in the textbook are provided except for answers to Concept Questions, History Questions, Projects, and some calculator exercises.

Feedback concerning errors, solution correctness, and manual style would be appreciated. These and any other comments can be sent directly to us at the address below or in care of the publisher.

The authors would like to thank David B. Johnson and Thomas A. Mowry for the opportunity to participate in their project. We hope that students and instructors will find this manual to be a useful instructional aid.

Deann Christianson
Elaine M. Werner
University of the Pacific
3601 Pacific Avenue
Stockton, CA  95211

# TO THE STUDENT

The purpose of this *Student Solutions Manual* is to assist you in successfully completing your mathematics course. Your class lectures are your best source of instruction and you should attend regularly. The key to success in mathematics is to regularly do the assignments given to you by your instructor.

This *Student Solutions Manual* will help you with the assignments. After you have tried and/or completed the assigned problems, verify the answers to odd problems using the answer key in the back of your textbook. You may also check your solutions against the selected exercises which have been solved for you in this manual. We have attempted to make these solutions as complete and as instructive as possible. Our solutions provide models for you in writing mathematics clearly and carefully.

We hope this manual will help you enjoy and successfully complete your mathematics course.

# Contents

# 1  Logic

1.  a)  Premise 1:  Form is "All A are B."

    Premise 2:  Ansel Adams is a specific master photographer which places his $x$ within the "master photographers" circle.

    Conclusion:  **VALID** ($x$ is within "artists" circle).

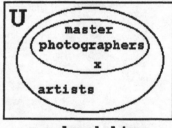

x = Ansel Adams

 

b)  Premise 1:  Form is "All A are B."

    Premise 2:  Ansel Adams is a specific artist which places his $x$ within the "artists" circle but not within the "master photographers" circle.

    Conclusion:  **INVALID** ($x$ is not within the "master photographers" circle).

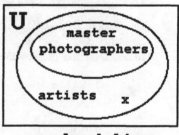

x = Ansel Adams

 

5.  Premise 1:  Form is "All A are B."

    Premise 2:  Fertilizer is outside "pesticide" circle, but not necessarily outside the "environment" circle.

    Conclusion:  **INVALID** ($x$ could be inside the "environment" circle).

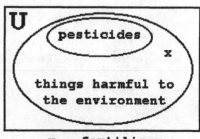

x = fertilizer

9.     Premise 1: Form is "All A are B."

       Premise 2: Form is "All B are C."

       Conclusion: **VALID** ("all poets" are within "taxi drivers" circle).

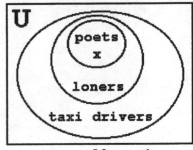

x = all poets

13.    Premise 1: Form is "All A are B."

       Premise 2: Route 66 is a specific road.

       Conclusion: **VALID** (*x* is within "Rome" circle).

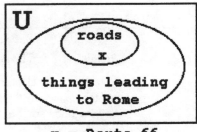

x = Route 66

17.    Premise 1: Form is "Some A are B."

       Premise 2: Form is "Some B are C," but nothing is indicated about women.

       Conclusion: **INVALID** (diagram can be constructed which meets the premises, but the conclusion does not follow).

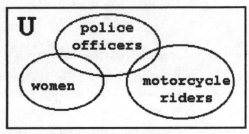

x = women who ride motorcycles

21.    a)    Inductive. The premises are specific cases which support the conclusion, but it might be the case that the electricity was off or the cable company or satellite disk was having problems.

       b)    Deductive. The premises are general cases that guarantee a specific conclusion.

25.    The numbers given are the sums of the even numbers beginning with 0; so the next number would be $0 + 2 + 4 + 6 + 8 = 20$.

       $\boxed{0}$, $0 + 2 = \boxed{2}$, $0 + 2 + 4 = \boxed{6}$, $0 + 2 + 4 + 6 = \boxed{12}$, $0 + 2 + 4 + 6 + 8 = \boxed{20}$

29.    The numbers given are the first five prime numbers, therefore the next number would be 13. The sequence is

       2, 3, 5, 7, 11, _13_ .

33.     If the letters given represent the first letter of the natural numbers, then the next two letters would be "F" for five and "S" for six.

One, Two, Three, Four, <u>F</u>ive, <u>S</u>ix

37.     The letters are the first letters of each of the given words. The next letter in the pattern would be "r".

41.     Notice that the letter given is the letter of the alphabet that follows the last letter that is in the word given. The letter that would fit the pattern would be "f."

45.     Substitute the fraction for $x$ in the equation:

$$ax^2 + bx + c = 0$$

$$a\left(\frac{-b+\sqrt{b^2-4ac}}{2a}\right)^2 + b\left(\frac{-b+\sqrt{b^2-4ac}}{2a}\right) + c = 0$$

$$a\left(\frac{b^2 + 2(-b)\sqrt{b^2-4ac} + \left(b^2-4ac\right)}{4a^2}\right) + \left(\frac{-b^2 + b\sqrt{b^2-4ac}}{2a}\right) + c = 0$$

$$\frac{2b^2 - 2b\sqrt{b^2-4ac} - 4ac}{4a} + \frac{-b^2 + b\sqrt{b^2-4ac}}{2a} + c = 0$$

$$\frac{2b^2 - 2b\sqrt{b^2-4ac} - 4ac}{4a} + \frac{-2b^2 + 2b\sqrt{b^2-4ac}}{4a} + c = 0$$

$$\frac{-4ac}{4a} + c = 0$$

$$-c + c = 0$$

---

## Section 1.2                                    Symbolic Logic

1.    a)    The given sentence "His name is George Washington" can be either true or false; therefore it is a statement.

b)    The given sentence is false; therefore it is a statement. (George Washington was the first president.)

c)    The given sentence is a question. As such, it is neither true nor false. It is not a statement.

d)    The given sentence is an opinion. As such, it is neither true nor false. It is not a statement.

5.    a)    Her dress is red.

b)    No computers are priced under $100.

c)    Some dogs are not four-legged animals.

d)    Some sleeping bags are waterproof.

9.    a)    $(p \vee q) \rightarrow r$ : Statement is a conditional (if...then...) of a disjunction (or) and a true
            statement.

      b)    $p \wedge q \wedge r$ : Statement is a conjunction (and) of three true statements.

      c)    $r \wedge \sim (p \vee q)$ : Statement is a conjunction (and) of a true statement and a negation (not) of a
            disjunction (or).

      d)    $q \rightarrow r$ : Rephrase as "If a person is a motorcycle rider, then the person wears leather
            jackets." Statement is conditional (if...then...).

13.   Translation 1:   $p$: It is a whole number.
                       $q$: It is an even number.
                       $r$: It is an odd number.
                       $p \rightarrow (q \vee r)$ :  (If $p$ , then $q$ or $r$)

      Translation 2:   $p$: It is a whole number.
                       $q$: It is not even.
                       $r$: It is not odd.
                       $p \rightarrow (\sim q \vee \sim r)$ :  (If $p$, then not $q$ or not $r$)

17.   Translation 1:   $p$: It is a muscle car.
                       $q$: It is from the Sixties.
                       $r$: It is a polluter.
                       $(p \wedge q) \rightarrow r$ :  (If $p$ and $q$, then $r$)

      Translation 2:   $p$: It is a muscle car.
                       $q$: It is from the Sixties.
                       $r$: It is a not polluter.
                       $(p \wedge q) \rightarrow \sim r$ :  (If $p$ and $q$, then not $p$)

21.   Translation 1:   $p$: You have a driver's license.
                       $q$: You have a credit card.
                       $r$: Your check is accepted.
                       $(\sim p \vee \sim q) \rightarrow \sim r$ :  (If not $p$ or not $q$, then not $r$)

      Translation 2:   $p$: You do not have a driver's license.
                       $q$: You do not have a credit card.
                       $r$: Your check is accepted.
                       $(p \vee q) \rightarrow \sim r$ :  (If $p$ or $q$, then not $r$)

25.   Translation 1:   $p$: You get a refund.
                       $q$: You get a store credit.
                       $r$: The product is defective.
                       $r \rightarrow (p \vee q)$ :  (If $r$, then $p$ or $q$)

      Translation 2:   $p$: You get a refund.
                       $q$: Your get a store credit.
                       $r$: The product works.
                       $\sim r \rightarrow (p \vee q)$ :  (If not $r$, then $p$ or $q$)

29.    a)    $(q \lor r) \to p$ : **If** I recycle my aluminum cans **or** my newspapers, **then** I am an
                environmentalist.

       b)    $\sim p \to \sim (q \lor r)$ : **If** I am **not** an environmentalist, **then** I do **not** recycle my aluminum cans
                **or** newspapers.

       c)    $(q \land r) \lor \sim p$ : I recycle both my aluminum cans and newspapers **or** I am **not** an
                environmentalist.

       d)    $(r \land \sim q \to \sim p$ : **If** I recycle my newspaper **and not** my aluminum cans, **then** I am **not** an
                environmentalist.

---

## Section 1.3                                                    Truth Tables

1.    There are $n = 2$ letters, so truth table rows $= 2^n = 2^2 = 4$.

| Label | Truth Value Assignment |
|---|---|
| $p$ | First half (2 rows) T, last half (2 rows) F |
| $q$ | Alternate T and F |
| $\sim q$ | Opposite of $q$ |
| $p \lor \sim q$ | If $p$ and $\sim q$ are F, then F (row 3), else T |

|  | $p$ | $q$ | $\sim q$ | $p \lor \sim q$ |
|---|---|---|---|---|
| 1. | T | T | F | T |
| 2. | T | F | T | T |
| 3. | F | T | F | F |
| 4. | F | F | T | T |

5.    There are $n = 2$ letters, so truth table rows $= 2^n = 2^2 = 4$.

| Label | Truth Value Assignment |
|---|---|
| $p$ | First half (2 rows) T, last half (2 rows) F |
| $q$ | Alternate T and F |
| $\sim q$ | Opposite of $q$ |
| $p \to \sim q$ | If $p$ is T and $\sim q$ is F, then F (row 1), else T |

|  | $p$ | $q$ | $\sim q$ | $p \to \sim q$ |
|---|---|---|---|---|
| 1. | T | T | F | F |
| 2. | T | F | T | T |
| 3. | F | T | F | T |
| 4. | F | F | T | T |

9.    There are $n = 2$ letters, so truth table rows $= 2^n = 2^2 = 4$.

| Label | Truth Value Assignment |
|---|---|
| $p$ | First half (2 rows) T, last half (2 rows) F |
| $q$ | Alternate T and F |
| $\sim p$ | Opposite of p |
| $p \vee q$ | If $p$ and $q$ are F, then F (row 4), else T |
| $(p \vee q) \to \sim p$ | If $(p \vee q)$ is T and $\sim p$ is F, then F (rows 1 and 2), otherwise T |

| | $p$ | $q$ | $\sim p$ | $p \vee q$ | $(p \vee q) \to \sim p$ |
|---|---|---|---|---|---|
| 1. | T | T | F | T | F |
| 2. | T | F | F | T | F |
| 3. | F | T | T | T | T |
| 4. | F | F | T | F | T |

13.    There are $n = 3$ letters, so truth table rows $= 2^n = 2^3 = 8$.

| Label | Truth Value Assignment |
|---|---|
| $p$ | First half (4 rows) T, last half (4 rows) F |
| $q$ | Alternate 2 Ts and 2 Fs |
| $r$ | Alternate T and F |
| $q \vee r$ | If $q$ and $r$ are F, then F (rows 4 and 8), else T |
| $\sim (q \vee r)$ | Opposite of $(q \vee r)$ |
| $p \wedge \sim (q \vee r)$ | If $p$ and $\sim (q \vee r)$ are T, then T (row 4), otherwise F |

| | $p$ | $q$ | $r$ | $q \vee r$ | $\sim (q \vee r)$ | $p \wedge \sim (q \vee r)$ |
|---|---|---|---|---|---|---|
| 1. | T | T | T | T | F | F |
| 2. | T | T | F | T | F | F |
| 3. | T | F | T | T | F | F |
| 4. | T | F | F | F | T | T |
| 5. | F | T | T | T | F | F |
| 6. | F | T | F | T | F | F |
| 7. | F | F | T | T | F | F |
| 8. | F | F | F | F | T | F |

17.    There are $n = 3$ letters, so truth table rows $= 2^n = 2^3 = 8$.

| Label | Truth Value Assignment |
|---|---|
| $p$ | First half (4 rows) T, last half (4 rows) F |
| $q$ | Alternate 2 Ts and 2 Fs |
| $r$ | Alternate T and F |
| $\sim r$ | Opposite of $r$ |
| $\sim r \vee p$ | If $\sim r$ and p are F, then F (rows 5 and 7), otherwise T |
| $q \wedge p$ | If $q$ and $p$ are T, then T (rows 1 and 2), else F |
| $(\sim r \vee p) \to (q \wedge p)$ | If $(\sim r \vee p)$ is T and $(q \wedge p)$ is F, then F (rows 3, 4, 6, and 8), else T |

|  | $p$ | $q$ | $r$ | $\sim r$ | $\sim r \vee p$ | $q \wedge p$ | $(\sim r \vee p) \to (q \wedge p)$ |
|---|---|---|---|---|---|---|---|
| 1. | T | T | T | F | T | T | T |
| 2. | T | T | F | T | T | T | T |
| 3. | T | F | T | F | T | F | F |
| 4. | T | F | F | T | T | F | F |
| 5. | F | T | T | F | F | F | T |
| 6. | F | T | F | T | T | F | F |
| 7. | F | F | T | F | F | F | T |
| 8. | F | F | F | T | T | F | F |

21.    $p$: It is raining.
       $q$: The streets are wet.
       $p \to q$ : If it is raining, then the streets are wet.

There are $n = 2$ letters, so truth table rows $= 2^n = 2^2 = 4$.

| Label | Truth Value Assignment |
|---|---|
| $p$ | First half (2 rows) T, last half (2 rows) F |
| $q$ | Alternate T and F |
| $p \to q$ | If $p$ is T and $q$ is F, then F (row 2), otherwise T |

|  | $p$ | $q$ | $p \to q$ |
|---|---|---|---|
| 1. | T | T | T |
| 2. | T | F | F |
| 3. | F | T | T |
| 4. | F | F | T |

25.    *p*: It is a square.

    *q*: It is a rectangle.

    $p \rightarrow q$ : If it is a square, then it is a rectangle.

There are $n = 2$ letters, so truth table rows $= 2^n = 2^2 = 4$.

| Label | Truth Value Assignment |
|---|---|
| *p* | First half (2 rows) T, last half (2 rows) F |
| *q* | Alternate T and F |
| $p \rightarrow q$ | If *p* is T and *q* is F, then F (row 2), otherwise T |

|    | *p* | *q* | $p \rightarrow q$ |
|---|---|---|---|
| 1. | T | T | T |
| 2. | T | F | F |
| 3. | F | T | T |
| 4. | F | F | T |

29.    *p*: You have a driver's license.

    *q*: You have a credit card.

    *r*: Your check is accepted.

    $(p \vee q) \rightarrow r$ : If you have a driver's license or a credit card, then your check is accepted.

There are $n = 3$ letters, so truth table rows $= 2^n = 2^3 = 8$.

| Label | Truth Value Assignment |
|---|---|
| *p* | First half (4 rows) T, last half (4 rows) F |
| *q* | Alternate 2 Ts and 2 Fs |
| *r* | Alternate T and F |
| $p \vee q$ | If p and *q* are F, then F (rows 7 and 8), otherwise T |
| $(p \vee q) \rightarrow r$ | If ( $p \vee q$ ) is T and *r* is F, then F (rows 2, ,4, and 6), otherwise T |

|    | *p* | *q* | *r* | $p \vee q$ | $(p \vee q) \rightarrow r$ |
|---|---|---|---|---|---|
| 1. | T | T | T | T | T |
| 2. | T | T | F | T | F |
| 3. | T | F | T | T | T |
| 4. | T | F | F | T | F |
| 5. | F | T | T | T | T |
| 6. | F | T | F | T | F |
| 7. | F | F | T | F | T |
| 8. | F | F | F | F | T |

33.    $p$: I have a college degree.

q: I have a job.

r: I own a house.

$p \wedge \sim (q \vee r)$ :  I have a college degree and I do not have a job or own a house.

There are $n = 3$ letters, so truth table rows $= 2^n = 2^3 = 8$.

| Label | Truth Value Assignment |
|---|---|
| $p$ | First half (4 rows) T, last half (4 rows) F |
| $q$ | Alternate 2 Ts and 2 Fs |
| $r$ | Alternate T and F |
| $q \vee r$ | If $q$ and $r$ are F, then F (rows 4 and 8), otherwise T |
| $\sim (q \vee r)$ | Opposite of $(q \vee r)$ |
| $p \wedge \sim (q \vee r)$ | If $p$ is T and $\sim (q \vee r)$ is T, then T (row 4), otherwise F |

|  | $p$ | $q$ | $r$ | $q \vee r$ | $\sim (q \vee r)$ | $p \wedge \sim (q \vee r)$ |
|---|---|---|---|---|---|---|
| 1. | T | T | T | T | F | F |
| 2. | T | T | F | T | F | F |
| 3. | T | F | T | T | F | F |
| 4. | T | F | F | F | T | T |
| 5. | F | T | T | T | F | F |
| 6. | F | T | F | T | F | F |
| 7. | F | F | T | T | F | F |
| 8. | F | F | F | F | T | F |

37.    $p$: The streets are wet.

q: It is raining.

$p \vee \sim q$ : The streets are wet or it is not raining.

$q \rightarrow p$ : If it is raining, then the streets are wet.

There are $n = 2$ letters, so truth table rows $= 2^n = 2^2 = 4$.

| Label | Truth Value Assignment |
|---|---|
| $p$ | First half (2 rows) T, last half (2 rows) F |
| $q$ | Alternate T and F |
| $\sim q$ | Opposite of $q$ |
| $p \vee \sim q$ | If $p$ and $\sim q$ are F, then F (row 3), else T |
| $q \rightarrow p$ | If $q$ is T and $p$ is F, then F (row 3), else T |

**37. Continued.**

| | $p$ | $q$ | $\sim q$ | $p \vee \sim q$ | $q \to p$ |
|---|---|---|---|---|---|
| 1. | T | T | F | T | T |
| 2. | T | F | T | T | T |
| 3. | F | T | F | F | F |
| 4. | F | F | T | T | T |

The two statements are equivalent because the values in the columns labeled $(p \vee \sim q)$ and $(q \to p)$ are the same.

**41.**    $p$: Handguns are outlawed.
$q$: Outlaws have handguns.
$p \to q$: If handguns are outlawed, then outlaws have handguns.
$q \to p$: If outlaws have handguns, then handguns are outlawed.

There are $n = 2$ letters, so truth table rows $= 2^n = 2^2 = 4$.

| Label | Truth Value Assignment |
|---|---|
| $p$ | First half (2 rows) T, last half (2 rows) F |
| $q$ | Alternate T and F |
| $p \to q$ | If $p$ is T and $q$ is F, then F (row 2), else T |
| $q \to p$ | If $q$ is T and $p$ is F, then F (row 3), else T |

| | $p$ | $q$ | $p \to q$ | $q \to p$ |
|---|---|---|---|---|
| 1. | T | T | T | T |
| 2. | T | F | F | T |
| 3. | F | T | T | F |
| 4. | F | F | T | T |

The two statements are **not equivalent** because the values in the columns labeled $(p \to q)$ and $(q \to p)$ are not the same.

45.    $p$: The plaintiff is innocent.

q: The insurance company settles out of court.

$p \vee \sim q$ : The plaintiff is innocent or the insurance company does not settle out of court.

$q \wedge \sim p$ : The insurance company settles out of court and the plaintiff is not innocent.

There are $n = 2$ letters, so truth table rows = $2^n = 2^2 = 4$.

| Label | Truth Value Assignment |
|---|---|
| $p$ | First half (2 rows) T, last half (2 rows) F |
| $q$ | Alternate T and F |
| $\sim p$ | Opposite of $p$ |
| $\sim q$ | Opposite of $q$ |
| $p \vee \sim q$ | If $p$ and $\sim q$ are F, then F (row 3), else T |
| $q \wedge \sim p$ | If $q$ and $\sim p$ are T, then T (row 3), else F |

|  | $p$ | $q$ | $\sim p$ | $\sim q$ | $p \vee \sim q$ | $q \wedge \sim p$ |
|---|---|---|---|---|---|---|
| 1. | T | T | F | F | T | F |
| 2. | T | F | F | T | T | F |
| 3. | F | T | T | F | F | T |
| 4. | F | F | T | T | T | F |

The two statements are **not equivalent** because the values in the columns labeled $(p \vee \sim q)$ and $(q \wedge \sim p)$ are not the same.

49.    $p$: I have a college degree.

q: I am not employed.

$p \wedge q$ : I have a college degree and I am not employed.

The negation is:  $\sim (p \wedge q) \equiv \sim p \vee \sim q$

$\sim p$: I do not have a college degree.

$\sim q$: I am employed.

$\sim p \vee \sim q$ : I do not have a college degree or I am employed.

53.    $p$: The building contains asbestos.

q: The original contractor is responsible.

$p \rightarrow q$ : If the building contains asbestos, the original contractor is responsible.

The negation is: $\sim (p \rightarrow q) \equiv p \wedge \sim q$

$p \wedge \sim q$ : The building contains asbestos and the original contractor is not responsible.

| **Section 1.4** | **More on Conditionals** |

1.    a)      $p \rightarrow q$ : If she is a police officer, then she carries a gun.

     b)      $q \rightarrow p$ : If she carries a gun, then she is a police officer.

     c)      $\sim p \rightarrow \sim q$ : If she is **not** a police officer, then she does **not** carry a gun.

     d)      $\sim q \rightarrow \sim p$ : If she does **not** carry a gun, then she is **not** a police officer.

     e)      Part (a) is equivalent to part (d) and part (b) is equivalent to part (c) because each equivalent pair was formed by negating and interchanging the premise and the conclusion.

5.    $p$: You will pass this mathematics course.
       $q$: You will fulfill a graduation requirement.

     a)      The inverse is   $\sim p \rightarrow \sim q$ : If you do **not** pass this mathematics course, then you will **not** fulfill a graduation requirement.

     b)      The converse is $q \rightarrow p$ : If you fulfill a graduation requirement, then you will pass this mathematics course.

     c)      The contrapositive is $\sim q \rightarrow \sim p$ : If you do **not** fulfill a graduation requirement, then you will **not** pass this mathematics course.

9.    $p$: You do not eat meat.
       $q$: You are a vegetarian.

     a)      The inverse is $\sim p \rightarrow \sim q$ : If you **do** eat meat, then you are **not** a vegetarian.

     b)      The converse is $q \rightarrow p$ : If you are a vegetarian, then you do not eat meat.

     c)      The contrapositive is $\sim q \rightarrow \sim p$ : If you are **not** a vegetarian, then you **do** eat meat.

13.    a)      In an "only if" compound statement the conclusion follows the "only if" so the premise is "I buy foreign products" and the conclusion is "domestic products are not available".

     b)      If I buy foreign products, then domestic products are not available.

     c)      A truth table needs to be constructed to find the conditions that make the statement false.

       $p$: I buy foreign products.
       $q$: Domestic products are available.
       $p \rightarrow \sim q$ : If I buy foreign products, then domestic products are not available.

       There are $n = 2$ letters, so truth table rows $= 2^n = 2^2 = 4$.

| Label | Truth Value Assignment |
|:---:|:---|
| $p$ | First half (2 rows) T, last half (2 rows) F |
| $q$ | Alternate T and F |
| $\sim q$ | Opposite of $q$ |
| $p \rightarrow \sim q$ | If $p$ is T and $\sim q$ is F, then F (row 1), else T |

13. c)  Continued.

|  | $p$ | $q$ | $\sim q$ | $p \to \sim q$ |
|---|---|---|---|---|
| 1. | T | T | F | F |
| 2. | T | F | T | T |
| 3. | F | T | F | T |
| 4. | F | F | T | T |

The conditions given in row 1 ($p$ and $q$ both true) would give a false expression. Therefore, the statement "I buy foreign products only if domestic products are not available," is false when I buy foreign products and domestic products are available.

17.  $p$: You obtain a refund.
  $q$: You have a receipt.
  $p \to q$: If you obtain a refund, then you have a receipt.
  $q \to p$: If you have a receipt, then you will obtain a refund.

The biconditional is $p \leftrightarrow q \equiv [(p \to q) \wedge (q \to p)]$: If you obtain a refund, then you have a receipt, **and** if you have a receipt, then you will obtain a refund.

21.  $p$: A polygon is a triangle.
  $q$: The polygon has three sides.
  $p \to q$: If a polygon is a triangle, then the polygon has three sides.
  $q \to p$: If the polygon has three sides, then the polygon is a triangle.

The biconditional is $p \leftrightarrow q \equiv [(p \to q) \wedge (q \to p)]$: If a polygon is a triangle, then the polygon has three sides, **and** if the polygon has three sides, then the polygon is a triangle.

25.  $p$: You earn less than \$12,000 per year.
  $q$: You are eligible for assistance.
  $p \to q$: If you earn less than \$12,000 per year, then you are eligible for assistance.
  $\sim q \to \sim p$: If you are **not** eligible for assistance, then you earn **at least** \$12,000 per year.

There are $n = 2$ letters, so truth table rows $= 2^n = 2^2 = 4$.

| Label | Truth Value Assignment |
|---|---|
| $p$ | First half (2 rows) T, last half (2 rows) F |
| $q$ | Alternate T and F |
| $p \to q$ | If $p$ is T and $q$ is F, then F (row 2), otherwise T |
| $\sim q$ | Opposite of $q$ |
| $\sim p$ | Opposite of $p$ |
| $\sim q \to \sim p$ | If $\sim q$ is T and $\sim p$ is F, then F (row 2), otherwise T |

25. Continued.

| | $p$ | $q$ | $p \rightarrow q$ | $\sim q$ | $\sim p$ | $\sim q \rightarrow \sim p$ |
|---|---|---|---|---|---|---|
| 1. | T | T | T | F | F | T |
| 2. | T | F | F | T | F | F |
| 3. | F | T | T | F | T | T |
| 4. | F | F | T | T | T | T |

$p \rightarrow q \equiv \sim q \rightarrow \sim p$. The two statements are **equivalent** because the values in columns $(p \rightarrow q)$ and $(\sim q \rightarrow \sim p)$ are the same.

29.    $p$: It is not raining.
       $q$: I walk to work.
       $p \rightarrow q$ : If it is not raining, then I walk to work.

       $(p \rightarrow q) \equiv (\sim q \rightarrow \sim p)$   (A conditional is equivalent to its contrapositive, therefore we need to negate and interchange the premise and the conclusion.)

       $\sim q \rightarrow \sim p$ : If I do **not** walk to work, then it is raining.

33.    $p$: You eat meat.
       $q$: You are not a vegetarian.
       $p \rightarrow q$ : If you eat meat, then you are not a vegetarian.

       $(p \rightarrow q) \equiv (\sim q \rightarrow \sim p)$   (A conditional is equivalent to its contrapositive, therefore we need to negate and interchange the premise and the conclusion.)

       $\sim q \rightarrow \sim p$ : If you **are** a vegetarian, then you do **not** eat meat.

37.    $p$: It is Sunday.
       $q$: I go to church.

       i)      $p \rightarrow q$ : If it is Sunday, then I go to church.
       ii)     $q \rightarrow p$ : I go to church **only if** it is Sunday.  (Conclusion follows "only if".)
       iii)    $\sim q \rightarrow \sim p$ : If I do **not** go to church, then it is **not** Sunday.
       iv)     $\sim p \rightarrow \sim q$ : If it is **not** Sunday, then I do **not** go to church.

       Using Figure 1.43:
               i $\equiv$ iii    conditional $\equiv$ contrapositive
               ii $\equiv$ iv         inverse $\equiv$ converse

| **Section 1.5** | **Analyzing Arguments** |
|---|---|

1.
    1. $p \rightarrow q$     ⎱ hypothesis
    2. $p$     ⎰
    ∴ $q$     } conclusion

The conditional representation is $[(p \rightarrow q) \wedge p] \rightarrow q$

5.
    1. $p \rightarrow q$     ⎱ hypothesis
    2. $\sim p$     ⎰
    ∴ $\sim q$     } conclusion

The conditional representation is $[(p \rightarrow q) \wedge \sim p] \rightarrow \sim q$

9.
    1. $p \rightarrow q$     ⎱ hypothesis
    2. $\sim q$     ⎰
    ∴ $\sim p$     } conclusion

The conditional representation is $[(p \rightarrow q) \wedge \sim q] \rightarrow \sim p$

There are $n = 2$ letters, so truth table rows $= 2^n = 2^2 = 4$.

| Label | Truth Value Assignment |
|---|---|
| $p$ | First half (2 rows) T, last half (2 rows) F |
| $q$ | Alternate T and F |
| Hypothesis 1: $p \rightarrow q$ | If $p$ is T and $q$ is F, then F (row 2), otherwise T |
| Hypothesis 2: $\sim q$ | Opposite of $q$ |
| $1 \wedge 2$: $(p \rightarrow q) \wedge \sim q$ | If $(p \rightarrow q)$ and $\sim q$ are T, then T (row 4), otherwise F |
| Conclusion: $\sim p$ | Opposite of $p$ |
| Conditional:<br>$[(p \rightarrow q) \wedge \sim q] \rightarrow \sim p$ | If $(1 \wedge 2)$ is T and the Conclusion is F, then F (no rows),<br>otherwise T |

|  | | | 1 | 2 | | $c$ | |
|---|---|---|---|---|---|---|---|
|  | $p$ | $q$ | $p \rightarrow q$ | $\sim q$ | $1 \wedge 2$ | $\sim p$ | $(1 \wedge 2) \rightarrow c$ |
| 1. | T | T | T | F | F | F | T |
| 2. | T | F | F | T | F | F | T |
| 3. | F | T | T | F | F | T | T |
| 4. | F | F | T | T | T | T | T |

The argument is **valid** as the conditional $[(p \rightarrow q) \wedge \sim q] \rightarrow \sim p$ is always true.

13.    $p$: The Democrats have a majority.
       $q$: Smith is appointed.
       $r$: Student loans are funded.

Symbolic form of the argument is:

$$
\left.\begin{array}{l} 1.\ p \rightarrow (q \wedge r) \\ 2.\ q \vee \sim r \end{array}\right\} \text{ hypothesis}
$$

$$
\left.\therefore\ \ \sim p \ \ \right\} \text{ conclusion}
$$

The conditional representation is $\{[\,p \rightarrow (q \wedge r)] \wedge (q \vee \sim r)\} \rightarrow \sim p$

There are $n = 3$ letters, so truth table rows $= 2^n = 2^3 = 8$.

| Label | Truth Value Assignment |
|---|---|
| $p$ | First half (4 rows) T, last half (4 rows) F |
| $q$ | Alternate 2 Ts and 2 Fs |
| $r$ | Alternate T and F |
| $\sim r$ | Opposite of $r$ |
| $q \wedge r$ | If $q$ and $r$ are T, then T (rows 1,5), else F |
| 1: $p \rightarrow (q \wedge r)$ | If $p$ is T and $(q \wedge r)$ is F, then F (rows 2, 3, and 4), otherwise T |
| 2: $q \vee \sim r$ | If $q$ is F and $\sim r$ is F then F (rows 3 and 7), otherwise T |
| $1 \wedge 2$ | If 1 is T and 2 is T, then T (rows 1, 5, 6, and 8), otherwise F |
| Conclusion: $\sim p$ | Opposite of $p$ |
| Conditional: (See Above) | If $(1 \wedge 2)$ is T and $c$ is F, then F (row 1), otherwise T |

| | $p$ | $q$ | $r$ | $\sim r$ | $q \wedge r$ | 1 | 2 | $1 \wedge 2$ | $c$ | $(1 \wedge 2) \rightarrow c$ |
|---|---|---|---|---|---|---|---|---|---|---|
| 1. | T | T | T | F | T | T | T | T | F | F |
| 2. | T | T | F | T | F | F | T | F | F | T |
| 3. | T | F | T | F | F | F | F | F | F | T |
| 4. | T | F | F | T | F | F | T | F | F | T |
| 5. | F | T | T | F | T | T | T | T | T | T |
| 6. | F | T | F | T | F | T | T | T | T | T |
| 7. | F | F | T | F | F | T | F | F | T | T |
| 8. | F | F | F | T | F | T | T | T | T | T |

The argument is **invalid** because row 1 of the conditional column of the truth table is false. If all three premises ($p$, $q$ and $r$) are true, then the Democrats must have a majority.

17.   $p$: It is a pesticide.
   $q$: It is harmful to the environment.
   $r$: It is a fertilizer.

Symbolic form of the argument is:

$$
\left.
\begin{array}{l}
1.\ p \to q \\
2.\ r \to\, \sim p
\end{array}
\right\} \text{ hypothesis}
$$

$$
\therefore\ r \to\, \sim q \quad \} \ \text{conclusion}
$$

The conditional representation is $[(p \to q) \wedge (r \to\, \sim p)] \to (r \to\, \sim q)$

There are $n = 3$ letters, so truth table rows $= 2^n = 2^3 = 8$.

| Label | Truth Value Assignment |
|---|---|
| $p$ | First half (4 rows) T, last half (4 rows) F |
| $q$ | Alternate 2 Ts and 2 Fs |
| $r$ | Alternate T and F |
| $\sim p$ | Opposite of $p$ |
| $\sim q$ | Opposite of $q$ |
| 1: $p \to q$ | If $p$ is T and $q$ is F, then F (rows 3 and 4), otherwise T |
| 2: $r \to\, \sim p$ | If $r$ is T and $\sim p$ is F then F (rows 1 and 3), otherwise T |
| $1 \wedge 2$ | If 1 is T and 2 is T, then T (rows 2, 5, 6, 7, and 8), otherwise F |
| Conclusion: $r \to\, \sim q$ | If $r$ is T and $\sim q$ is F, then F (rows 1 and 5), otherwise T |
| Conditional: (See Above) | If $(1 \wedge 2)$ is T and $c$ is F, then F (row 5), otherwise T |

| | $p$ | $q$ | $r$ | $\sim p$ | $\sim q$ | 1 | 2 | $1 \wedge 2$ | $c$ | $(1 \wedge 2) \to c$ |
|---|---|---|---|---|---|---|---|---|---|---|
| 1. | T | T | T | F | F | T | F | F | F | T |
| 2. | T | T | F | F | F | T | T | T | T | T |
| 3. | T | F | T | F | T | F | F | F | T | T |
| 4. | T | F | F | F | T | F | T | F | T | T |
| 5. | F | T | T | T | F | T | T | T | F | F |
| 6. | F | T | F | T | F | T | T | T | T | T |
| 7. | F | F | T | T | T | T | T | T | T | T |
| 8. | F | F | F | T | T | T | T | T | T | T |

The argument is **invalid** because row 5 of the conditional column of the truth table is false. Since both $q$ and $r$ are true, it is possible for a fertilizer to be harmful to the environment.

21.  $p$: It is a professor.
$q$: It is a millionaire.
$r$: It is illiterate.

Symbolic form of the argument is:

$$
\left.
\begin{array}{l}
1.\ p \to \sim q \\
2.\ q \to \sim r
\end{array}
\right\} \text{ hypothesis}
$$

$$
\therefore\ p \to \sim r \quad \} \text{ conclusion}
$$

The conditional representation is  $[(p \to \sim q) \wedge (q \to \sim r)] \to (p \to \sim r)$

There are $n = 3$ letters, so truth table rows $= 2^n = 2^3 = 8$.

| Label | Truth Value Assignment |
|---|---|
| $p$ | First half (4 rows) T, last half (4 rows) F |
| $q$ | Alternate 2 Ts and 2 Fs |
| $r$ | Alternate T and F |
| $\sim q$ | Opposite of $q$ |
| $\sim r$ | Opposite of $r$ |
| 1: $p \to \sim q$ | If $p$ is T and $\sim q$ is F, then F (rows 1 and 2), otherwise T |
| 2: $q \to \sim r$ | If $q$ is T and $\sim r$ is F then F (rows 1 and 5), otherwise T |
| $1 \wedge 2$ | If 1 is T and 2 is T, then T (rows 3, 4, 6, 7, and 8), otherwise F |
| Conclusion: $p \to \sim r$ | If $p$ is T and $\sim r$ is F, then F (rows 1 and 3), otherwise T |
| Conditional: (See Above) | If $(1 \wedge 2)$ is T and $c$ is F, then F (row 3), otherwise T |

|  | $p$ | $q$ | $r$ | $\sim q$ | $\sim r$ | 1 | 2 | $1 \wedge 2$ | $c$ | $(1 \wedge 2) \to c$ |
|---|---|---|---|---|---|---|---|---|---|---|
| 1. | T | T | T | F | F | F | F | F | F | T |
| 2. | T | T | F | F | T | F | T | F | T | T |
| 3. | T | F | T | T | F | T | T | T | F | F |
| 4. | T | F | F | T | T | T | T | T | T | T |
| 5. | F | T | T | F | F | T | F | F | T | T |
| 6. | F | T | F | F | T | T | T | T | T | T |
| 7. | F | F | T | T | F | T | T | T | T | T |
| 8. | F | F | F | T | T | T | T | T | T | T |

The argument is **invalid** because row 3 of the conditional column of the truth table is false. Since both $p$ and $r$ are true, it is possible for a professor to be illiterate.

25.   *p*: The defendant is innocent.
      *q*: The defendant goes to jail.

Symbolic form of the argument is:

$$\left.\begin{array}{l} 1.\ \ p \rightarrow\, \sim q \\ 2.\ \ q \end{array}\right\} \text{hypothesis}$$

$$\therefore\ \ \sim p \qquad \} \text{ conclusion}$$

The conditional representation is  $[(p \rightarrow\, \sim q) \wedge q] \rightarrow\, \sim p$

There are *n* = 2 letters, so truth table rows = $2^n = 2^2 = 4$.

| Label | Truth Value Assignment |
|---|---|
| *p* | First half (2 rows) T, last half (2 rows) F |
| 2: *q* | Alternate T and F |
| $\sim q$ | Opposite of *q* |
| 1: $p \rightarrow\, \sim q$ | If *p* is T and $\sim q$ is F, then F (row 1), otherwise T |
| $1 \wedge 2$ | If $(p \rightarrow\, \sim q)$ and *q* are T, then T (row 3), otherwise F |
| Conclusion: $\sim p$ | Opposite of *p* |
| Conditional: $[(p \rightarrow\, \sim q) \wedge q] \rightarrow\, \sim p$ | If $(1 \wedge 2)$ is T and the Conclusion is F, then F (no rows), otherwise T |

| | | 2 | | 1 | | *c* | |
|---|---|---|---|---|---|---|---|
| | *p* | *q* | $\sim q$ | $p \rightarrow\, \sim q$ | $1 \wedge 2$ | $\sim p$ | $(1 \wedge 2) \rightarrow c$ |
| 1. | T | T | F | F | F | F | T |
| 2. | T | F | T | T | F | F | T |
| 3. | F | T | F | T | T | T | T |
| 4. | F | F | T | T | F | T | T |

The argument is **valid** as the conditional  $[(p \rightarrow\, \sim q) \wedge q] \rightarrow\, \sim p$  is always true.

29.   *p*: You listen to rock and roll.
      *q*: You go to heaven.
      *r*: You are a moral person.

Symbolic form of the argument is:

$$\left.\begin{array}{l} 1.\ \ p \rightarrow\, \sim q \\ 2.\ \ r \rightarrow q \end{array}\right\} \text{hypothesis}$$

$$\therefore\ \ p \rightarrow\, \sim r \qquad \} \text{ conclusion}$$

The conditional representation is  $[(p \rightarrow\, \sim q) \wedge (r \rightarrow q)] \rightarrow (p \rightarrow\, \sim r)$

29. Continued.

There are $n = 3$ letters, so truth table rows $= 2^n = 2^3 = 8$.

| Label | Truth Value Assignment |
|---|---|
| $p$ | First half (4 rows) T, last half (4 rows) F |
| $q$ | Alternate 2 Ts and 2 Fs |
| $r$ | Alternate T and F |
| $\sim q$ | Opposite of $q$ |
| $\sim r$ | Opposite of $r$ |
| 1: $p \rightarrow \sim q$ | If $p$ is T and $\sim q$ is F, then F (rows 1 and 2), otherwise T |
| 2: $r \rightarrow q$ | If $r$ is T and $q$ is F then F (rows 3 and 7), otherwise T |
| $1 \wedge 2$ | If 1 is T and 2 is T, then T (rows 4, 5, 6, and 8), otherwise F |
| Conclusion: $p \rightarrow \sim r$ | If $p$ is T and $\sim r$ is F, then F (rows 1 and 3), otherwise T |
| Conditional: (See Above) | If $(1 \wedge 2)$ is T and $c$ is F, then F (no rows), otherwise T |

| | $p$ | $q$ | $r$ | $\sim q$ | $\sim r$ | 1 | 2 | $1 \wedge 2$ | $c$ | $(1 \wedge 2) \rightarrow c$ |
|---|---|---|---|---|---|---|---|---|---|---|
| 1. | T | T | T | F | F | F | T | F | F | T |
| 2. | T | T | F | F | T | F | T | F | T | T |
| 3. | T | F | T | T | F | T | F | F | F | T |
| 4. | T | F | F | T | T | T | T | T | T | T |
| 5. | F | T | T | F | F | T | T | T | T | T |
| 6. | F | T | F | F | T | T | T | T | T | T |
| 7. | F | F | T | T | F | T | F | F | T | T |
| 8. | F | F | F | T | T | T | T | T | T | T |

The argument is **valid** because the conditional $[(p \rightarrow \sim q) \wedge (r \rightarrow q)] \rightarrow (p \rightarrow \sim r)$ is always true.

33.    $p$: It is intelligible.
     $q$: It puzzles me.
     $r$: It is logic.

Symbolic form of the argument is:

       1. $p \rightarrow \sim q$    ⎫
                   ⎬ hypothesis
      2. $r \rightarrow q$      ⎭

      ∴ $r \rightarrow \sim p$    ⎬ conclusion

The conditional representation is $[(p \rightarrow \sim q) \wedge (r \rightarrow q)] \rightarrow (r \rightarrow \sim p)$

33. Continued.

There are $n = 3$ letters, so truth table rows $= 2^n = 2^3 = 8$.

| Label | Truth Value Assignment |
|---|---|
| $p$ | First half (4 rows) T, last half (4 rows) F |
| $q$ | Alternate 2 Ts and 2 Fs |
| $r$ | Alternate T and F |
| $\sim p$ | Opposite of $p$ |
| $\sim q$ | Opposite of $q$ |
| 1: $p \rightarrow \sim q$ | If $p$ is T and $\sim q$ is F, then F (rows 1 and 2), otherwise T |
| 2: $r \rightarrow q$ | If $r$ is T and $q$ is F then F (rows 3 and 7), otherwise T |
| $1 \wedge 2$ | If 1 is T and 2 is T, then T (rows 4, 5, 6, and 8), otherwise F |
| Conclusion: $r \rightarrow \sim p$ | If $r$ is T and $\sim p$ is F, then F (rows 1 and 3), otherwise T |
| Conditional: (See Above) | If $(1 \wedge 2)$ is T and $c$ is F, then F (no rows), otherwise T |

| | $p$ | $q$ | $r$ | $\sim p$ | $\sim q$ | 1 | 2 | $1 \wedge 2$ | $c$ | $(1 \wedge 2) \rightarrow c$ |
|---|---|---|---|---|---|---|---|---|---|---|
| 1. | T | T | T | F | F | F | T | F | F | T |
| 2. | T | T | F | F | F | F | T | F | T | T |
| 3. | T | F | T | F | T | T | F | F | F | T |
| 4. | T | F | F | F | T | T | T | T | T | T |
| 5. | F | T | T | T | F | T | T | T | T | T |
| 6. | F | T | F | T | F | T | T | T | T | T |
| 7. | F | F | T | T | T | T | F | F | T | T |
| 8. | F | F | F | T | T | T | T | T | T | T |

The argument is **valid** because the conditional $[(p \rightarrow \sim q) \wedge (r \rightarrow q)] \rightarrow (r \rightarrow \sim p)$ is always true.

37.  $p$: It is a wasp.
  $q$: It is friendly.
  $r$: It is a puppy.

  Symbolic form of the argument is:

  1. $p \rightarrow \sim q$
  2. $r \rightarrow \sim (\sim q)$ or $r \rightarrow q$      } hypothesis

  $\therefore \ r \rightarrow \sim p$      } conclusion

  The conditional representation is $\{[\, p \rightarrow \sim (\sim q)] \wedge (r \rightarrow q)\} \rightarrow (r \rightarrow \sim p)$

**37. Continued.**

There are $n = 3$ letters, so truth table rows $= 2^n = 2^3 = 8$.

| Label | Truth Value Assignment |
|-------|------------------------|
| $p$ | First half (4 rows) T, last half (4 rows) F |
| $q$ | Alternate 2 Ts and 2 Fs |
| $r$ | Alternate T and F |
| $\sim p$ | Opposite of $p$ |
| $\sim q$ | Opposite of $q$ |
| 1: $p \to \sim q$ | If $p$ is T and $\sim q$ is F, then F (rows 1 and 2), otherwise T |
| 2: $r \to q$ | If $r$ is T and $q$ is F then F (rows 3 and 7), otherwise T |
| $1 \wedge 2$ | If 1 is T and 2 is T, then T (rows 4, 5, 6, and 8), otherwise F |
| Conclusion: $r \to \sim p$ | If $r$ is T and $\sim p$ is F, then F (rows 1 and 3), otherwise T |
| Conditional: (See Above) | If $(1 \wedge 2)$ is T and $c$ is F, then F (no rows), otherwise T |

| | $p$ | $q$ | $r$ | $\sim p$ | $\sim q$ | 1 | 2 | $1 \wedge 2$ | $c$ | $(1 \wedge 2) \to c$ |
|----|---|---|---|---|---|---|---|---|---|---|
| 1. | T | T | T | F | F | F | T | F | F | T |
| 2. | T | T | F | F | F | F | T | F | T | T |
| 3. | T | F | T | F | T | T | F | F | F | T |
| 4. | T | F | F | F | T | T | T | T | T | T |
| 5. | F | T | T | T | F | T | T | T | T | T |
| 6. | F | T | F | T | F | T | T | T | T | T |
| 7. | F | F | T | T | T | T | F | F | T | T |
| 8. | F | F | F | T | T | T | T | T | T | T |

The argument is **valid** because the conditional $[(p \to \sim q) \wedge (r \to q)] \to (p \to \sim r)$ is always true.

---

| **Chapter 1** | **Review** |
|---|---|

1.     a)     Inductive. The premises are specific cases which support the conclusion.
        b)     Deductive. The premises are general cases that support a specific conclusion.

5. Premise 1: Form is "All A are B."

   Premise 2: Rocky is outside the "truck drivers" circle, but not necessarily outside the "union members" circle.

   Conclusion: **INVALID** ($x$ could be inside the "union members" circle).

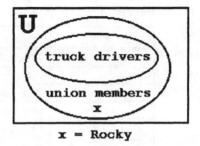

   x = Rocky

9. Premise 1: Form is "Some A are B."

   Premise 2: Form is "All A are B."

   Conclusion: **VALID** (Some electricians would be in the "carpenters" circle

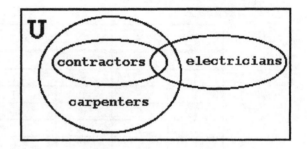

13. a) $p \wedge q$ : Statement is a conjunction (and).

    b) $\sim q \rightarrow \sim p$ : Statement is a conditional (if...then...) of two negations (not).

    c) $p \wedge \sim q$ : Statement is a conjunction (and) of a true statement and a negation (not).

    d) $\sim (p \vee q)$ : Statement is a negation (not) of a disjunction (or).

17. There are $n = 2$ letters, so truth table rows $= 2^n = 2^2 = 4$.

| Label | Truth Value Assignment |
|-------|------------------------|
| $p$ | First half (2 rows) T, last half (2 rows) F |
| $q$ | Alternate T and F |
| $\sim q$ | Opposite of $q$ |
| $p \vee \sim q$ | If $p$ and $\sim q$ are F, then F (row 3), otherwise T |

|     | $p$ | $q$ | $\sim q$ | $p \vee \sim q$ |
|-----|-----|-----|----------|-----------------|
| 1.  | T   | T   | F        | T               |
| 2.  | T   | F   | T        | T               |
| 3.  | F   | T   | F        | F               |
| 4.  | F   | F   | T        | T               |

21.     There are $n = 3$ letters, so truth table rows $= 2^n = 2^3 = 8$.

| Label | Truth Value Assignment |
|---|---|
| $p$ | First half (4 rows) T, last half (4 rows) F |
| $q$ | Alternate 2 Ts and 2 Fs |
| $r$ | Alternate T and F |
| $p \vee r$ | If $p$ and $r$ are both F, then F (rows 6 and 8), otherwise T |
| $\sim (p \vee r)$ | Opposite of $(p \vee r)$ |
| $q \vee \sim (p \vee r)$ | If $q$ and $\sim (p \vee r)$ are both F, then F (rows 3, 4, and 7), otherwise T |

|  | $p$ | $q$ | $r$ | $p \vee r$ | $\sim (p \vee r)$ | $q \vee \sim (p \vee r)$ |
|---|---|---|---|---|---|---|
| 1. | T | T | T | T | F | T |
| 2. | T | T | F | T | F | T |
| 3. | T | F | T | T | F | F |
| 4. | T | F | F | T | F | F |
| 5. | F | T | T | T | F | T |
| 6. | F | T | F | F | T | T |
| 7. | F | F | T | T | F | F |
| 8. | F | F | F | F | T | T |

25.     $p$: The car is reliable.

$q$: The car is expensive.

$\sim p \vee q$ : The car is unreliable or expensive.

$p \rightarrow q$ : If the car is reliable, then it is expensive.

There are $n = 2$ letters, so truth table rows $= 2^n = 2^2 = 4$.

| Label | Truth Value Assignment |
|---|---|
| $p$ | First half (2 rows) T, last half (2 rows) F |
| $q$ | Alternate T and F |
| $\sim p$ | Opposite of $p$ |
| $\sim p \vee q$ | If $\sim p$ and $q$ are both F, then F (row 2), otherwise T |
| $p \rightarrow q$ | If $p$ is T and $q$ if F, then F (row 2), otherwise T |

|  | $p$ | $q$ | $\sim p$ | $\sim p \vee q$ | $p \rightarrow q$ |
|---|---|---|---|---|---|
| 1. | T | T | F | T | T |
| 2. | T | F | F | F | F |
| 3. | F | T | T | T | T |
| 4. | F | F | T | T | T |

$\sim p \vee q \equiv p \rightarrow q$ . The two statements are **equivalent** because the values in the columns labeled $(\sim p \vee q)$ and $(p \rightarrow q)$ are the same.

29.   $p$: Jesse had a party.
      $q$: Nobody came to the party.
      $p \wedge q$ : Jesse had a party and nobody came.

      By De Morgan's Law the negation is:   $\sim(p \wedge q) \equiv \ \sim p \ \vee \sim q$

            $\sim p$: Jesse did not have a party.
            $\sim q$: Somebody came to the party.
            $\sim p \vee \sim q$ : Jesse did not have a party or somebody came to the party.

33.   $p$: His application is ignored.
      $q$: The selection procedure has been violated.
      $p \rightarrow q$ : If his application is ignored, the selection procedure has been violated.

      The negation is:   $\sim(p \rightarrow q) \equiv p \wedge \sim q$

            $\sim q$: The selection procedure has not been violated.
            $p \wedge \sim q$ : His application is ignored and the selection procedure has not been violated.

37.   a)     In an "only if" compound statement the conclusion follows the "only if" so the premise is
             "The economy improves" and the conclusion is "unemployment goes down."
      b)     If the economy improves, then unemployment goes down.

41.   $p$: You are allergic to dairy products.
      $q$: You cannot eat cheese.
      $p \rightarrow q$ : If you are allergic to dairy products, you cannot eat cheese.

      $q \rightarrow p$ : If you cannot eat cheese, then you are allergic to dairy products.

      There are $n = 2$ letters, so truth table rows $= 2^n = 2^2 = 4$.

| Label | Truth Value Assignment |
|-------|------------------------|
| $p$ | First half (2 rows) T, last half (2 rows) F |
| $q$ | Alternate T and F |
| $p \rightarrow q$ | If $p$ is T and $q$ is F, then F (row 2), else T |
| $q \rightarrow p$ | If $q$ is T and $p$ is F, then F (row 3), else T |

|   | $p$ | $q$ | $p \rightarrow q$ | $q \rightarrow p$ |
|---|-----|-----|-------------------|-------------------|
| 1. | T | T | T | T |
| 2. | T | F | F | T |
| 3. | F | T | T | F |
| 4. | F | F | T | T |

      The two statements are **not equivalent** because the values in the columns labeled $(p \rightarrow q)$ and
      $(q \rightarrow p)$ are not the same.

45.    $p$: You do pay attention.
       $q$: You do learn the new method.

Symbolic form of the argument is:

$$1. \sim p \rightarrow \sim q$$
$$2. \ q$$

hypothesis

$$\therefore \ p$$

} conclusion

The conditional representation is $[(\sim p \rightarrow \sim q) \wedge q] \rightarrow p$

There are $n = 2$ letters, so truth table rows $= 2^n = 2^2 = 4$.

| Label | Truth Value Assignment |
|---|---|
| $p$ | First half (2 rows) T, last half (2 rows) F |
| 2: $q$ | Alternate T and F |
| $\sim p$ | Opposite of $p$ |
| $\sim q$ | Opposite of $q$ |
| 1: $\sim p \rightarrow \sim q$ | If $\sim p$ is T and $\sim q$ is F, then F (row 3), otherwise T |
| $1 \wedge 2$ | If $(\sim p \rightarrow \sim q)$ and $q$ are T, then T (row 1), otherwise F |
| Conclusion: $p$ | $p$ |
| Conditional: $[(\sim p \rightarrow \sim q) \wedge q] \rightarrow p$ | If $(1 \wedge 2)$ is T and the Conclusion is F, then F (no rows), otherwise T |

| | | 2 | | | 1 | | $c$ | |
|---|---|---|---|---|---|---|---|---|
| | $p$ | $q$ | $\sim p$ | $\sim q$ | $\sim p \rightarrow \sim q$ | $1 \wedge 2$ | $p$ | $(1 \wedge 2) \rightarrow c$ |
| 1. | T | T | F | F | T | T | T | T |
| 2. | T | F | F | T | T | F | T | T |
| 3. | F | T | T | F | F | F | F | T |
| 4. | F | F | T | T | T | F | F | T |

The argument is **valid** as the conditional $[(\sim p \rightarrow \sim q) \wedge q] \rightarrow p$ is always true.

49.  *p*: I will go to the concert.
     *q*: You buy me a ticket.

Symbolic form of the argument is:

$$\left.\begin{array}{l} 1.\ p \to q \\ \underline{2.\ q} \end{array}\right\} \text{hypothesis}$$

$$\therefore\ p \quad \left.\right\} \text{ conclusion}$$

The conditional representation is $[(p \to q) \wedge q] \to p$

There are $n = 2$ letters, so truth table rows $= 2^n = 2^2 = 4$.

| Label | Truth Value Assignment |
|---|---|
| *p* | First half (2 rows) T, last half (2 rows) F |
| 2: *q* | Alternate T and F |
| 1: $p \to q$ | If *p* is T and *q* is F, then F (row 2), otherwise T |
| $1 \wedge 2$ | If $(p \to q)$ and *q* are T, then T (rows 1 and 3), otherwise F |
| Conclusion: *p* | *p* |
| Conditional: $[(p \to q) \wedge q] \to p$ | If $(1 \wedge 2)$ is T and the Conclusion is F, then F (row 3), otherwise T |

|  |  |  | 2 | 1 |  | *c* |  |
|---|---|---|---|---|---|---|---|
|  | *p* | *q* | $p \to q$ | $1 \wedge 2$ | *p* | $(1 \wedge 2) \to c$ |
| 1. | T | T | T | T | T | T |
| 2. | T | F | F | F | T | T |
| 3. | F | T | T | T | F | F |
| 4. | F | F | T | F | F | T |

The argument is **invalid** because row 3 of the conditional $[(p \to q) \wedge q] \to p$ is false.

53.    $p$: It is a professor.
       $q$: It is educated.
       $r$: It is a monkey.

Symbolic form of the argument is:

$$
\left.
\begin{array}{l}
1.\ p \to \sim (\sim q)\ \text{or}\ p \to q \\
2.\ \underline{r \to \sim p}
\end{array}
\right\} \text{hypothesis}
$$

$$
\therefore\ r \to \sim q \qquad \left.\right\} \text{conclusion}
$$

The conditional representation is  $[(p \to q) \wedge (r \to \sim p)] \to (r \to \sim q)$

There are $n = 3$ letters, so truth table rows $= 2^n = 2^3 = 8$.

| Label | Truth Value Assignment |
|---|---|
| $p$ | First half (4 rows) T, last half (4 rows) F |
| $q$ | Alternate 2 Ts and 2 Fs |
| $r$ | Alternate T and F |
| $\sim p$ | Opposite of $p$ |
| $\sim q$ | Opposite of $q$ |
| 1: $p \to q$ | If $p$ is T and $q$ is F, then F (rows 3 and 4), otherwise T |
| 2: $r \to \sim p$ | If $r$ is T and $\sim p$ is F then F (rows 1 and 3), otherwise T |
| $1 \wedge 2$ | If 1 is T and 2 is T, then T (rows 2, 5, 6, 7, and 8), otherwise F |
| Conclusion: $r \to \sim q$ | If $r$ is T and $\sim q$ is F, then F (rows 1 and 5), otherwise T |
| Conditional: (See Above) | If $(1 \wedge 2)$ is T and $c$ is F, then F (row 5), otherwise T |

|  | $p$ | $q$ | $r$ | $\sim p$ | $\sim q$ | 1 | 2 | $1 \wedge 2$ | $c$ | $(1 \wedge 2) \to c$ |
|---|---|---|---|---|---|---|---|---|---|---|
| 1. | T | T | T | F | F | T | F | F | F | T |
| 2. | T | T | F | F | F | T | T | T | T | T |
| 3. | T | F | T | F | T | F | F | F | T | T |
| 4. | T | F | F | F | T | F | T | F | T | T |
| 5. | F | T | T | T | F | T | T | T | F | F |
| 6. | F | T | F | T | F | T | T | T | T | T |
| 7. | F | F | T | T | T | T | T | T | T | T |
| 8. | F | F | F | T | T | T | T | T | T | T |

The argument is **invalid** because row 5 of the conditional column of the truth table is false.

# 2  Sets and Counting

**Section 2.1**                                                    **Sets and Set Operations**

1.  a)  "The set of all black automobiles" **is** well-defined because an automobile is either black or
        not black.
    b)  "The set of all inexpensive automobiles" **is not** well-defined because we need a level of
        cost (e.g., under $8,000) specified to determine whether an automobile is expensive or
        inexpensive.
    c)  "The set of all prime numbers" **is** well-defined as it can be determined whether or not a
        number is prime or not prime.
    d)  "The set of all large numbers" **is not** well-defined because different people will have
        different definitions for large.

5.  Proper subsets:       No elements: ∅

                          One element: {yes}, {no}, {undecided}

                          Two elements: {yes, no}, {yes, undecided}, {no, undecided}

    Improper subset:      Three elements: {yes, no, undecided}

9.  a)  $A \cap B$ = set of elements common to both $A$ and $B$ = ∅

    b)  $A \cup B$ = set of elements in either $A$ or $B$ or both = {0, 1, 2, 3, 4, 5, 6, 7, 8, 9} = $U$

    c)  $A'$ = set of elements in $U$ but not in $A$ = {0, 2, 4, 6, 8} = $B$

    d)  $B'$ = set of elements in $U$ but not in $B$ = {1, 3, 5, 7, 9} = $A$

13. $B'$ = set of all elements in $U$ but not in $B$ = {Monday, Tuesday, Wednesday, Thursday}

17. $A \cap B$ ($A$ intersect $B$) is the region that is
    common to both $A$ and $B$.

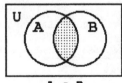

A ∩ B

21.    $A \cup B'$ ($A$ union the complement of $B$) is the region that is in $A$ together with all the region that is not in $B$.

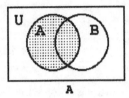 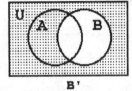 

25.    $A' \cup B'$ (the complement of $A$ union the complement of $B$) is the region that is not in $A$ together with the region that is not in $B$.

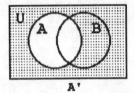 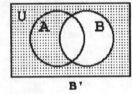 

29.    $U$ = {students | the student is a high school senior}
$A$ = {students | the student owns an automobile}
$M$ = {students | the student owns a motorcycle}
$A \cap M$ = {students | the student owns both}

$n(U) = 500,\ n(A) = 91,\ n(M) = 123,\ n(A \cap M) = 29$

a)    $n(A \text{ only}) = 91 - 29 = 62$
$n(M \text{ only}) = 123 - 29 = 94$
$n(A \cup M) = n(A) + n(M) - n(A \cap M)$
$= 91 + 123 - 29 = 185$
$n(\text{None}) = n(U) - n(A \cup M) = 500 - 185 = 315$

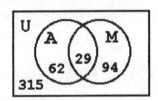

b)    The percent of students owning an automobile or a motorcycle is

$$\frac{n(A \cup M)}{n(U)} = \frac{185}{500} = 0.37 = 37\%$$

33.    $M$ = {Maine, Maryland, Massachusetts, Michigan, Minnesota, Mississippi, Missouri, Montana}
(Hint: Use an atlas.)

$n(U) = 50,\ n(M) = 8$

$n(M') = n(U) - n(M) = 50 - 8 = 42$

37.    $R$ = {September, October, November, December}.

$n(U) = 12, n(R) = 4$

$n(R\,') = n(U) - n(R) = 12 - 4 = 8$

41.    $U$ = {cards | the card is from an ordinary deck}
$S$ = {cards | the card is a spade}
$A$ = {cards | the card is an ace}
$S \cap A$ = {cards | the card is both a spade and an ace}

$n(U) = 52, n(S) = 13, n(A) = 4, n(S \cap A) = 1$

Find "spades **or** aces" which is the union of set $S$ and set $A$.

$n(S \cup A) = n(S) + n(A) - n(S \cap A) = 13 + 4 - 1 = 16$ spades or aces

45.    $U$ = {cards | the card is from an ordinary deck}
$F$ = {cards | the card is a face card (jack, queen, king)}
$B$ = {cards | the card is black (spade, club)}

Find "face cards **and** black" which is the intersection of set $F$ and set $B$.

$F \cap B$ = {club jack, club queen, club king, spade jack, spade queen, spade king}

$n(F \cap B) = 6$ face cards that are black

49.    $U$ = {cards | the card is from an ordinary deck}
$A$ = {cards | the card is an ace}
$E$ = {cards | the card is an eight}

Find "aces **and** eights" which is the intersection of set $A$ and set $E$.

$A \cap E = \varnothing$  There are no cards that are both an ace and an eight.

$n(A \cap E) = 0$

53.    a)    $A = \{a\}, n(A) = 1.$
$S$ = the subsets of $A = \{\varnothing, \{a\}\}$
$n(S) = 2$

b)    $A = \{a, b\}, n(A) = 2.$
$S$ = the subsets of $A = \{\varnothing, \{a\}, \{b\}, \{a, b\}\}$
$n(S) = 4$

c)    $A = \{a, b, c\}, n(A) = 3$
$S$ = the subsets of $A = \{\varnothing, \{a\}, \{b\}, \{c\}, \{a, b\}, \{a, c\}, \{b, c\}, \{a, b, c,\}\}$
$n(S) = 8$

53. Continued.

   d)   $A = \{a, b, c, d\}, n(A) = 4.$
      $S = $ the subsets of $A = \{\varnothing, \{a\}, \{b\}, \{c\}, \{d\}, \{a, b\}, \{a, c\}, \{a, d\}, \{b, c\}, \{b, d\}, \{c, d\},$
               $\{a, b, c,\}, \{a, b, d\}, \{a, c, d\}, \{b, c, d\}, \{a, b, c, d\}\}$
      $n(S) = 16$

   e)   Yes, each time an element is added to the set the number of subsets doubles.  In other
        words, the number of subsets is two raised to the cardinal number ($2^{n(A)}$).

   f)   $A = \{a, b, c, d, e, f\}, n(A) = 6$
        The number of subsets would be $2^{n(A)} = 2^6 = 64.$

57.   Omitted.

61.   Omitted.

65.   Comparing the possible answers against the given restrictions leaves (d) as the only answer that
      conforms to all the restrictions.

| Answer | John | Juneko | Restriction Violated |
|--------|------|--------|----------------------|
| a. | J, K, L | M, N, O | M and O cannot be in the same group. |
| b. | J, K, P | L, M, N | J and P cannot be in the same group. |
| c. | K, N, P | J, M, O | If N is in John's group, P must be in Juneko's group. |
| d. | L, M, N | K, O, P | No violation of the restrictions |
| e. | M, O, P | J, K, N | M and O cannot be in the same group |

69.   The answer must be (e).  By the restrictions:  If K is in John's group, M must be in Juneko's group.
      Also, M and O cannot be in the same group, therefore O cannot be in Juneko's group.  O could be
      in John's group, but it could also be the poster that is not used.  Nothing is known about J, L, or N
      which means answer (e) is the only one that can be true.

---

| **Section 2.2** | **Applications of Venn Diagrams** |

1.   $U = \{$people $|$ the person was interviewed in the survey$\}$
     $V = \{$people$|$ the person owned a VCR$\}$
     $M = \{$people $|$ the person owned a microwave oven$\}$

     $n(U) = 200, n(V) = 94, n(M) = 127, n(V \cap M) = 78$

1. Continued.

| Region | Set | $n$(Set) | Calculation |
|--------|-----|----------|-------------|
| Both | $V \cap M$ | 78 | Given |
| $V$ Only | $V \cap M'$ | 16 | $n(V) - n(V \cap M) = 94 - 78 = 16$ |
| $M$ Only | $M \cap V'$ | 49 | $n(M) - n(V \cap M) = 127 - 78 = 49$ |
| Neither | $(V \cup M)'$ | 57 | $n(U) - n(V \cap M) - n(V \cap M') - n(M \cap V')$<br>$= 200 - 78 - 16 - 49 = 57$ |

Construct a Venn diagram from the table.

a)   $n(V \cup M) = 16 + 78 + 49 = 143$ people who own **either** a VCR **or** a microwave oven.

b)   $n(V \cap M') = 16$ people who own **only** a VCR but **not** a microwave oven.

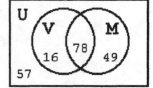

c)   $n(M \cap V') = 49$ people who own **only** a microwave oven but **not** a VCR.

d)   $n((V \cup M)') = 57$ people who own **neither** a VCR **nor** a microwave oven.

5.   $U = \{$adults | the adult was in the survey$\}$
$B = \{$adults | the adult watched the Big Game$\}$
$M = \{$adults | the adult watched the New Movie$\}$
$n(U) = 674,\ n((B \cup M)') = 226,\ n(M) = 289,\ n(M \cap B') = 183$

| Region | Set | $n$(Set) | Calculation |
|--------|-----|----------|-------------|
| Neither | $(B \cup M)'$ | 226 | Given |
| $M$ Only | $M \cap B'$ | 183 | Given |
| Both | $B \cap M$ | 106 | $n(M) - n(M \cap B') = 289 - 183 = 106$ |
| $B$ Only | $B \cap M'$ | 159 | $n(U) - n(M) - n((B \cup M)')$<br>$= 674 - 289 - 226 = 159$ |

Construct a Venn diagram from the table.

**5.** Continued.

a) $n(B \cap M) = 106$ adults watched **both** programs.

b) $n(B \cup M) = 159 + 106 + 183 = 448$ adults watched **at least one** program.

c) $n(B) = 159 + 106 = 265$ adults watched the Big Game.

d) $n(B \cap M') = 159$ adults watched **only** the Big Game but **not** the New Movie.

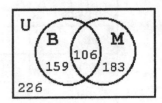

**9.** $U = \{\text{people} \mid \text{the person was interviewed in the survey}\}$
$C = \{\text{people} \mid \text{the person owned a personal computer}\}$
$T = \{\text{people} \mid \text{the person owned an cellular telephone}\}$
$V = \{\text{people} \mid \text{the person owned a VCR}\}$

$n(C) = 313,$
$n(T) = 232,$
$n(V) = 269,$

$n(\text{had all three}) = n(C \cap T \cap V) = 69,$

$n(\text{had none}) = n(C' \cap T' \cap V') = 64,$

$n(\text{had } T \text{ and } V) = n(T \cap V) = 98,$

$n(\text{had } T \text{ only}) = n(C' \cap T \cap V') = 57,$

$n(\text{had } C \text{ and } V \text{ but not } T) = n(C \cap T' \cap V) = 104$

| Region | Set | $n$(Set) | Calculation |
|--------|-----|----------|-------------|
| All | $C \cap T \cap V$ | 69 | Given |
| None | $C' \cap T' \cap V'$ | 64 | Given |
| $T$ Only | $C' \cap T \cap V'$ | 57 | Given |
| $T$ and $V$ and not $C$ | $C' \cap T \cap V$ | 29 | $n(T \cap V) - n(C \cap T \cap V) = 98 - 69 = 29$ |
| $T$ and $C$ and not $V$ | $C \cap T \cap V'$ | 77 | $n(T) - n(C \cap T \cap V) - n(C' \cap T \cap V)$ $- n(C' \cap T \cap V') = 232 - 69 - 29 - 57 = 77$ |
| $C$ and $V$ and not $T$ | $C \cap T' \cap V$ | 104 | Given |
| $C$ only | $C \cap T' \cap V'$ | 63 | $n(C) - n(C \cap T \cap V) - n(C \cap T \cap V')$ $- n(C \cap T' \cap V) = 313 - 69 - 77 - 104 = 63$ |
| $V$ only | $C' \cap T' \cap V$ | 67 | $n(V) - n(C \cap T \cap V) - n(C' \cap T \cap V)$ $- n(C \cap T' \cap V) = 269 - 69 - 29 - 104 = 67$ |

9. Continued.

Construct a Venn diagram from the table.

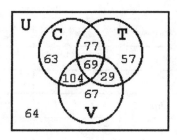

$n(U) = 69 + 64 + 57 + 29 + 77 + 104 + 63 + 67 = 530$

a)    $n$(cellular telephone) = $n(T)$ = 232

In percent, $\dfrac{n(T)}{n(U)} = \dfrac{232}{530} = 0.437735 \approx 43.8\%$

b)    $n$(cellular phone only) = $n(C' \cap T \cap V')$

In percent, $\dfrac{n(C' \cap T \cap V')}{n(U)} = \dfrac{57}{530} = 0.107547 \approx 10.8\%$

13.    $U$ = {members | the member belongs to the Eye and I Photo Club}
       $B$ = {members | the member used black and white film}
       $C$ = {members | the member used color film}
       $I$ = {members | the member used infrared film}

$n(B) = 77,$

$n(B \text{ only}) = n(B \cap C' \cap I') = 24,$

$n(C) = 65,$

$n(C \text{ only}) = n(C \cap B' \cap I') = 18,$

$n(B \text{ or } C) = n(B \cup C) = 101,$

$n(I) = 27,$

$n(\text{all}) = n(B \cap C \cap I) = 9,$

$n(\text{none}) = n(B' \cap C' \cap I') = 8$

13. Continued.

| Region | Set | $n$(Set) | Calculation |
|---|---|---|---|
| None | $B' \cap C' \cap I'$ | 8 | Given |
| All | $B \cap C \cap I$ | 9 | Given |
| $B$ Only | $B \cap C' \cap I'$ | 24 | Given |
| $C$ Only | $B' \cap C \cap I'$ | 18 | Given |
| $B$ and $C$ and not $I$ | $B \cap C \cap I'$ | 32 | Step 1: <br> $\quad n(B \cup C) = n(B) + n(C) - n(B \cap C)$ <br> $\quad\quad 101 = 77 + 65 - n(B \cap C)$ <br> $\quad n(B \cap C) = 142 - 101 = 41$ <br> Step 2: <br> $\quad n(B \cap C) - n(B \cap C \cap I) = 41 - 9 = 32$ |
| $B$ and $I$ and not $C$ | $B \cap C' \cap I$ | 12 | $n(B) - n(B \cap C' \cap I') - n(B \cap C)$ <br> $= 77 - 24 - 41 = 12$ |
| $C$ and $I$ and not $B$ | $B' \cap C \cap I$ | 6 | $n(C) - n(C \cap B' \cap I') - n(B \cap C)$ <br> $= 65 - 18 - 41 = 6$ |
| $I$ Only | $B' \cap C' \cap I$ | 0 | $n(I) - n(B \cap I \cap C') - n(B \cap C \cap I)$ <br> $\quad - n(C \cap I \cap B') = 27 - 12 - 9 - 6 = 0$ |

Construct a Venn diagram from the table.

$n(U) = 8 + 9 + 24 + 18 + 32 + 12 + 6 + 0 = 109$

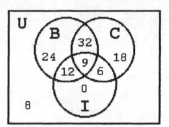

a)    $n$(infrared film only) $= n(B' \cap C' \cap I) = 0$

In percent, $\dfrac{n(B' \cap C' \cap I)}{n(U)} = \dfrac{0}{109} = 0.0 = 0\%$

13. Continued.

b)    $T$ = {members | the member used at least two types of film}

$$n(T) = n(B \cap C \cap I) + n(B \cap C \cap I') + n(B \cap I \cap C') + n(C \cap I \cap B')$$
$$= 9 + 32 + 12 + 6$$
$$= 59 \text{ members used at least two types of film}$$

In percent, $\dfrac{n(T)}{n(U)} = \dfrac{59}{109} = 0.541284 \approx 54.1\%$ .

17.    $U$ = {people | the person was in the survey}
       $D$ = {people | the person listened to WOLD (oldies)}
       $J$ = {people | the person listened to WJZZ (jazz)}
       $T$ = {people | the person listened to WTLK (talk show news)}

$n(D) = 140,$
$n(J) = 95,$
$\text{n}(T) = 134,$
$n(D \text{ or } J) = n(D \cup J) = 235,$
$n(D \text{ and } T) = n(D \cap T) = 48,$
$n(T \text{ or } J) = n(T \cup J) = 208,$
$n(\text{none}) = n(D' \cap J' \cap T') = 25$

| Region | Set | $n$(Set) | Calculation |
|---|---|---|---|
| None | $D' \cap J' \cap T'$ | 25 | Given |
| $D$ and $J$ and not $T$ | $D \cap J \cap T'$ | 0 | Step 1: $n(D \cup J) = n(D) + n(J) - n(D \cap J)$ $235 = 140 + 95 - n(D \cap J)$ $n(D \cap J) = 235 - 235 = 0$ Step 2: $D$ and $J$ are mutually exclusive.  Draw a Venn diagram with no intersection for $D$ and $J$. |
| $D$ and $T$ | $D \cap J' \cap T$ | 48 | Given |
| $D$ Only | $D \cap J' \cap T'$ | 92 | $n(D) - n(D \cap J' \cap T) = 140 - 48 = 92$ |
| $J$ and $T$ and not $D$ | $D' \cap J \cap T$ | 21 | $n(J \cup T) = n(J) + n(T) - n(J \cap T)$ $208 = 95 + 134 - n(J \cap T)$ $n(J \cap T) = 229 - 208 = 21$ |
| $J$ Only | $D' \cap J \cap T'$ | 74 | $n(J) - n(D' \cap J \cap T) = 95 - 21 = 74$ |
| $T$ Only | $D' \cap J' \cap T$ | 65 | $n(T) - n(D \cap J' \cap T) - n(D' \cap J \cap T)$ $= 134 - 48 - 21 = 65$ |

17.  Continued.

Construct a Venn diagram from the table.

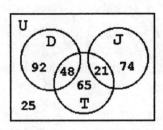

$n(U) = 25 + 48 + 92 + 21 + 74 + 65 = 325$

a)      $n(\text{WTLK only}) = n(D' \cap J' \cap T) = 65$

In percent, $\dfrac{n(D' \cap J' \cap T)}{n(U)} = \dfrac{65}{325} = 0.20 = 20\%$ .

b)      $T' = \{\text{people} \mid \text{the person did not listen to WTLK}\}$

$n(T') = n(U) - n(T) = 325 - 134 = 191$

In percent, $\dfrac{n(T')}{n(U)} = \dfrac{191}{325} = 0.587692 \approx 58.8\%.$

21.      $A = \{0, 2, 4, 5, 9\}, B' = \{0, 3, 4, 5, 6\}$

$(A' \cup B)' = (A')' \cap B' = A \cap B' = \{0, 4, 5\}$

25.      Fill in the Venn diagram as shown.

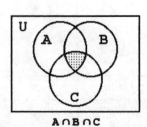

$A \cap B \cap C$

29.    Fill in the Venn diagram as shown.

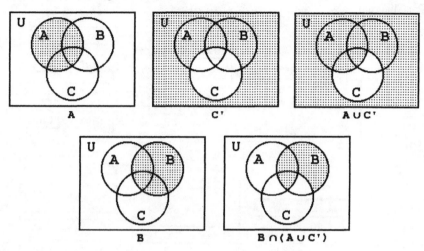

33.    Omitted.

37.    The correct answer is (d).  The second restriction excludes (a) and (b).  The third restriction excludes (c).  The fourth restriction excludes (e) because both Ed and Grant would not attend.

41.    The correct answer is (e).  If the meeting is held on Friday, then by the second restriction Carmine cannot attend.  If Carmine cannot attend, then by the last restriction Frank cannot attend.  Nothing is known as to whether or not the other board members given as choices will attend so it could be assumed that they might attend.

| **Section 2.3** | **Introduction to Combinatorics** |

1.    There are two possible outcomes (heads or tails) on each coin.

    a)    Create a box for each coin and enter possible outcomes and then multiply.

| possible different outcomes | = | outcomes from nickel | x | outcomes from dime | x | outcomes from quarter |

                    =    2    x    2    x    2    =    8

1. Continued.

    b)     Tree diagram:

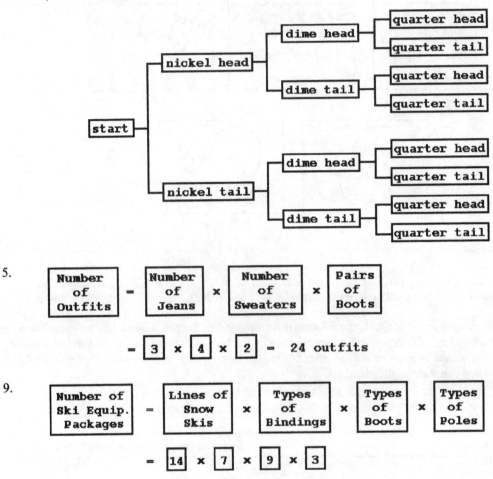

5.

$$\boxed{\begin{array}{c}\text{Number}\\\text{of}\\\text{Outfits}\end{array}} = \boxed{\begin{array}{c}\text{Number}\\\text{of}\\\text{Jeans}\end{array}} \times \boxed{\begin{array}{c}\text{Number}\\\text{of}\\\text{Sweaters}\end{array}} \times \boxed{\begin{array}{c}\text{Pairs}\\\text{of}\\\text{Boots}\end{array}}$$

$$= \boxed{3} \times \boxed{4} \times \boxed{2} = \text{24 outfits}$$

9.

$$\boxed{\begin{array}{c}\text{Number of}\\\text{Ski Equip.}\\\text{Packages}\end{array}} = \boxed{\begin{array}{c}\text{Lines of}\\\text{Snow}\\\text{Skis}\end{array}} \times \boxed{\begin{array}{c}\text{Types}\\\text{of}\\\text{Bindings}\end{array}} \times \boxed{\begin{array}{c}\text{Types}\\\text{of}\\\text{Boots}\end{array}} \times \boxed{\begin{array}{c}\text{Types}\\\text{of}\\\text{Poles}\end{array}}$$

$$= \boxed{14} \times \boxed{7} \times \boxed{9} \times \boxed{3}$$

= **2646 ski equipment packages**

13.

$$\boxed{\begin{array}{c}\text{Number of}\\\text{Social Security}\\\text{Numbers}\end{array}} = \boxed{\text{Ten choices for each of nine digits}}$$

$= 10 \cdot 10 \cdot 10 \cdot 10 \cdot 10 \cdot 10 \cdot 10 \cdot 10 \cdot 10 = 10^9$
$= 1{,}000{,}000{,}000$
$= 1$ billion possible Social Security numbers

17.  a)

| Number of telephone area codes | = | first digit (not 0 or 1) | x | second digit (only 0 or 1) | x | third digit (not 0 or 1) |
|---|---|---|---|---|---|---|

= $\boxed{8}$ x $\boxed{2}$ x $\boxed{8}$ = 128 area codes

b)

| Number of telephone area codes | = | first digit (not 0 or 1) | x | second digit | x | third digit |
|---|---|---|---|---|---|---|

= $\boxed{8}$ x $\boxed{10}$ x $\boxed{10}$ = 800 area codes

c)  Omitted.

21.

| Number of student I.D. numbers | = | first digit (nonzero) | x | three digits (may repeat) | x | three letters from A,B,C,D,E (not repeated) |
|---|---|---|---|---|---|---|

= $\boxed{9}$ x $\boxed{10}$ x $\boxed{10}$ x $\boxed{10}$ x $\boxed{5}$ x $\boxed{4}$ x $\boxed{3}$ = 540,000

25.  $10! = 10 \cdot 9 \cdot 8 \cdot 7 \cdot \ldots \cdot 2 \cdot 1 = 3,628,800$  (Use calculator)

29.  $6! \cdot 4! = (6 \cdot 5 \cdot 4 \cdot 3 \cdot 2 \cdot 1) \cdot (4 \cdot 3 \cdot 2 \cdot 1) = (720) \cdot (24) = 17,280$

33.  $\dfrac{8!}{5! \cdot 3!} = \dfrac{8 \cdot 7 \cdot 6 \cdot 5!}{5! \cdot 3 \cdot 2 \cdot 1} = \dfrac{8 \cdot 7 \cdot 6}{3 \cdot 2 \cdot 1} = \dfrac{336}{6} = 56$

When dividing factorials, cancel the largest common factorial in the denominator and numerator and then multiply or divide the remaining numbers. In this case, the largest factorial was 5!.

37.  $\dfrac{82!}{80! \cdot 2!} = \dfrac{82 \cdot 81 \cdot 80!}{80! \cdot 2 \cdot 1} = \dfrac{82 \cdot 81}{2 \cdot 1} = \dfrac{6642}{2} = 3321$

41.  When $n = 5, r = 5, \dfrac{n!}{(n-r)!} = \dfrac{5!}{(5-5)!} = \dfrac{5!}{0!} = \dfrac{5 \cdot 4 \cdot 3 \cdot 2 \cdot 1}{1} = 120$ .

45.  When $n = 5, r = 5, \dfrac{n!}{(n-r)!r!} = \dfrac{5!}{(5-5)! \cdot 5!} = \dfrac{5!}{0! \cdot 5!} = \dfrac{1}{1 \cdot 1} = 1$ .

49.  Omitted.

53.   The correct choice is (d). From the first restriction, the pink car must be in space #3. From the second restriction, the black and yellow cars could be on either side of the pink car as long as they are together. From the third restriction, the white car could park in either space #1 or space #2 if the gray, black, and yellow cars occupy spaces #4, #5, and #6. However, if the black and yellow cars are in space #1 and space #2, the white car could not be in space #5 without being next to the gray car.

---

| Section 2.4 | Permutations and Combinations |
| --- | --- |

1.   a)   When $n = 7, r = 3$, $\,_nP_r = \dfrac{n!}{(n-r)!} = \,_7P_3 = \dfrac{7!}{(7-3)!} = \dfrac{7!}{4!} = \dfrac{7\cdot6\cdot5\cdot4!}{4!} = 7\cdot6\cdot5 = 210$

   b)   When $n = 7, r = 3$, $\,_nC_r = \dfrac{n!}{(n-r)!r!} = \,_7C_3 = \dfrac{7!}{(7-3)!\cdot3!} = \dfrac{7\cdot6\cdot5\cdot4!}{4!\cdot3!} = \dfrac{7\cdot6\cdot5}{3\cdot2\cdot1} = \dfrac{210}{6} = 35$

5.   a)   When $n = 14, r = 1$, $\,_nP_r = \dfrac{n!}{(n-r)!} = \,_{14}P_1 = \dfrac{14!}{(14-1)!} = \dfrac{14!}{13!} = \dfrac{14\cdot13!}{13!} = 14$

   b)   When $n = 14, r = 1$, $\,_nC_r = \dfrac{n!}{(n-r)!r!} = \,_{14}C_1 = \dfrac{14!}{(14-1)!\cdot1!} = \dfrac{14\cdot13!}{13!\cdot1!} = \dfrac{14}{1} = 14$

9.   a)   When $n = x, r = x - 1$, $\,_nP_r = \dfrac{n!}{(n-r)!} = \,_xP_{x-1} = \dfrac{x!}{[x-(x-1)]!} = \dfrac{x!}{(x-x+1)!} = \dfrac{x!}{1!} = x!$

   b)   When $n = x, r = x - 1$, $\,_nC_r = \dfrac{n!}{(n-r)!r!} = \,_xC_{x-1} = \dfrac{x!}{[x-(x-1)]!\cdot(x-1)!} = \dfrac{x(x-1)!}{1!\cdot(x-1)!} = \dfrac{x}{1} = x$

13.   a)   When $n = 3, r = 2$, $\,_nP_r = \dfrac{n!}{(n-r)!} = \,_3P_2 = \dfrac{3!}{(3-2)!} = \dfrac{3!}{1!} = \dfrac{3\cdot2\cdot1}{1} = 6$

   b)   {a, b}, {a, c}, {b, a}, {b, c}, {c, a}, {c, b}

17.   12 students, each to give a presentation.

   a)   The order **is** important so use a **permutation** where $n = 12$ and $r = 12$.

$$\,_nP_r = \dfrac{n!}{(n-r)!} = \,_{12}P_{12} = \dfrac{12!}{(12-12)!} = \dfrac{12!}{0!} = \dfrac{12\cdot11\cdot10\cdot\cdots\cdot1}{1} = 479,001,600$$

   b)   There is only one way in which the presentation order will be alphabetical.

21.  There are 14 teams, each team playing every other team once. The order in which they play is not important so combinations are used where $n = 14$ and $r = 2$.

$$_nC_r = \frac{n!}{(n-r)!r!} = {}_{14}C_2 = \frac{14!}{(14-2)!\cdot2!} = \frac{14\cdot13\cdot12!}{12!\cdot2\cdot1} = \frac{14\cdot13}{2\cdot1} = \frac{182}{2} = 91$$

25.  Select a four-person committee from 8 women and 6 men.  The Fundamental Principle of Counting tells us to multiply the number from each category (women and men) together.  In a committee the order is not important so combinations will be used.

a)  Must have 2 women and 2 men.

| Number of Possible Committees | = | Select 2 women from 8 (n = 8, r = 2) | × | Select 2 men from 6 (n = 6, r = 2) |
|---|---|---|---|---|

$$\boxed{\text{women}} \times \boxed{\text{men}} = \boxed{_8C_2} \times \boxed{_6C_2}$$

$$(_8C_2)\cdot(_6C_2) = \frac{8!}{(8-2)!\cdot2!} \cdot \frac{6!}{(6-2)!\cdot2!} = \frac{8!}{6!\cdot2!} \cdot \frac{6!}{4!\cdot2!} = \frac{8\cdot7}{2\cdot1} \cdot \frac{6\cdot5}{2\cdot1} = 28\cdot15 = 420$$

b)  For any mixture we add women and men together to obtain 4 possible members.  The order is not important so we will use combinations where $n = 14$ and $r = 4$.

$$_{14}C_4 = \frac{14!}{(14-4)!\cdot4!} = \frac{14\cdot13\cdot12\cdot11\cdot10!}{10!\cdot4\cdot3\cdot2\cdot1} = \frac{14\cdot13\cdot12\cdot11}{4\cdot3\cdot2\cdot1} = \frac{24,024}{24} = 1001$$

c)  To have a majority of women there must be more women than men.  It is possible to meet this condition either by having 3 women and 1 man or by having 4 women and no men.  Calculate the possible outcomes from each case and then add the results together.

Case 1: 3 women and 1 man

| Number of Possible Committees | = | Select 3 women from 8 (n = 8, r = 3) | × | Select 1 man from 6 (n = 6, r = 1) |
|---|---|---|---|---|

$$\boxed{\text{women}} \times \boxed{\text{men}} = \boxed{_8C_3} \times \boxed{_6C_1}$$

$$(_8C_3)\cdot(_6C_1) = \frac{8!}{(8-3)!\cdot3!} \cdot \frac{6!}{(6-1)!\cdot1!} = \frac{8!}{5!\cdot3!} \cdot \frac{6!}{5!\cdot1!} = \frac{8\cdot7\cdot6\cdot5!}{5!\cdot3\cdot2\cdot1} \cdot \frac{6\cdot5!}{5!\cdot1} = \frac{336}{6} \cdot \frac{6}{1} = 336$$

25. c) Continued.

Case 2: 4 women and no men

| Number of Possible Committees | = | Select 4 women from 8 (n = 8, r = 4) | × | Select no men from 6 (n = 6, r = 0) |
|---|---|---|---|---|

$$\boxed{\text{women}} \times \boxed{\text{men}} = \boxed{_8C_4} \times \boxed{_6C_0}$$

$$(_8C_4) \cdot (_6C_0) = \frac{8!}{(8-4)! \cdot 4!} \cdot \frac{6!}{(6-0)! \cdot 0!} = \frac{8!}{4! \cdot 4!} \cdot \frac{6!}{6! \cdot 0!} = \frac{8 \cdot 7 \cdot 6 \cdot 5 \cdot 4!}{4 \cdot 3 \cdot 2 \cdot 1 \cdot 4!} \cdot \frac{1}{1} = \frac{1680}{24} \cdot \frac{1}{1} = 70$$

Adding Case 1 and Case 2 together, the total possible committees will be:

Case 1 + Case 2 = 336 + 70 = 406.

29.   Select a five-card poker hand from a deck of cards consisting of four suits with 13 different cards per suit. The Fundamental Principle of Counting tells us to multiply the number from each category (suits and cards) together. In a poker hand order is not important so combinations will be used.

a)   Must have 3 aces out of 4 possible and then have 2 cards from the remaining 48 cards.

| Number of Possible Poker Hands | = | Select 3 aces from 4 (n = 4, r = 3) | × | Select 2 cards from 48 (n = 48, r = 2) |
|---|---|---|---|---|

$$\boxed{\text{aces}} \times \boxed{\text{non–aces}} = \boxed{_4C_3} \times \boxed{_{48}C_2}$$

$$(_4C_3) \cdot (_{48}C_2) = \frac{4!}{(4-3)! \cdot 3!} \cdot \frac{48!}{(48-2)! \cdot 2!} = \frac{4 \cdot 3!}{1! \cdot 3!} \cdot \frac{48 \cdot 47 \cdot 46!}{46! \cdot 2 \cdot 1} = \frac{4}{1} \cdot \frac{2256}{2} = 4(1128) = 4512$$

There are 4512 possible hands that would contain exactly three aces

29. Continued.

    b)    Must have 1 card out of 13 and then have 3 different suits of that card and then any 2 cards from the remaining 48.  There are 13 ways to get the first card.

| Number of Possible Poker Hands | = | 13 ways to select 3 cards from 4 ($n = 4$, $r = 3$) | × | Select 2 cards from 48 ($n = 48$, $r = 2$) |
|---|---|---|---|---|

$$\boxed{\text{same kind}} \times \boxed{\text{non-same}} = 13 \times \boxed{_4C_3} \times \boxed{_{48}C_2}$$

$$13(_4C_3)\cdot(_{48}C_2) = 13 \cdot \frac{4!}{(4-3)!\cdot 3!}\cdot\frac{48!}{(48-2)!\cdot 2!}$$

$$= 13 \cdot \frac{4\cdot 3!}{1!\cdot 3!}\cdot\frac{48\cdot 47\cdot 46!}{46!\cdot 2\cdot 1} = 13(4)(1128) = 58,656$$

There are 58,656 possible hands that would contain three-of-a-kind.

33.    Choose 6 numbers out of 53 numbers.  The order is not important so use a combination where $n = 53$ and $r = 6$.

$$_{53}C_6 = \frac{53!}{(53-6)!\cdot 6!} = \frac{53\cdot 52\cdot 51\cdot 50\cdot 49\cdot 48\cdot 47!}{47!\cdot 6\cdot 5\cdot 4\cdot 3\cdot 2\cdot 1} = \frac{1.65293856\times 10^{10}}{720} = 22,957,480$$

There are 22,957,480 possible lottery tickets

37.    The 5/36 lottery would be easier to win because there are fewer combinations of numbers that can be chosen.

41.    Pascal's Triangle is Figure 2.32.
    a)    If the first row is counted as the $0^{th}$ row, then $_4C_2$ would be found in the $4^{th}$ row.
    b)    Counting the first row as the $0^{th}$ row, $_nC_r$ would be found in the $n^{th}$ row.
    c)    No, the second number in the fourth row is not $_4C_2$.
    d)    Yes, the third number in the fourth row is $_4C_2$.
    e)    The location of $_nC_r$ would be in the $n^{th}$ row (counting the first row as the $0^{th}$ row) and the $(r + 1)$ number.

45.    Each team will play the 5 other teams in the league one time each, so (c) is the correct choice.

49.    The correct choice is e) by elimination of the other choices.  Choices (a), (b) and (c) are eliminated because Team A could lose all five games or it could win the other four games, but it definitely lost it's first game to team D.  Choice (d) is eliminated because Team B had to lose at least one game to team D, so it could not win five games.

| **Section 2.5** | **Voting Systems** |

1.   a)   The total votes cast were $314 + 155 + 1052 + 479 = 2000$ votes.

   b)   Cruz wins the election with 1052 votes.

   c)   Yes, the winner received $\dfrac{1052}{2000} = 53\%$ of the vote.

5.   a)   The total votes cast were $6 + 8 + 11 + 5 = 30$ votes.

   b)   Looking at the row of $1^{st}$ choices, we see that
   W received 5 votes,
   P received $6 + 8 = 14$ votes, and
   B received 11 votes.
   The winner was the Park which received the most votes.

   c)   The Park received $\dfrac{14}{30} = 47\%$ of the vote.

   d)   The majority of the votes would be $\dfrac{30}{2} + 1 = 15 + 1 = 16$ votes.

   Eliminate the Warehouse because it received the fewest $1^{st}$ choice votes.

   The new table would be:

|  | Number of ballots cast | | | |
|---|---|---|---|---|
|  | 6 | 8 | 11 | 5 |
| $1^{st}$ choice | P | P | B | B |
| $2^{nd}$ choice | B | B | P | P |

   From the new row of $1^{st}$ choice votes,
   P received $6 + 8 = 14$ votes, and
   B received $11 + 5 = 16$ votes.
   The Beach would be the winner.

   e)   The Beach received $\dfrac{16}{30} = 53\%$ of the votes.

5. Continued.

f)    Tally the votes for each location and then multiply by 3 points for a $1^{st}$ choice vote, 2 points for a $2^{nd}$ choice vote, and 1 point for a $3^{rd}$ choice vote.

|  | Warehouse | Park | Beach |
|---|---|---|---|
| $1^{st}$ choice (3 points each) | 5 votes $5 \times 3 = 15$ points | $6 + 8 = 14$ votes $14 \times 3 = 42$ points | 11 votes $11 \times 3 = 33$ points |
| $2^{nd}$ choice (2 points each) | $8 + 11 = 19$ votes $19 \times 2 = 38$ points | 0 votes $0 \times 2 = 0$ points | $6 + 5 = 11$ votes $11 \times 2 = 22$ points |
| $3^{rd}$ choice (1 point each) | 6 votes $6 \times 1 = 6$ points | $11 + 5 = 16$ votes $16 \times 1 = 16$ points | 8 votes $8 \times 1 = 8$ points |
| Total points | $15 + 38 + 6$ $= 59$ points | $42 + 0 + 16$ $= 58$ points | $33 + 22 + 8$ $= 63$ points |

The Beach would be the winner, because it has the most total points.

g)    The Beach received 63 points.

h)    3 possible locations used 2 at a time gives $_3C_2 = \dfrac{3!}{(3-2)! \cdot 2!} = \dfrac{3 \cdot 2!}{1! \cdot 2!} = 3$ pairwise comparisons.

The pairs, the preferences by column, and the votes received are as follows:

|  | Columns Won | Votes Received | Winner | Points Received |
|---|---|---|---|---|
| W vs P | W wins 3 and 4 P wins 1 and 2 | $11 + 5 = 16$ $6 + 8 = 14$ | W | 1 |
| W vs B | W wins 2 and 4 B wins 1 and 3 | $8 + 5 = 13$ $6 + 11 = 17$ | B | 1 |
| B vs P | B wins 3 and 4 P wins 1 and 2 | $11 + 5 = 16$ $6 + 8 = 14$ | B | 1 |

Tally the points to show that B received 2 points, W received 1 point, and P received 0 points. The Beach is the winner.

i)    The Beach received 2 points.

9.  a)    The total votes cast were $25 + 13 + 19 + 27 + 30 + 26 = 140$ votes.

b)    Looking at the row of $1^{st}$ choices, we see that
          B received $25 + 13 = 38$ votes,
          N received $19 + 27 = 46$ votes, and
          S received $30 + 26 = 56$ votes.
The winner was Shattuck who received the most votes.

c)    Shattuck received $\dfrac{56}{140} = 40\%$ of the vote.

9. Continued

d)    The majority of the votes would be $\dfrac{140}{2}+1=70+1=71$ votes.

Eliminate Budd because he received the fewest 1$^{st}$ choice votes.

The new table would be:

| | Number of ballots cast | | | | | |
|---|---|---|---|---|---|---|
| | 25 | 13 | 19 | 27 | 30 | 26 |
| 1$^{st}$ choice | N | S | N | N | S | S |
| 2$^{nd}$ choice | S | N | S | S | N | N |

From the new row of 1$^{st}$ choice votes,
    N received 25 + 19 + 27 = 71 votes, and
    S received 13 + 30 + 26 = 69 votes.
Nirgiotis would be the winner.

e)    Nirgiotis received $\dfrac{71}{140}=0.507142...\approx 51\%$ of the votes.

f)    Tally the votes for each location and then multiply by 3 points for a 1$^{st}$ choice vote, 2 points for a 2$^{nd}$ choice vote, and 1 point for a 3$^{rd}$ choice vote.

| | Budd | Nirgiotis | Shattuck |
|---|---|---|---|
| 1$^{st}$ choice (3 points each) | 25 + 13 = 38 votes<br>38 × 3 = 114 points | 19 + 27 = 46 votes<br>46 × 3 = 138 points | 30 + 26 = 56 votes<br>56 × 3 = 168 points |
| 2$^{nd}$ choice (2 points each) | 27 + 30 = 57 votes<br>57 × 2 = 114 points | 25 + 26 = 51 votes<br>51 × 2 = 102 points | 13 + 19 = 32 votes<br>32 × 2 = 64 points |
| 3$^{rd}$ choice (1 point each) | 19 + 26 = 45 votes<br>45 × 1 = 45 points | 13 + 30 = 43 votes<br>43 × 1 = 43 points | 25 + 27 = 52 votes<br>52 × 1 = 52 points |
| Total points | 114 + 114 + 45<br>= 273 points | 138 + 102 + 43<br>= 283 points | 168 + 64 + 52<br>= 284 points |

Shattuck would be the winner because he has the most total points.

g)    Shattuck received 284 points.

9. Continued.

   h)     3 possible candidates used 2 at a time gives $_3C_2 = \dfrac{3!}{(3-2)! \cdot 2!} = \dfrac{3 \cdot 2!}{1! \cdot 2!} = 3$ pairwise comparisons.

   The pairs, the preferences by column, and the votes received are as follows:

| | Columns Won | Votes Received | Winner | Points Received |
|---|---|---|---|---|
| B vs N | B wins 1, 2, and 5 | $25 + 13 + 30 = 68$ | N | 1 |
| | N wins 3, 4 and 6 | $19 + 27 + 26 = 72$ | | |
| B vs S | B wins 1, 2 and 4 | $25 + 13 + 27 = 65$ | S | 1 |
| | S wins 3, 5 and 6 | $19 + 30 + 26 = 75$ | | |
| N vs S | N wins 1, 3 and 4 | $25 + 19 + 27 = 71$ | N | 1 |
| | S wins 2, 5 and 6 | $13 + 30 + 26 = 69$ | | |

   Tally the points to show that B received 0 points, N received 2 points, and S received 1 point. Nirgiotis is the winner.

   i)     Nirgiotis received 2 points.

13.  a)     Total votes cast are $1897 + 1025 + 4368 + 2790 + 6897 + 9571 + 5206 = 31,754$ votes.

     b)     Looking at the row of 1$^{st}$ choices, we see that
                 A received $1897 + 1025 = 2922$ votes,
                 B received 4368 votes,
                 C received 2790 votes,
                 D received $6897 + 9571 = 16,468$ votes, and
                 E received 5206 votes.
     The winner was Darter who received the most votes.

     c)     Darter received $\dfrac{16,468}{31,754} = 0.51861\ldots \approx 52\%$ of the vote.

     d)     The majority of the votes would be $\dfrac{31,754}{2} + 1 = 15,877 + 1 = 15,878$ votes.

     Since Darter already has more than 15,878 votes, there is no need to complete the run–off process.

     e)     Darter received $\dfrac{16,468}{31,754} = 0.51861\ldots \approx 52\%$ of the votes.

     f)     Tally the votes for each candidate and then multiply by 5 points for a 1$^{st}$ choice vote, 4 points for a 2$^{nd}$ choice vote, 3 points for a 3rd choice vote, 2 points for a 4$^{th}$ choice vote, and 1 point for a 5th choice vote.

13. Continued.

| | Addley | Burke | Ciento | Darter | Epp |
|---|---|---|---|---|---|
| 1st choice votes | $1897 + 1025$ $= 2922$ | 4368 | 2790 | $6897 + 9571$ $= 16,468$ | 5206 |
| Points (5 per vote) | $2922 \times 5$ $= 14,610$ | $4368 \times 5$ $= 21,840$ | $2790 \times 5$ $= 13,950$ | $16,468 \times 5$ $= 82,340$ | $5206 \times 5$ $= 26,030$ |
| 2nd choice votes | 0 | 6897 | 1897 | $1025 + 4368$ $+ 5206$ $= 10,599$ | $2790 + 9571$ $= 12,361$ |
| Points (4 per vote) | $0 \times 4$ $= 0$ | $6897 \times 4$ $= 27,588$ | $1897 \times 4$ $= 7588$ | $10,599 \times 4$ $= 42,396$ | $12,361 \times 4$ $= 49,444$ |
| 3rd choice votes | 0 | 9571 | $1025 + 4368$ $+ 6897$ $+ 5206$ $= 17,496$ | $1897 + 2790$ $= 4687$ | 0 |
| Points (3 per vote) | $0 \times 3$ $= 0$ | $9571 \times 3$ $= 28,713$ | $17,496 \times 3$ $= 52,488$ | $4687 \times 3$ $= 14,061$ | $0 \times 3$ $= 0$ |
| 4th choice votes | 9571 | $1897 + 2790$ $+ 5206$ $= 9893$ | 0 | 0 | $1025 + 4368$ $+ 6897$ $= 12,290$ |
| Points (2 per vote) | $9571 \times 2$ $= 19,142$ | $9893 \times 2$ $= 19,786$ | $0 \times 2$ $= 0$ | $0 \times 2$ $= 0$ | $12,290 \times 2$ $= 24,580$ |
| 5th choice votes | $4368 + 2790$ $+ 6897$ $+ 5206$ $= 19,261$ | 1025 | 9571 | 0 | 1897 |
| Points (1 per vote) | $19,261 \times 1$ $= 19,261$ | $1025 \times 1$ $= 1025$ | $9571 \times 1$ $= 9571$ | $0 \times 1$ $= 0$ | $1897 \times 1$ $= 1897$ |
| Total points | 53,013 | 98,952 | 83,597 | 138,797 | 101,951 |

Darter would be the winner, because he has the most total points.

g)      Darter received 138,797 points.

13. Continued.

h)     5 candidates, 2 at a time gives $_5C_2 = \dfrac{5!}{(5-2)! \cdot 2!} = \dfrac{5 \cdot 4 \cdot 3!}{3! \cdot 2!} = 10$ pairwise comparisons.

The pairs, the preferences by column, and the votes received are as follows:

| | Columns Won | Votes Received | Winner | Points Received |
|---|---|---|---|---|
| A vs B | A wins 1 and 2 | $1897 + 1025 = \mathbf{2922}$ | B | 1 |
| | B wins 3, 4, 5, 6, and 7 | $4368 + 2790 + 6897 + 9571 + 5206 = \mathbf{28{,}832}$ | | |
| A vs C | A wins 1, 2, and 6 | $1897 + 1025 + 9571 = \mathbf{12{,}493}$ | C | 1 |
| | C wins 3, 4, 5, and 7 | $4368 + 2790 + 6897 + 5206 = \mathbf{19{,}261}$ | | |
| A vs D | A wins 1 and 2 | $1897 + 1025 = \mathbf{2922}$ | D | 1 |
| | D wins 3, 4, 5, 6, and 7 | $4368 + 2790 + 6897 + 9571 + 5206 = \mathbf{28{,}832}$ | | |
| A vs E | A wins 1 and 2 | $1897 + 1025 = \mathbf{2922}$ | E | 1 |
| | E wins 3, 4, 5, 6, and 7 | $4368 + 2790 + 6897 + 9571 + 5206 = \mathbf{28{,}832}$ | | |
| B vs C | B wins 3, 5, and 6 | $4368 + 6897 + 9571 = \mathbf{20{,}836}$ | B | 1 |
| | C wins 1, 2, 4, and 7 | $1897 + 1025 + 2790 + 5206 = \mathbf{10{,}918}$ | | |
| B vs D | B wins 3 | $\mathbf{4368}$ | D | 1 |
| | D wins 1, 2, 4, 5, 6, and 7 | $1897 + 1025 + 2790 + 6897 + 9571 + 5206 = \mathbf{27{,}386}$ | | |
| B vs E | B wins 1, 3, and 5 | $1897 + 4368 + 6897 = \mathbf{13{,}162}$ | E | 1 |
| | E wins 2, 4, 6, and 7 | $1025 + 2790 + 9571 + 5206 = \mathbf{18{,}592}$ | | |
| C vs D | C wins 1 and 4 | $1897 + 2790 = \mathbf{4687}$ | D | 1 |
| | D wins 2, 3, 5, 6, and 7 | $1025 + 4368 + 6897 + 9571 + 5206 = \mathbf{27{,}067}$ | | |
| C vs E | C wins 1, 2, 3, 4, and 5 | $1897 + 1025 + 4368 + 2790 + 6897 = \mathbf{16{,}977}$ | C | 1 |
| | E wins 6 and 7 | $9571 + 5206 = \mathbf{14{,}777}$ | | |
| D vs E | D wins 1, 2, 3, 5, and 6 | $1897 + 1025 + 4368 + 6897 + 9571 = \mathbf{23{,}758}$ | D | 1 |
| | E wins 4 and 7 | $2790 + 5206 = \mathbf{7996}$ | | |

**13. h)  Continued.**

Tally the points to show that
A received 0 points,
B received 2 points,
C received 2 points,
D received 4 points, and
E received 2 points.
Darter is the winner.

i)      Darter received 4 points.

17.      6 possible candidates, 2 at a time gives $_6C_2 = \dfrac{6!}{(6-2)! \cdot 2!} = \dfrac{6 \cdot 5 \cdot 4!}{4! \cdot 2!} = 15$ pairwise comparisons.

21.      a)      7 possible candidates used 2 at a time gives $_7C_2 = \dfrac{7!}{(7-2)! \cdot 2!} = \dfrac{7 \cdot 6 \cdot 5!}{5! \cdot 2!} = \dfrac{7 \cdot 6}{2 \cdot 1} = 21$ pairwise

comparisons.  Each candidate would be compared with each of the other 6 candidates one time, therefore the maximum number of points possible for a candidate would be 6.

b)      It is possible that a candidate would not win any of the pairwise comparisons and therefore, would receive a minimum of 0 points.  (See Example 8 in the text.)

---

## Section 2.6 — Infinite Sets

1.      $S$ is equivalent to $C$ ($S \sim C$) because they both contain 4 elements, $n(S) = n(C) = 4$.

Possible one-to-one correspondence (capital city $\leftrightarrow$ state):

Sacramento $\leftrightarrow$ California

Lansing $\leftrightarrow$ Michigan

Richmond $\leftrightarrow$ Virginia

Topeka $\leftrightarrow$ Kansas

5.      $C$ contains 22 multiples of 3, so $n(C) = 22$ and $D$ contains 22 multiples of 4, so $n(D) = 22$. Therefore, $C$ and $D$ have the same cardinal number, $C \sim D$.

Possible one-to-one correspondence ($3n \leftrightarrow 4n$):
$$C = \{3, 6, 9, 12, ..., 3n, ..., 63, 66\}$$
$$\updownarrow \updownarrow \updownarrow \updownarrow \ \cdots \ \updownarrow \ \cdots \ \updownarrow \ \updownarrow$$
$$D = \{4, 8, 12, 16, ..., 4n, ..., 84, 88\}$$

9.    $A$ is the set of odd numbers from 1 through 123 which can be represented by $2n - 1$. The last number, 123, would be the $62^{nd}$ number in the set, $2(62) - 1 = 124 - 1 = 123$. So $n(A) = 62$.

$B$ is the set of odd numbers from 125 through 247 which can be represented by $2n + 123$. The last number, 247 would be the $62^{nd}$ number in the set, $2(62) + 123 = 124 + 123 = 247$. So $n(B) = 62$.

The cardinal numbers of $A$ and $B$ are the same, so $A \sim B$.

Possible one-to-one correspondence ($2n - 1 \leftrightarrow 2n + 123$):

$$A = \{\ 1,\quad 3,\quad 5,\ ...,\quad 2n - 1,\quad ...,\ 121, 123\}$$
$$\updownarrow\ \ \updownarrow\ \ \updownarrow\ \ ...\quad \updownarrow\quad ...\ \updownarrow\ \ \updownarrow$$
$$B = \{125, 127, 129,\ ...,\ 2n + 123,\ ...,\ 245, 247\}$$

13.   a)   The elements of $N$ and $T$ can be paired up as follows:

$$N = \{1, 2, 3, 4,\ ...,\ n,\quad n + 1,\ ...\}$$
$$\updownarrow\ \updownarrow\ \updownarrow\ \updownarrow\ ...\ \updownarrow\quad \updownarrow\quad ...$$
$$T = \{3, 6, 9, 12,\ ...,\ 3n, 3(n + 1), ...\}$$

Any natural number, $n \in N$, corresponds with the multiple of three, $3n \in T$. There exists a one-to-one correspondence between the elements of $N$ and $T$, therefore the sets are equivalent; that is, $T \sim N$.

b)   $936 = 3n \in T$, so $n = \dfrac{936}{3} = 312 \in N$

c)   $x = 3n \in T$, so $n = \dfrac{x}{3} \in N$

d)   $936 = n \in N$, so $3n = 3(936) = 2808 \in T$

e)   $n \in N$, so $3n \in T$

17.    Let $A'B$ = line segment [0, 1] and $AC$ = line segment [0, 3]. Line segment $AC$ is three times as long as line segment $A'B$. Draw $A'B$ above $AC$ and extend line segments $AA'$ and $CB$ so that they meet at point $D$.

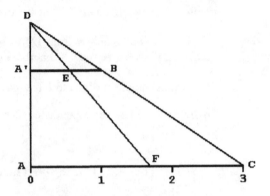

Any point $E$ on $A'B$ can be paired up with the unique point $F$ on $AC$ formed by the intersection of lines $DE$ and $AC$. Conversely, any point $F$ on segment $AC$ can be paired up with the unique point $E$ on $A'B$ formed by the intersection of lines $DF$ and $A'B$.

Therefore, a one-to-one correspondence exists between the two segments and the interval [0, 1] contains exactly the same number of points as the interval [0, 3]. Thus, the intervals are equivalent.

21.    Draw the circle of radius 1 cm inside the circle of radius 5 cm so that they have the same center $C$.

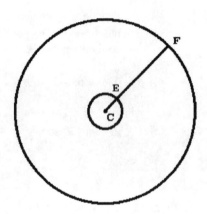

Any point $E$ on the smaller circle can be paired with the unique point $F$ on the larger circle by drawing a line from the center of the circle through $E$. Conversely, any point $F$ on the larger circle can be paired with the unique point $E$ on the inner circle by drawing a line from the center of the circle through $F$.

Therefore, a one-to-one correspondence exists between the two circles and they are equivalent.

## Chapter 2                                                              Review

1.    a)    "The set of all multiples of 5" **is** well-defined because a number is a multiple of 5 or it is not.
       b)    "The set of all difficult math problems" **is not** well-defined because different people will have different definitions for difficult.
       c)    "The set of all great movies" **is not** well-defined because different people will have different definitions of great.
       d)    "The set of all Oscar-winning movies" **is** well-defined because a movie is either an Oscar winner or it is not.

5. a) $n(U) = 61, n(A) = 32, n(B) = 26, n(A \cup B) = 40$

$n(A \cup B) = n(A) + n(B) - n(A \cap B)$

$40 = 32 + 26 - n(A \cap B)$

$40 = 58 - n(A \cap B)$

$n(A \cap B) = 58 - 40 = 18$

b) $n(A \text{ only}) = n(A) - n(A \cap B) = 32 - 18 = 14$

$n(B \text{ only}) = n(B) - n(A \cap B) = 26 - 18 = 8$

$n(\text{None}) = n(U) - n(A \cup B) = 61 - 40 = 21$

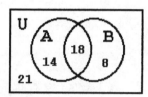

9. a) Create a box for each event, enter each possible outcome and then multiply.

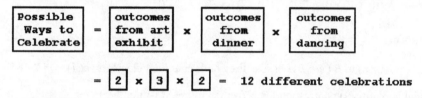

b) Tree diagram:

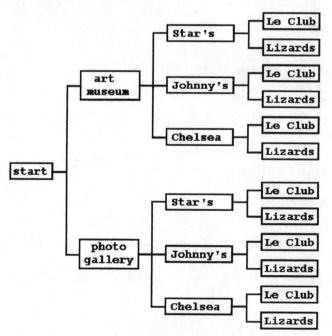

13.    Select 3 items at a time from 11 items.

a)    The order **is not** important so use a **combination** where $n = 11$ and $r = 3$.

$$_nC_r = \frac{n!}{(n-r)!\,r!} = {_{11}C_3} = \frac{11!}{(11-3)!\cdot 3!} = \frac{11\cdot 10\cdot 9\cdot 8!}{8!\cdot 3!} = \frac{11\cdot 10\cdot 9}{3\cdot 2\cdot 1} = \frac{990}{6} = 165$$

b)    The order **is** important so use a **permutation** where $n = 11$ and $r = 3$.

$$_nP_r = \frac{n!}{(n-r)!} = {_{11}P_3} = \frac{11!}{(11-3)!} = \frac{11!}{8!} = \frac{11\cdot 10\cdot 9\cdot 8!}{8!} = 11\cdot 10\cdot 9 = 990$$

17.    There are 10 teams with 3 possible end-of-the-season rankings.  The order **is** important so use a **permutation** where $n = 10$ and $r = 3$.

$$_nP_r = \frac{n!}{(n-r)!} = {_{10}P_3} = \frac{10!}{(10-3)!} = \frac{10!}{7!} = \frac{10\cdot 9\cdot 8\cdot 7!}{7!} = 10\cdot 9\cdot 8 = 720$$

21.    Omitted.

25.    $A$ is equivalent to $B$ $(A \sim B)$ because they both contain 4 elements, $n(A) = n(B) = 5$.

Possible one-to-one correspondence (roman numeral $\leftrightarrow$ number):
$$A = \{\ \text{I,}\quad \text{II,}\quad \text{III,}\quad \text{IV,}\quad \text{V}\}$$
$$\updownarrow\quad \updownarrow\quad \updownarrow\quad \updownarrow\quad \updownarrow$$
$$B = \{\text{one, two, three, four, five}\}$$

29.    a)    $396 \in N$ is even, so it corresponds to $\dfrac{3}{2}(396) = \dfrac{1188}{2} = 594$.

b)    $396 \in A$ is positive, so

$$396 = \frac{3}{2}n$$
$$792 = 3n$$
$$n = 264$$

c)    $-153 \in A$ is negative so

$$-153 = \frac{3}{2}(1-n)$$
$$-306 = 3(1-n)$$
$$-102 = 1-n$$
$$n = 103$$

d)    $n(A) = \aleph_0$.

# 3  Probability

1.  Individual solutions will vary depending on what happens when the single die is rolled.  The following table records the results of a trial of this experiment.

| Result<br>(Number of 6s rolled in four rolls of die) | Number of times in 10 trials that<br>result occurred |
|:---:|:---:|
| 0 | 3 |
| 1 | 4 |
| 2 | 3 |
| 3 | 0 |
| 4 | 0 |

Betting that you could roll at least one six, you would have won 7 times ($70) and lost three times ($30).  Thus, you would have won $40 (70 − 30 = 40).

5.  You bet $5 on a 17-20 split.  The house odds are 17 to 1.
    a)   You win $85.  (5•17 = 85)
    b)   You win $85.  (5•17 = 85)
    c)   You lose $5.

9.  You bet $10 on 31-32-33-34-35-36 line.  The house odds are 5 to 1.
    a)   You lose $10.
    b)   You win $50. (10•5 = 50)

13.  You bet $50 on the odd numbers.  The house odds are 1 to 1.
    a)   You lose $50.
    b)   You win $50.

17.  You bet $30 on the 1-2 split (house odds are 17 to 1), and also bet $15 on the even numbers (house odds are 1 to 1).
    a)   You win $495.  (30•17 − 15•1 = 510 − 15 = 495)
    b)   You win $525.  (30•17 + 15•1 = 510 + 15 = 525)
    c)   You lose $45.  (−30 − 15 = −45)
    d)   You lose $15.  (−30 + 15•1 = −30 + 15 = −15)

21.  Since the house odds are 2 to 1 you must bet $1000/2 = $500 to win at least $1000.  You could lose $500.

25.    a)    There are 12 face cards in a deck of cards. A jack, queen and king from each of 4 suits.
       b)    There are 52 cards in a deck of cards. The fraction that are face cards is 12/52 = 3/13.

---

## Section 3.2                                                    Basic Terms of Probability

---

1.    The experiment is the picking of a jellybean.

5.    a)    Event $A$ = picking a red or yellow jelly bean
            $n(S) = 12 + 8 + 10 + 5 = 35$, $n(A) = 8 + 10 = 18$

            $$p(A) = \frac{n(A)}{n(S)} = \frac{18}{35}$$

      b)    This means that 18 out of every 35 possible outcomes are a success.

9.    a)    Event $B$ = picking a white jelly bean
            $n(S) = 12 + 8 + 10 + 5 = 35$, $n(B) = 0$

            $$p(B) = \frac{n(B)}{n(S)} = \frac{0}{35} = 0$$

      b)    This means that there is no possible way to have a successful outcome.

13.   a)    Event $A$ = picking a red or yellow jelly bean
            $n(S) = 12 + 8 + 10 + 5 = 35$, $n(A) = 8 + 10 = 18$
            $n(A') = n(S) - n(A) = 35 - 18 = 17$

            $o(B) = n(A){:}n(A') = 18{:}17$

      b)    This means that there are 18 successful outcomes for every 17 possible failures.

17.   $E$ = drawing a black card
      $n(S) = 52$, $n(E) = 26$, $n(E') = n(S) - n(E) = 52 - 26 = 26$

      a)    $$p(E) = \frac{n(E)}{n(S)} = \frac{26}{52} = \frac{1}{2}$$

      b)    $o(E) = n(E){:}n(E') = 26{:}26 = 1{:}1$

      c)    The probability means that one out of every two possible outcomes is a success (that is, a black card). The odds mean that there is one possible success for every one possible failure.

21.    $E$ = drawing a queen of spades
       $n(S) = 52, n(E) = 1, n(E') = n(S) - n(E) = 52 - 1 = 51$

       a)    $p(E) = \dfrac{n(E)}{n(S)} = \dfrac{1}{52}$

       b)    $o(E) = n(E):n(E') = 1:51$

       c)    The probability means that there is one possible success for every 52 possible outcomes.
             The odds mean that there is one possible success for every 51 possible failures.

25.    $E$ = drawing a card above a 4 = {5, 6, 7, 8, 9, 10, J, Q, K, A}
       $E' = \{2, 3, 4\}$
       $n(S) = 52, n(E) = 10(4 \text{ suits}) = 40, n(E') = 3(4 \text{ suits}) = 12$

       a)    $p(E) = \dfrac{n(E)}{n(S)} = \dfrac{40}{52} = \dfrac{10}{13}$

       b)    $o(E) = n(E):n(E') = 40:12 = 10:3$

       c)    The probability means that there are ten possible successes for every 13 possible outcomes.
             The odds mean that there are ten possible successes for every 3 possible failures.

29.    $E$ = ball lands on a single number
       $n(S) = 38, n(E) = 1, n(E') = n(S) - n(E) = 38 - 1 = 37$

       a)    $p(E) = \dfrac{n(E)}{n(S)} = \dfrac{1}{38}$

       b)    $o(E) = n(E):n(E') = 1:37$

       c)    The probability means that there is one possible success for every 38 possible outcomes.
             The odds means that there is one possible success for every 37 possible failures.

33.    $E$ = ball lands on one of five numbers
       $n(S) = 38, n(E) = 5, n(E') = n(S) - n(E) = 38 - 5 = 33$

       a)    $p(E) = \dfrac{n(E)}{n(S)} = \dfrac{5}{38}$

       b)    $o(E) = n(E):n(E') = 5:33$

       c)    The probability means that there are 5 possible successes for every 38 possible outcomes.
             This means that for every five possible successes there are 33 possible failures.

37.   $E$ = ball lands on an even number
      $n(S) = 38$, $n(E) = 18$, $n(E') = n(S) - n(E) = 38 - 18 = 20$

   a)   $p(E) = \dfrac{n(E)}{n(S)} = \dfrac{18}{38} = \dfrac{9}{19}$

   b)   $o(E) = n(E){:}n(E') = 18{:}20 = 9{:}10$

   c)   The probability means that there are 9 possible successes for every 19 possible outcomes.
        This means that for every 9 possible successes there are 10 possible failures.

41.   $o(E) = n(E){:}n(E') = 3{:}2$, so $n(E) = 3$, $n(E') = 2$.
      $n(S) = n(E) + n(E') = 3 + 2 = 5$.

      $p(E) = \dfrac{n(E)}{n(S)} = \dfrac{3}{5}$

45.   $E$ = ball lands on an odd number
      $n(S) = 38$, $n(E) = 18$, $n(E') = n(S) - n(E) = 38 - 18 = 20$

   a)   $p(E) = \dfrac{n(E)}{n(S)} = \dfrac{18}{38} = \dfrac{9}{19} = \dfrac{a}{b}$

   b)   $o(E) = a{:}(b - a) = 9{:}(19 - 9) = 9{:}10$

   c)   The probability means that there are $a = 9$ possible successes for every $b = 19$ possible
        outcomes. The odds mean that for every $a = 9$ possible successes there are $(b - a) = 10$
        possible failures.

49.   a)   $S = \{(b, b, b), (b, b, g), (b, g, b), (b, g, g), (g, b, b), (g, b, g), (g, g, b), (g, g, g)\}$

   b)   $E = \{(b, g, g), (g, b, g), (g, g, b)\}$

   c)   $F = \{(b, g, g), (g, b, g), (g, g, b), (g, g, g)\}$

   d)   $G = \{(g, g, g)\}$

      For e) through j): $n(S) = 8$, $n(E) = 3$, $n(F) = 4$, $n(G) = 1$

   e)   $p(E) = \dfrac{n(E)}{n(S)} = \dfrac{3}{8}$

   f)   $p(F) = \dfrac{n(F)}{n(S)} = \dfrac{4}{8} = \dfrac{1}{2}$

   g)   $p(G) = \dfrac{n(G)}{n(S)} = \dfrac{1}{8}$

   h)   $o(E) = n(E){:}n(E') = n(E){:}[n(S) - n(E)] = 3{:}(8 - 3) = 3{:}5$

   i)   $o(F) = n(F){:}n(F') = n(F){:}[n(S) - n(F)] = 4{:}(8 - 4) = 4{:}4 = 1{:}1$

   j)   $o(G) = n(G){:}n(G') = n(G){:}[n(S) - n(G)] = 1{:}(8 - 1) = 1{:}7$

53. The couple would be more likely to have children of different sexes because there are only 2 ways of having the same sex, {b, b, b} or {g, g, g}, while there are six ways of having three children of opposite sexes. (See Exercise 49.)

57. Pink snapdragons have one red gene (r) and one white gene (w). The Punnett square would look like the following:

|   | r | w |
|---|---|---|
| r | rr | wr |
| w | rw | ww |

← first parent's genes
← offspring
← offspring

↑
second parent's genes

a)  $R = \{rr\}$  (red offspring)

$$p(R) = \frac{n(R)}{n(S)} = \frac{1}{4}$$

b)  $W = \{ww\}$  (white offspring)

$$p(W) = \frac{n(W)}{n(S)} = \frac{1}{4}$$

c)  $P = \{rw, wr\}$ (pink offspring)

$$p(P) = \frac{n(P)}{n(S)} = \frac{2}{4} = \frac{1}{2}$$

61. One parent is a carrier of Tay-Sachs disease and one parent is not. Let T denote the disease-free gene and t denote the recessive Tay-Sachs disease gene. The Punnett square would look like the following:

|   | T | t |
|---|---|---|
| T | TT | tT |
| T | TT | tT |

← carrier parent's genes
← offspring
← offspring

↑
non-carrier parent's genes

a)  $D = \{tt\} = \varnothing$  (child has the disease)

$$p(D) = \frac{n(D)}{n(S)} = \frac{0}{4} = 0$$

**61.** Continued.

b)   $C = \{tT, tT\}$   (child is a carrier)

$$p(C) = \frac{n(C)}{n(S)} = \frac{2}{4} = \frac{1}{2}$$

c)   $H = \{TT, TT, tT, tT\}$   (child is healthy---includes carriers)

$$p(H) = \frac{n(H)}{n(S)} = \frac{4}{4} = 1$$

**65.** Omitted.

---

## Section 3.3                                    Basic Rules of Probability

1.   Events $E$ and $F$ **are not** mutually exclusive since a person could be a woman and a doctor.

5.   Events $E$ and $F$ **are not** mutually exclusive since a person could have both brown hair and gray hair mixed together or a person could have gray hair dyed brown.

9.   Events $E$ and $F$ **are** mutually exclusive since a "four" is not an "odd number."

13.   $A$ = card is a ten, $B$ = card is a spade
$n(S) = 52$, $n(A) = 4$, $n(B) = 13$, $n(A \cap B) = 1$

a)   a ten and a spade $(A \cap B)$:

$$p(A \cap B) = \frac{n(A \cap B)}{n(S)} = \frac{1}{52}$$

b)   a ten or a spade $(A \cup B)$:

$$p(A \cup B) = p(A) + p(B) - p(A \cap B) = \frac{n(A)}{n(S)} + \frac{n(B)}{n(S)} - \frac{n(A \cap B)}{n(S)} = \frac{4}{52} + \frac{13}{52} - \frac{1}{52} = \frac{16}{52} = \frac{4}{13}$$

c)   not a ten of spades $((A \cap B)')$:

$$p((A \cap B)') = 1 - p(A \cap B) = 1 - \frac{1}{52} = \frac{52}{52} - \frac{1}{52} = \frac{51}{52}$$

17.    $A$ = card is above a five
       $A'$ = card is five or below = $\{2, 3, 4, 5\}$
       $B$ = card is below a ten
       $B'$ = card is ten or above = $\{10, J, Q, K, A\}$

a)      $n(A') = (4 \text{ suits})(4 \text{ cards}) = 16$, $n(S) = 52$

$$p(A) = 1 - p(A') = 1 - \frac{n(A')}{n(S)} = 1 - \frac{16}{52} = \frac{36}{52} = \frac{9}{13}$$

b)      $n(B') = (4 \text{ suits})(5 \text{ cards}) = 20$

$$p(B) = 1 - p(B') = 1 - \frac{n(B')}{n(S)} = 1 - \frac{20}{52} = \frac{32}{52} = \frac{8}{13}$$

c)      $(A \cap B)$ = card is above a five and below a ten = $\{6, 7, 8, 9\}$
        $n(A \cap B) = (4 \text{ suits})(4 \text{ cards}) = 16$

$$p(A \cap B) = \frac{n(A \cap B)}{n(S)} = \frac{16}{52} = \frac{4}{13}$$

d)      $(A \cup B)$ = card is above a five or below a ten = card is any card = $S$

$$p(A \cup B) = p(A) + p(B) - p(A \cap B) = \frac{9}{13} + \frac{8}{13} - \frac{4}{13} = \frac{13}{13} = 1$$

        or, $p(A \cup B) = p(S) = 1$

21.    $F$ = card is a face card = $\{J, Q, K\}$
       $n(S) = 52$, $n(F) = (4 \text{ suits})(3 \text{ cards}) = 12$

$$p(F') = 1 - p(F) = 1 - \frac{n(F)}{n(S)} = 1 - \frac{12}{52} = \frac{40}{52} = \frac{10}{13}$$

25.    $F$ = card is a jack or higher = $\{J, Q, K, A\}$
       $n(S) = 52$, $n(F) = (4 \text{ suits})(4 \text{ cards}) = 16$

$$p(F') = 1 - p(F) = 1 - \frac{n(F)}{n(S)} = 1 - \frac{16}{52} = \frac{36}{52} = \frac{9}{13}$$

29.    If $p(E) = \frac{2}{7} = \frac{n(E)}{n(S)}$, then $n(E) = 2$ and $n(S) = 7$.

       $n(E') = n(S) - n(E) = 7 - 2 = 5$

       $o(E) = n(E):n(E') = 2:5$

       $o(E') = n(E'):n(E) = 5:2$

33.    From Exercise 32, if $p(E) = \dfrac{a}{b}$, then $o(E') = (b-a):a$.

$K$ = card is a king

$n(S) = 52$, $n(K) = (4 \text{ suits})(1 \text{ card}) = 4$

$$p(K) = \frac{n(K)}{n(S)} = \frac{4}{52} = \frac{1}{13} = \frac{a}{b}$$

$o(K') = (b-a):a = (13-1):1 = 12:1$

37.    From Exercise 32, if $p(E) = \dfrac{a}{b}$, then $o(E') = (b-a):a$.

$F$ = card is a four or below = $\{2, 3, 4\}$

$n(S) = 52$, $n(F) = (4 \text{ suits})(3 \text{ cards}) = 12$

$$p(F) = \frac{n(F)}{n(S)} = \frac{12}{52} = \frac{3}{13} = \frac{a}{b}$$

$o(F') = (b-a):a = (13-3):3 = 10:3$

41.    $S$ = shopper who was polled
$A$ = shopper made a purchase and was happy
$B$ = shopper made a purchase and was not happy
$C$ = shopper did not make a purchase and was happy
$D$ = shopper did not make a purchase and was not happy

$n(S) = 700$, $n(A) = 151$, $n(B) = 133$, $n(C) = 201$, $n(D) = 215$

a)     $p(A) = \dfrac{n(A)}{n(S)} = \dfrac{151}{700}$

b)     $A \cup B \cup C$ = shopper made a purchase or was happy

$$p(A \cup B \cup C) = p(A) + p(B) + p(C) = \frac{n(A)}{n(S)} + \frac{n(B)}{n(S)} + \frac{n(C)}{n(S)}$$

$$= \frac{151}{700} + \frac{133}{700} + \frac{201}{700} = \frac{485}{700} = \frac{97}{140}$$

45.    $S$ = customer who was polled
$E$ = size of bill is between \$40.00 and \$79.99
$n(S) = 1{,}000$, $n(E) = 183 + 177 = 360$

a)     $p(E) = \dfrac{n(E)}{n(S)} = \dfrac{360}{1000} = \dfrac{9}{25}$

b)     $p(E') = 1 - p(E) = 1 - \dfrac{n(E)}{n(S)} = 1 - \dfrac{9}{25} = \dfrac{16}{25}$

49.    $D$ = rolling doubles = $\{(1, 1), (2, 2), (3, 3), (4, 4), (5, 5), (6, 6)\}$
       $E$ = rolling a 7 = $\{(1, 6), (2, 5), (3, 4), (4, 3), (5, 2), (6, 1)\}$
       $F$ = rolling an 11 = $\{(5, 6), (6, 5)\}$
       $n(S) = 36$, $n(D) = 6$, $n(E) = 6$, $n(F) = 2$

   a)    rolling a 7 or 11, use Rule 5 ($E$ and $F$ are mutually exclusive):

$$p(E \cup F) = p(E) + p(F) = \frac{n(E)}{n(S)} + \frac{n(F)}{n(S)} = \frac{6}{36} + \frac{2}{36} = \frac{8}{36} = \frac{2}{9}$$

   b)    rolling a 7 or 11 or doubles, use Rule 5 ($D$, $E$ and $F$ are mutually exclusive):

$$p(D \cup E \cup F) = p(D) + p(E) + p(F) = \frac{n(D)}{n(S)} + \frac{n(E)}{n(S)} + \frac{n(F)}{n(S)}$$

$$= \frac{6}{36} + \frac{6}{36} + \frac{2}{36} = \frac{14}{36} = \frac{7}{18}$$

53.    $D$ = rolling doubles = $\{(1, 1), (2, 2), (3, 3), (4, 4), (5, 5), (6, 6)\}$
       $E$ = rolling even sums = $\{(1, 1), (1, 3), (2, 2), (3, 1), ..., (6, 6)\}$

       $n(S) = 36$, $n(D) = 6$, $n(E) = 18$ (see figure 3.6)

   a)    rolling even and doubles, $(E \cap D) = D$ (All doubles have even sums):

$$p(E \cap D) = \frac{n(E \cap D)}{n(S)} = \frac{n(D)}{n(S)} = \frac{6}{36} = \frac{1}{6}$$

   b)    rolling even or doubles, $(E \cup D)$:

$$p(E \cup D) = p(E) + p(D) - p(E \cap D) = p(E) + p(D) - p(D) = p(E) = \frac{n(E)}{n(S)} = \frac{18}{36} = \frac{1}{2}$$

57.    Let T denote the disease-free gene and t denote the recessive Tay-Sachs disease gene.  Create the Punnett squares as indicated.

       $D$ = $\{tt\}$  (child has the disease)
       $C$ = $\{tT, Tt\}$  (child is a carrier)
       $D \cup C$ = $\{tt, tT, Tt\}$  (child either has the disease or is a carrier)

   a)    Each parent is a Tay-Sachs carrier.

|   | T | t |   |
|---|---|---|---|
|   | **T** | **t** | ← carrier parent's genes |
| **T** | TT | tT | ← offspring |
| **t** | Tt | tt | ← offspring |

↑
carrier parent's genes

57. a). Continued.

$$n(S) = 4, n(D) = 1, n(C) = 2$$

$$p((D \cup C)') = 1 - p(D \cup C) = 1 - (p(D) + p(C)) = 1 - \left(\frac{1}{4} + \frac{2}{4}\right) = 1 - \frac{3}{4} = \frac{1}{4}$$

b) One parent is a Tay-Sachs carrier and the other parent has no Tay-Sachs gene.

| | **T** | **t** | |
|---|---|---|---|
| **T** | TT | tT | ← offspring |
| **T** | TT | tT | ← offspring |

← carrier parent's genes

↑

no Tay-Sachs gene

$$n(S) = 4, n(D) = 0, n(C) = 2$$

$$p((D \cup C)') = 1 - p(D \cup C) = 1 - (p(D) + p(C)) = 1 - \left(\frac{0}{4} + \frac{2}{4}\right) = 1 - \frac{2}{4} = \frac{2}{4} = \frac{1}{2}$$

c) One parent has Tay-Sachs and the other parent has no Tay-Sachs gene.

| | **t** | **t** | |
|---|---|---|---|
| **T** | tT | tT | ← offspring |
| **T** | tT | tT | ← offspring |

← Tay-Sachs parent's genes

↑

no Tay-Sachs gene

$$n(S) = 4, n(D) = 0, n(C) = 4$$

$$p((D \cup C)') = 1 - p(D \cup C) = 1 - [p(D) + p(C)] = 1 - \left(\frac{0}{4} + \frac{4}{4}\right) = 1 - 1 = 0$$

61. $S$ = flashlights in a large shipment
$A$ = flashlight has a defective bulb
$B$ = flashlight has a defective battery
$(A \cap B)$ = flashlight has both defects

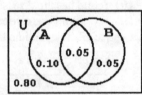

$$p(S) = 1, p(A) = 0.15, p(B) = 0.10, p(A \cap B) = 0.05$$

a) $p(A \cup B) = p(A) + p(B) - p(A \cap B) = 0.15 + 0.10 - 0.05 = 0.20$

b) $p(A' \cup B') = p((A \cap B)') = 1 - p(A \cap B) = 1 - 0.05 = 0.95$

c) $p(A' \cap B') = p((A \cup B)') = 1 - p(A \cup B) = 1 - 0.20 = 0.80$

65.   $p(E)+p(E')=\dfrac{n(E)}{n(S)}+\dfrac{n(E')}{n(S)}=\dfrac{n(E)}{n(S)}+\dfrac{n(S)-n(E)}{n(S)}=\dfrac{n(E)+n(S)-n(E)}{n(S)}=\dfrac{n(S)}{n(S)}=1$

69.   a)     $\dfrac{18}{33}=\dfrac{3\cdot6}{3\cdot11}=\dfrac{6}{11}$

      b)     Check these results on your calculator.

73.   a)     $\dfrac{6}{15}+\dfrac{10}{21}=\dfrac{6}{3\cdot5}\left(\dfrac{7}{7}\right)+\dfrac{10}{3\cdot7}\left(\dfrac{5}{5}\right)=\dfrac{42}{105}+\dfrac{50}{105}=\dfrac{92}{105}$

      b)     Check these results on your calculator.

---

| **Section 3.4** | **Combinatorics and Probability** |
|---|---|

1.    $S$ = all possible lists of 30 birthdays
      $E$ = lists of 30 birthdays in which at least 2 of those birthdays are the same
      $E'$ = lists of 30 birthdays in which no two birthdays are the same

      The birthdays may not be repeated so we use **permutations**.

      $_nP_r=\dfrac{n!}{(n-r)!}$ , permutation where $n=365$, $r=30$

      $n(E')=_{365}P_{30}=\dfrac{365!}{(365-30)!}=\dfrac{365!}{335!}=\dfrac{365\cdot364\cdot363\cdot\ldots\cdot336\cdot335!}{335!}=2.1710302\times10^{76}$

      $n(S)=365\cdot365\cdot\ldots\cdot365=365^{30}$

      $p(E)=1-p(E')=1-\dfrac{n(E')}{n(S)}=1-\dfrac{2.1710302\times10^{76}}{365^{30}}=0.706316...\approx0.7063$

5.    Winning second prize requires 2 categories:
           5 of the 6 winning numbers ($n=6$, $r=5$), and
           one that is not a winning number ($n=38$, $r=1$).

      $_6C_5=\dfrac{6!}{5!\cdot(6-5)!}=\dfrac{6!}{5!\cdot1!}=\dfrac{6}{1}=6$

      $_{38}C_1=\dfrac{38!}{1!\cdot(38-1)!}=\dfrac{38!}{1!\cdot37!}=\dfrac{38}{1}=38$

      $p\left(\begin{array}{c}\text{second prize in}\\\text{a 6/44 lottery}\end{array}\right)=\dfrac{n(E)}{n(S)}=\dfrac{_6C_5\cdot_{38}C_1}{_{44}C_6}=\dfrac{6\cdot38}{7,059,052}=\dfrac{228}{7,059,052}\approx\dfrac{1}{30,961}\approx0.00003$

      or, approximately 1 in 31 thousand

9.    a)    The order is **not** important so use **combinations** where $n = 39$, $r = 5$

$$_{39}C_5 = \frac{39!}{5! \cdot (39-5)!} = \frac{39!}{5! \cdot 34!} = \frac{39 \cdot 38 \cdot 37 \cdot 36 \cdot 35}{5 \cdot 4 \cdot 3 \cdot 2 \cdot 1} = 575,757$$

$E$ = winning first prize
$n(E) = 1$, $n(S) = {}_{39}C_5 = 575,757$

$$p(\text{winning 5/39 lottery}) = \frac{n(E)}{n(S)} = \frac{1}{575,757} \approx 0.000001737$$

or, approximately 1 in 600 thousand.

b)    Winning second prize requires 2 categories:
        4 of the 5 winning numbers ($n = 5$, $r = 4$), and
        one that is not a winning number ($n = 34$, $r = 1$).

$$_5C_4 = \frac{5!}{4! \cdot (5-4)!} = \frac{5!}{4! \cdot 1!} = \frac{5}{1} = 5$$

$$_{34}C_1 = \frac{34!}{1! \cdot (34-1)!} = \frac{34!}{1! \cdot 33!} = \frac{34}{1} = 34$$

$$p\left(\begin{array}{c}\text{second prize in}\\ \text{a 5/39 lottery}\end{array}\right) = \frac{n(E)}{n(S)} = \frac{_5C_4 \cdot {}_{34}C_1}{_{39}C_5} = \frac{5 \cdot 34}{575,757} = \frac{170}{575,757} \approx \frac{1}{3387} \approx 0.000295$$

or, approximately 1 in 3 thousand

13.    The sample space consists of 8 numbers from a possible 80 ($n = 80$, $r = 8$). The order is **not** important so use **combinations**.

$$_{80}C_8 = \frac{80!}{8! \cdot (80-8)!} = \frac{80!}{8! \cdot 72!} = 2.899 \times 10^{10}$$

Selecting all eight winning spots in eight-spot keno requires having 8 of a possible 20 winning numbers ($n = 20$, $r = 8$) combined with none that are not winning ($n \doteq 60$, $r = 0$). The order is **not** important so use **combinations**.

$$_{20}C_8 = \frac{20!}{8! \cdot (20-8)!} = \frac{20!}{8! \cdot 12!} = 125,970$$

$$_{60}C_0 = \frac{60!}{0! \cdot (60-0)!} = \frac{60!}{0! \cdot 60!} = 1$$

$$p(\text{8 winning spots}) = \frac{_{20}C_8 \cdot {}_{60}C_0}{_{80}C_8} = \frac{(125,970)(1)}{2.899 \times 10^{10}} \approx 0.000004$$

Similar calculations are necessary for 7, 6, 5, and 4 winning spots.

13. Continued.

Seven of a possible 20 winning numbers ($n = 20$, $r = 7$) combined with 1 that is not winning ($n = 60$, $r = 1$).

$$p(7 \text{ winning spots}) = \frac{_{20}C_7 \cdot _{60}C_1}{_{80}C_8} = \frac{(77,520)(60)}{2.899 \times 10^{10}} \approx 0.000160$$

Six of a possible 20 winning numbers ($n = 20$, $r = 6$) combined with 2 that are not winning ($n = 60$, $r = 2$).

$$p(6 \text{ winning spots}) = \frac{_{20}C_6 \cdot _{60}C_2}{_{80}C_8} = \frac{(38,760)(1770)}{2.899 \times 10^{10}} \approx 0.002367$$

Five of a possible 20 winning numbers ($n = 20$, $r = 5$) combined with 3 that are not winning ($n = 60$, $r = 3$).

$$p(5 \text{ winning spots}) = \frac{_{20}C_5 \cdot _{60}C_3}{_{80}C_8} = \frac{(15,504)(34,220)}{2.899 \times 10^{10}} \approx 0.018303$$

Four of a possible 20 winning numbers ($n = 20$, $r = 4$) combined with 4 that are not winning ($n = 60$, $r = 4$).

$$p(4 \text{ winning spots}) = \frac{_{20}C_4 \cdot _{60}C_4}{_{80}C_8} = \frac{(4845)(487,635)}{2.899 \times 10^{10}} \approx 0.081504$$

Selecting less than four winning spots in eight-spot keno requires having 1, 2, or 3 of a possible 20 winning numbers. Or, using the Laws of Probabilities, add the above probabilities and subtract from one.

$$p(\text{less than 4 winning spots}) = 1 - [p(8 \text{ winning}) + p(7 \text{ winning}) + \ldots + p(4 \text{ winning})]$$
$$= 1 - (0.000004 + 0.000160 + 0.002367 + 0.018303 + 0.081504)$$
$$= 1 - 0.102338$$
$$= 0.897662$$

| Outcome | Probability |
|---|---|
| 8 winning spots | 0.000004 |
| 7 winning spots | 0.000160 |
| 6 winning spots | 0.002367 |
| 5 winning spots | 0.018303 |
| 4 winning spots | 0.081504 |
| fewer than 4 winning spots | 0.897662 |

17.    12 burritos to go, 5 with hot peppers, 7 without hot peppers.  Pick three burritos at random.
$E$ = all have hot peppers, ($n = 5$, $r = 3$) implies none without hot peppers ($n = 7$, $r = 0$)

$$_5C_3 = \frac{5!}{3! \cdot (5-3)!} = \frac{5!}{3! \cdot 2!} = \frac{5 \cdot 4}{2 \cdot 1} = 10$$

$$_7C_0 = \frac{7!}{0! \cdot (7-0)!} = \frac{7!}{7!} = 1$$

$$_{12}C_3 = \frac{12!}{3! \cdot (12-3)!} = \frac{12!}{3! \cdot 9!} = \frac{12 \cdot 11 \cdot 10}{3 \cdot 2 \cdot 1} = \frac{1320}{6} = 220$$

$$p(E) = \frac{_5C_3 \cdot {_7C_0}}{_{12}C_3} = \frac{(10)(1)}{220} = 0.04545\ldots \approx 0.05$$

21.    12 burritos to go, 5 with hot peppers, 7 without hot peppers.  Pick three burritos at random.  At
most one with hot peppers implies either exactly one with hot peppers or none with hot peppers.
$E$ = exactly one with hot peppers ($n = 5$, $r = 1$) implies two without hot peppers ($n = 7$, $r = 2$)

$$_5C_1 = \frac{5!}{1! \cdot (5-1)!} = \frac{5!}{1! \cdot 4!} = \frac{5}{1} = 5$$

$$_7C_2 = \frac{7!}{2! \cdot (7-2)!} = \frac{7!}{2! \cdot 5!} = \frac{7 \cdot 6}{2 \cdot 1} = 21$$

$$_{12}C_3 = \frac{12!}{3! \cdot (12-3)!} = \frac{12!}{3! \cdot 9!} = \frac{12 \cdot 11 \cdot 10}{3 \cdot 2 \cdot 1} = \frac{1320}{6} = 220$$

$$p(E) = \frac{_5C_1 \cdot {_7C_2}}{_{12}C_3} = \frac{(5)(21)}{220} = \frac{105}{220}$$

$F$ = none with hot peppers ($n = 5$, $r = 0$) implies three without hot peppers ($n = 7$, $r = 3$)

$$_5C_0 = \frac{5!}{0! \cdot (5-0)!} = \frac{5!}{0! \cdot 5!} = 1$$

$$_7C_3 = \frac{7!}{3! \cdot (7-3)!} = \frac{7!}{3! \cdot 4!} = \frac{7 \cdot 6 \cdot 5}{3 \cdot 2 \cdot 1} = 35$$

$$_{12}C_3 = \frac{12!}{3! \cdot (12-3)!} = \frac{12!}{3! \cdot 9!} = \frac{12 \cdot 11 \cdot 10}{3 \cdot 2 \cdot 1} = \frac{1320}{6} = 220$$

$$p(F) = \frac{_5C_3 \cdot {_7C_0}}{_{12}C_3} = \frac{(1)(35)}{220} = \frac{35}{220}$$

$$p\binom{\text{at most one}}{\text{hot pepper}} = p(E) + p(F) = \frac{105}{220} + \frac{35}{220} = \frac{140}{220} \approx 0.64$$

25.    The sample space consists of 2 applicants selected from a possible 200 ($n = 200$, $r = 2$).  The order
is **not** important so use **combinations**.

$$_{200}C_2 = \frac{200!}{2! \cdot (200-2)!} = \frac{200!}{2! \cdot 198!} = \frac{200 \cdot 199}{2 \cdot 1} = \frac{39,800}{2} = 19,900$$

a)    $E$ = exactly 2 women, ($n = 60$, $r = 2$) implies no men ($n = 140$, $r = 0$)

$$_{60}C_2 = \frac{60!}{2! \cdot (60-2)!} = \frac{60!}{2! \cdot 58!} = \frac{60 \cdot 59}{2 \cdot 1} = 1770$$

$$_{140}C_0 = \frac{140!}{0! \cdot (140-0)!} = \frac{140!}{0! \cdot 140!} = \frac{1}{1} = 1$$

$$p(E) = \frac{_{60}C_2 \cdot _{140}C_0}{_{200}C_2} = \frac{(1770)(1)}{19,900} = 0.0889... \approx 0.09$$

b)    $E$ = exactly 1 woman, ($n = 60$, $r = 1$) implies one man ($n = 140$, $r = 1$)

$$_{60}C_1 = \frac{60!}{1! \cdot (60-1)!} = \frac{60!}{1! \cdot 59!} = \frac{60}{1} = 60$$

$$_{140}C_1 = \frac{140!}{1! \cdot (140-1)!} = \frac{140!}{1! \cdot 139!} = \frac{140}{1} = 140$$

$$p(E) = \frac{_{60}C_1 \cdot _{140}C_1}{_{200}C_2} = \frac{(60)(140)}{19,900} = 0.4221... \approx 0.42$$

c)    $E$ = exactly 2 men, ($n = 140$, $r = 2$) implies no women ($n = 60$, $r = 0$)

$$_{140}C_2 = \frac{140!}{2! \cdot (140-2)!} = \frac{140!}{2! \cdot 138!} = \frac{140 \cdot 139}{2 \cdot 1} = 9730$$

$$_{60}C_0 = \frac{60!}{0! \cdot (60-0)!} = \frac{60!}{0! \cdot 60!} = \frac{1}{1} = 1$$

$$p(E) = \frac{_{140}C_2 \cdot _{60}C_0}{_{200}C_2} = \frac{(9730)(1)}{19,900} = 0.4889... \approx 0.49$$

d)    Omitted.

## Section 3.5                                                        Expected Value

1.   a)   $E$ = ball lands on one of two numbers.  House odds are 17 to 1.  (See Figure 3.1.)

$$p(\text{win } E) = \frac{2}{38} = \frac{1}{19}, p(\text{lose } E) = 1 - \frac{1}{19} = \frac{18}{19}$$

expected value = $17 \cdot p(\text{win } E) + (-\$1) \cdot p(\text{lose } E)$

$$= 17 \cdot \frac{1}{19} + (-1) \cdot \frac{18}{19} = \frac{17}{19} - \frac{18}{19} = -\frac{1}{19} \approx -\$0.053$$

     b)   On average, $0.053 would be lost on a $1 two-number bet.

5.   a)   $E$ = ball lands on one of six numbers.  House odds are 5 to 1.  (See Figure 3.1.)

$$p(\text{win } E) = \frac{6}{38} = \frac{3}{19}, p(\text{lose } E) = 1 - \frac{3}{19} = \frac{16}{19}$$

expected value = $5 \cdot p(\text{win } E) + (-\$1) \cdot p(\text{lose } E)$

$$= 5 \cdot \frac{3}{19} + (-1) \cdot \frac{16}{19} = \frac{15}{19} - \frac{16}{19} = \frac{-1}{19} \approx -\$0.053$$

     b)   On average, $0.053 would be lost on a $1 six-number bet.

9.   a)   $E$ = ball lands on one of the red numbers.  House odds are 1 to 1.  (See Figure 3.1.)

$$p(\text{win } E) = \frac{18}{38} = \frac{9}{19}, p(\text{lose } E) = 1 - \frac{9}{19} = \frac{10}{19}$$

expected value = $1 \cdot p(\text{win } E) + (-\$1) \cdot p(\text{lose } E)$

$$= 1\left(\frac{9}{19}\right) + (-1)\frac{10}{19} = \frac{9}{19} + \frac{-10}{19} = \frac{-1}{19} \approx -\$0.053$$

     b)   On average, $0.053 would be lost on a $1 red-number bet.

13.   Multiply each possible number of books by its probability to calculate the expected number of books checked out.

expected number = $0(0.15) + 1(0.35) + 2(0.25) + 3(0.15) + 4(0.05) + 5(0.05)$

$$= 0 + 0.35 + 0.50 + 0.45 + 0.20 + 0.25$$

$$= 1.75 \text{ books checked out of the library.}$$

17.   $35\left(\dfrac{1}{38}\right) + (-1)\left(\dfrac{37}{38}\right) = \dfrac{35 + (-1)(37)}{38} = \dfrac{35 - 37}{38} = \dfrac{-2}{38} \approx -0.053$

21.  The pharmaceuticals stock would be the better choice when the expected value of the stock is greater than the expected value of the bank account which is 0.045. Let $p$ = the probability of success for the pharmaceutical stocks, then $(1 - p)$ = the probability of failure for the stocks.

expected value of stocks = $p(0.5) - (1 - p)(0.6)$
expected value of bank account = 0.045

expected value of stocks > expected value of bank account

$$(0.5)p - (0.6)(1 - p) > 0.045$$
$$0.5p - 0.6 + 0.6p > 0.045$$
$$-0.6 + 1.1p > 0.045$$
$$1.1p > 0.045 + 0.6 = 0.645$$
$$p > \frac{0.645}{1.1} = 0.586364 \approx 0.59$$

When the probability of success for the pharmaceutical stocks is 0.59 or more it would be the better choice for Erica.

25.  Multiply the probability of each winning bet by the expected profit to calculate the expected value of the bet.

| Number of winning spots | Calculation | Probability | Profits |
|---|---|---|---|
| 4 | $\dfrac{_{20}C_4 \cdot {_{60}}C_4}{_{80}C_8}$ | 0.081504 | $0 |
| 5 | $\dfrac{_{20}C_5 \cdot {_{60}}C_3}{_{80}C_8}$ | 0.018303 | $4 |
| 6 | $\dfrac{_{20}C_6 \cdot {_{60}}C_2}{_{80}C_8}$ | 0.002367 | $99 |
| 7 | $\dfrac{_{20}C_7 \cdot {_{60}}C_1}{_{80}C_8}$ | 0.000160 | $1479 |
| 8 | $\dfrac{_{20}C_8 \cdot {_{60}}C_0}{_{80}C_8}$ | 0.000004 | $18,999 |

$p$(winning) = sum of probability column = $0.081504 + 0.018303 + ... + 0.000004 = 0.102338$

$p$(not winning) = $1 - p$(winning) = $1 - 0.102338 = 0.897662$ with a profit = $-$1$

expected value of bet = $0(0.081504) + 4(0.018303) + 99(0.002367) + 1479(0.000160)$
$$+ 18,999(0.000004) - 1(0.897662)$$
$$= 0.620181 - 0.897662$$
$$= -\$0.277481 \approx -\$0.28$$

29.    $p$(burning down) $= p = 0.01$
       $p$(not burning down) $= 1 - p = 1 - 0.01 = 0.99$

       expected value of loss $= 120{,}000 \cdot p + (0) \cdot (1 - p)$
       $\phantom{expected value of loss} = 120{,}000(0.01) + (0)(0.99) = \$1200$

       The annual premium should be more than $1200.

33.    Multiply the retail value of each prize less the cost of the ticket by the probability of winning.

   a)  $n(S) = 1000$, $n$(win any prize) $= 26$, $n$(not winning a prize) $= 974$.

   $$\text{expected value} = \$21{,}565\left(\frac{1}{1000}\right) + \$925\left(\frac{1}{1000}\right) + \$485\left(\frac{2}{1000}\right) + \$85\left(\frac{2}{1000}\right)$$
   $$+ \$165\left(\frac{20}{1000}\right) - \$15\left(\frac{974}{1000}\right)$$

   $$= 21.565 + 0.925 + 0.970 + 0.170 + 3.300 - 14.610 = \$12.32$$

   b)  $n(S) = 2000$, $n$(win any prize) $= 26$, $n$(not winning a prize) $= 1974$.

   $$\text{expected value} = \$21{,}565\left(\frac{1}{2000}\right) + \$925\left(\frac{1}{2000}\right) + \$485\left(\frac{2}{2000}\right) + \$85\left(\frac{2}{2000}\right)$$
   $$+ \$165\left(\frac{20}{2000}\right) - \$15\left(\frac{1974}{2000}\right)$$

   $$= 10.7825 + 0.4625 + 0.4850 + 0.0850 + 1.6500 - 14.8050 = -\$1.34$$

   c)  $n(S) = 3000$, $n$(win any prize) $= 26$, $n$(not winning a prize) $= 2974$.

   $$\text{expected value} = \$21{,}565\left(\frac{1}{3000}\right) + \$925\left(\frac{1}{3000}\right) + \$485\left(\frac{2}{3000}\right) + \$85\left(\frac{2}{3000}\right)$$
   $$+ \$165\left(\frac{20}{3000}\right) - \$15\left(\frac{2974}{3000}\right)$$

   $$= 7.1883 + 0.3083 + 0.3233 + 0.0567 + 1.1000 - 14.8700 \approx -\$5.89$$

37.    $\text{management expenses} = \dfrac{\$13,000,000}{2} = \$6,500,000$

cost of strategy = cost of tickets purchases + management expenses
        = $5,000,000 + $6,500,000 = $11,500,000

winnings = jackpot − (cost) = $27,000,000 − $11,500,000 = $15,500,000

$$p(\text{win}) = \frac{n(\text{tickets purchased})}{n(\text{possible combinations})} = \frac{5,000,000}{7,000,000} = \frac{5}{7}$$

$$p(\text{lose}) = \frac{n(\text{tickets not purchased})}{n(\text{possible combinations})} = \frac{2,000,000}{7,000,000} = \frac{2}{7}$$

expected value = winnings • p(win) − cost • p(lose)

$$= \$15,500,000\left(\frac{5}{7}\right) - \$11,500,000\left(\frac{2}{7}\right) \approx \$11,071429 - 3,285,714 = \$7,785,714$$

41.    Complete a table as follows:

| Bet Number | Bet | Result | Winnings/ Losses | Total Winnings/ Losses | Cash Left |
|---|---|---|---|---|---|
| 1 | $1 | lose | −$1 | −$1 | $99 |
| 2 | $2 | lose | −$2 | −$3 | $97 |
| 3 | $4 | lose | −$4 | −$7 | $93 |
| 4 | $8 | lose | −$8 | −$15 | $85 |
| 5 | $16 | lose | −$16 | −$31 | $69 |
| 6 | $32 | lose or win | −$32 or +$32 | −$63 or +$1 | $37 or $101 |

The gambler could afford six successive losses before there would not be enough money left to place another bet. If the gambler lost each bet except the last one, the winnings would be $1.

## Section 3.6                    Conditional Probability

1.    a)    $p(N) = p(\text{response was no}) = \dfrac{n(\text{response was no})}{n(\text{total responses})} = \dfrac{n(N)}{n(S)} = \dfrac{140}{600} \approx 0.23 = 23\%$

This means that of the people responding to the survey 23% felt there was not too much violence on television.

**1. Continued**

b)  $p(W) = p(\text{response from women}) = \dfrac{n(\text{response from women})}{n(\text{total responses})} = \dfrac{n(W)}{n(S)} = \dfrac{320}{600}$

$$\approx 0.53 = 53\%$$

This means that 53% of the people responding to the survey were women.

c)  $p(N|W) = p(\text{response was no, given by a woman}) = \dfrac{n(N \cap W)}{n(W)} = \dfrac{45}{320} \approx 0.14 = 14\%$

This means that 14% of the women responding to the survey felt there was not too much violence on television.

d)  $p(W|N) = p(\text{response from a woman, given it was no}) = \dfrac{n(W \cap N)}{n(N)} = \dfrac{45}{140} \approx 0.32 = 32\%$

This means that 32% of the people who responded that there was not too much violence on television were women.

e)  $p(N \cap W) = p(\text{response was no and from a woman})$

$$= \dfrac{n(N \cap W)}{n(S)} = \dfrac{45}{600} \approx = 0.08 = 8\%$$

This means that 8% of the total people who responded to the survey were people who said there was not too much violence on television and they were women.

f)  $p(W \cap N) = p(\text{response from a woman and response was no})$

$$= \dfrac{n(W \cap N)}{n(S)} = \dfrac{45}{600} = 0.075 \approx 0.08 = 8\%$$

This means that 8% of the total people who responded to the survey were women who said there was not too much violence on television.

5.  $A$ = driver had accident
$B$ = driver was 20 – 24

$n(B) = 16,900,000, \ n(A \cap B) = 5,800,000$

$$p(A \mid B) = \dfrac{n(A \cap B)}{n(B)} = \dfrac{5,800,000}{16,900,000} = 0.343195 \approx 0.34$$

9.    $A$ = card dealt first is a diamond
      $B$ = card dealt second is a spade

      $n(S) = 52,\ n(A) = 13,\ n(B) = 13$

   a)    $p(A) = \dfrac{n(A)}{n(S)} = \dfrac{13}{52} = \dfrac{1}{4} = 0.25$

   b)    $p(B \mid A) = \dfrac{n(B)}{n(\text{cards left})} = \dfrac{13}{51} = 0.255$

   c)    $p(A \cap B) = p(B \mid A) \cdot p(A) = \dfrac{13}{51} \cdot \dfrac{1}{4} = \dfrac{13}{204} \approx 0.064$

   d)
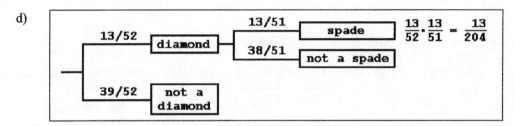

13.   a)    $p(\text{rolling a 6}) = \dfrac{n(\text{rolling a 6})}{n(S)} = \dfrac{1}{6}$

   b)    $p(\text{rolling a 6} \mid \text{even}) = \dfrac{n(\text{rolling a 6 and even})}{n(\text{even})} = \dfrac{1}{3}$

   c)    $p(\text{rolling a 6} \mid \text{odd}) = \dfrac{n(\text{rolling a 6 and odd})}{n(\text{odd})} = \dfrac{0}{3} = 0$

   d)    $p(\text{even} \mid \text{rolling a 6}) = \dfrac{n(\text{even and rolling a 6})}{n(\text{rolling a 6})} = \dfrac{1}{1} = 1$

17.   Use Figure 3.6 to calculate the cardinal numbers.

   a)    $p(\text{sum} = 4) = \dfrac{n(\text{sum} = 4)}{n(S)} = \dfrac{3}{36} = \dfrac{1}{12}$

   b)    $p(\text{sum} = 4 \mid \text{sum} < 6) = \dfrac{n(\text{sum} = 4 \text{ and sum} < 6)}{n(\text{sum} < 6)} = \dfrac{3}{10}$

   c)    $p(\text{sum} < 6 \mid \text{sum} = 4) = \dfrac{n(\text{sum} < 6 \text{ and sum} = 4)}{n(\text{sum} = 4)} = \dfrac{3}{3} = 1$

21. $S$ = shopper who was polled
    $A$ = shopper made a purchase and was happy with service
    $B$ = shopper made a purchase and was unhappy with service
    $C$ = shopper made no purchase and was happy with service
    $D$ = shopper made no purchase and was unhappy with service

    $n(S) = 700$, $n(A) = 125$, $n(B) = 111$, $n(C) = 148$, $n(D) = 316$

    Given that a shopper was happy, find the probability of making a purchase.

| | Purchase | No Purchase | Total |
|---|---|---|---|
| **Happy** | $n(A) = 125$ | $n(C) = 148$ | 273 |

$$p(\text{purchase} \mid \text{happy}) = \frac{n(\text{purchase})}{n(\text{happy})} = \frac{125}{273} = 0.4579 \approx 0.46$$

25. $A$ = the next four cards after the first card are spades
    $B$ = the first card is a spade
    After each spade is drawn there is one less for the numerator and one less for the denominator.

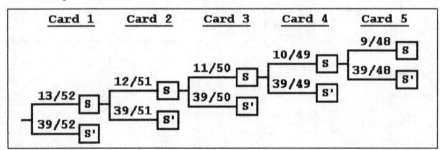

$$p(B) = \frac{13}{52}, \; p(A \cap B) = \frac{13}{52} \cdot \frac{12}{51} \cdot \frac{11}{50} \cdot \frac{10}{49} \cdot \frac{9}{48}$$

$$p(A \mid B) = \frac{p(A \cap B)}{p(B)} = \frac{\frac{13}{52} \cdot \frac{12}{51} \cdot \frac{11}{50} \cdot \frac{10}{49} \cdot \frac{9}{48}}{\frac{13}{52}} = \frac{12}{51} \cdot \frac{11}{50} \cdot \frac{10}{49} \cdot \frac{9}{48} = \frac{11,880}{5,997,600} \approx 0.0020$$

29.   To calculate probabilities reduce the numerator and denominator by one for each branch of the tree.

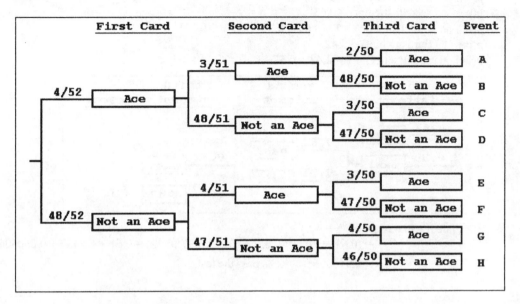

$p(\text{exactly one ace}) = p(D) + p(F) + p(G) =$

$$= \frac{4}{52} \cdot \frac{48}{51} \cdot \frac{47}{50} + \frac{48}{52} \cdot \frac{4}{51} \cdot \frac{47}{50} + \frac{48}{52} \cdot \frac{47}{51} \cdot \frac{4}{50} = 3 \cdot \frac{9024}{132,600}$$

$$= 3(0.0680542) \approx 0.20$$

33.

| Probabilities Given | Complements of These Probabilities |
|---|---|
| $p(\text{Japan}) = 0.38$ | $p(\text{America}) = p(\text{not Japan}) = 1 - 0.38 = 0.62$ |
| $p(\text{defective} \mid \text{Japan}) = 0.017$ | $p(\text{not defective} \mid \text{Japan}) = 1 - 0.017 = 0.983$ |
| $p(\text{defective} \mid \text{America}) = 0.011$ | $p(\text{not defective} \mid \text{America}) = 1 - 0.011 = 0.989$ |

$p(\text{defective and Japan}) = p(\text{defective} \mid \text{Japan}) \cdot p(\text{Japan}) = (0.017)(0.38) = .00646 \approx 0.6\%$

37.   a)   $p(\text{female} \mid \text{Democrat}) = \dfrac{n(\text{female and Democrat})}{n(\text{Democrats})} = \dfrac{10}{50} = 0.20$

   b)   $p(\text{female} \mid \text{Republican}) = \dfrac{n(\text{female and Republican})}{n(\text{Republicans})} = \dfrac{3}{49} \approx 0.6$

   c)   If a woman is a senator, it is more likely that she is a Democrat than a Republican.

41.   $p$(pass on first attempt) = 0.61
      $p$(fail on first attempt) = 1 – 0.61 = 0.39
      $p$(pass on second attempt) = 0.63
      $p$(fail on second attempt) = 1 – 0.63 = 0.37
      $p$(pass on third attempt) = 0.42
      $p$(fail on third attempt) = 1 – 0.42 = 0.58

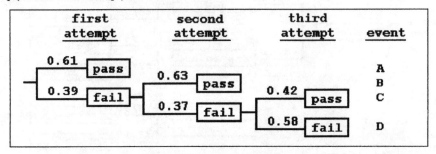

$p$(a student will pass) = $p(A) + p(B) + p(C)$
$= (0.61) + (0.39)(0.63) + (0.39)(0.37)(0.42) = 0.61 + 0.2457 + 0.060606$
$= 0.916306 \approx$ 92% pass the test

45.   $A$ = {ace}
      $B$ = {ten, jack, queen, or king}

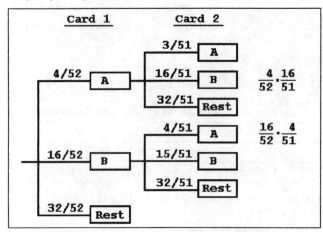

$p(B \mid A) = \dfrac{4}{52} \cdot \dfrac{16}{51} = \dfrac{64}{2652}$, $p(A \mid B) = \dfrac{16}{52} \cdot \dfrac{4}{51} = \dfrac{64}{2652}$

$p(A \cup B) = p(B \mid A) + p(A \mid B) = \dfrac{64}{2652} + \dfrac{64}{2652} = \dfrac{128}{2652} = 0.048265 \approx 0.05$

49.   The complement of $A \mid B = A' \mid B$.

## Section 3.7                          Independence; Trees in Genetics

1.    a)    $E$ and $F$ are **dependent** since knowing that a person is a woman affects the probability that a person is a doctor.  There are fewer women doctors than men doctors.  $p(E \mid F) < p(E)$.

      b)    $E$ and $F$ are **not mutually exclusive** since you could be both a woman and a doctor simultaneously.  $E \cap F \neq \varnothing$

5.    a)    Knowing a person has gray hair **does not affect** whether or not that person has brown hair as gray hair could either be dyed brown or some other color or left natural.  Therefore, the two events are **independent**.  $p(E \mid F) = p(E)$.

      b)    $E$ and $F$ are **not mutually exclusive** since a gray-haired person could dye her hair brown or a person can have both gray and brown hair mixed together.  $p(E \cap F) \neq \varnothing$.

9.    $p(E) = \dfrac{1}{6}$ , $p(F) = \dfrac{3}{6}$,

      $p(E \cap F) = 0$ since you cannot get a four and an odd number at the same time.

      $$p(E \mid F) = \frac{p(E \cap F)}{p(F)} = \frac{0}{\frac{3}{6}} = 0$$

      a)    $E$ and $F$ are **dependent** since $p(E \mid F) \neq p(E)$.

      b)    $E$ and $F$ are **mutually exclusive** since $E \cap F = \varnothing$.

      c)    The events $E$ and $F$ can be **dependent** and **mutually exclusive** at the same time.

13.   $S$ = shopper who was polled
      $A$ = being happy with the service
      $B$ = making a purchase

|             | Happy        | Not Happy | Total         |
|-------------|--------------|-----------|---------------|
| **Purchase**    | 125          | 111       | n(B) = 236    |
| **No Purchase** | 148          | 316       | 464           |
| **Total**       | n(A) = 273   | 427       | n(S) = 700    |

$$p(B) = \frac{n(B)}{n(S)} = \frac{236}{700} \approx 0.337$$

$$p(B \mid A) = \frac{n(A \cap B)}{n(A)} = \frac{125}{273} \approx 0.458$$

Since $p(B \mid A) \neq p(B)$, being happy with the service and making a purchase are **dependent** events. We can conclude that people who are happy with the service are more likely to make a purchase.

17.

| Probabilities Given | Complements of These Probabilities |
|---|---|
| $p$(buying HAL computer) = 0.60 | $p$(purchased different brand) = $p$(not HAL) = 1 − 0.60 = 0.40 |
| $p$(quit using computer) = 0.04 | $p$(still using computer) = 1 − 0.04 = 0.96 |
| $p$(quit │ used HAL computer) = 0.03 | $p$(still using computer │ used HAL computer) = 1 − 0.03 = 0.97 |

$p$(quit using computer) ≠ $p$(quit │ used HAL computer), so the events are not independent. The conclusion is that owners of HAL computers are less likely to quit using computers.

21.  a)  $p$(failure) = 0.01 for a computer **independent** of another computer. Use the product rule for independent events.

$$p(A \cap B \cap C) = p(A) \cdot p(B) \cdot p(C)$$

$$p(\text{all 3 fail}) = p(\text{failure}) \cdot p(\text{failure}) \cdot p(\text{failure})$$
$$= (0.01)(0.01)(0.01) = 0.000001 \text{ or } 1.0 \times 10^{-6}$$

b)  3 computers:

$$p(3 \text{ fail}) = (0.01)^3 = 0.000001 = \frac{1}{1,000,000} \text{ or 1 in a million}$$

4 computers:

$$p(4 \text{ fail}) = (0.01)^4 = 0.00000001 = \frac{1}{100,000,000} \text{ or 1 in a hundred million}$$

5 computers:

$$p(5 \text{ fail}) = (0.01)^5 = 0.0000000001 = \frac{1}{10,000,000,000} \text{ or 1 in 10 billion}$$

5 computers would be needed if a $\dfrac{1}{\text{billion}}$ chance of failure is required. Therefore, 4 backup computers are needed.

25.  $p(\text{HIV-positive}) = \dfrac{1,000,000}{261,000,000} = 0.003831$

$p(\text{non-HIV-positive}) = 1 - 0.003831 = 0.996169$
$p(\text{SUDS positive} \mid \text{AIDS/HIV}) = 0.999$
$p(\text{SUDS negative} \mid \text{non-AIDS/HIV}) = 0.996$

Construct a tree diagram using the above information.

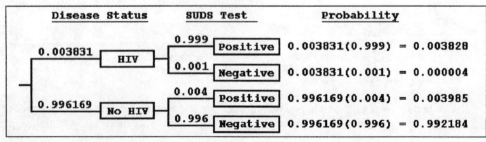

25.  Continued.

a)   $p(\text{HIV} \mid +) = \dfrac{0.003828}{0.003828 + 0.003985} = 0.489953 \approx 0.490$

b)   $p(\text{no HIV} \mid -) = \dfrac{0.992184}{0.992184 + 0.000004} = 0.999996 \approx 1.000$

c)   $p(\text{no HIV} \mid +) = \dfrac{0.003985}{0.003985 + 0.003828} = 0.510047 \approx 0.510$

d)   $p(\text{HIV} \mid -) = \dfrac{0.000004}{0.000004 + 0.992184} = 0.000004 \approx 0.0$

e)   (a) and (c).  (If the test was positive, you would want to know how accurate the test is in predicting the presence of HIV virus.)

f)   The false positive is answer (c).  The false negative is answer (d).

g)   Omitted.

h)   Omitted.

29.  a)   Three Punnett squares are used.

| Type A | | |
|---|---|---|
| Both Parents are Carriers | | |
|  | **C** | **c** |
| **C** | CC | Cc |
| **c** | Cc | cc |

| Type B | | |
|---|---|---|
| Only One Parent is a Carrier | | |
|  | **C** | **c** |
| **C** | CC | Cc |
| **C** | CC | Cc |

| Type C | | |
|---|---|---|
| Neither Parent is a Carrier | | |
|  | **C** | **C** |
| **C** | CC | CC |
| **C** | CC | CC |

The grandfather comes from Type A but does not have cystic fibrosis.  Therefore, his probability of being a carrier is 2/3 and of not being a carrier is 1/3.  (cc possibility is eliminated.)

The grandfather's children (parents of the cousins) each have a probability of being a carrier of 1/2 if the grandfather is a carrier (Type B) and a probability of 0 if the grandfather is not a carrier (Type C).

Likewise, the cousins each have a probability of being a carrier of 1/2 if the parent is a carrier  (Type B) and a probability of 0 if the parent is not a carrier (Type C).

29. a). Continued.

Construct a tree diagram to find the probability of a cousin being a carrier.

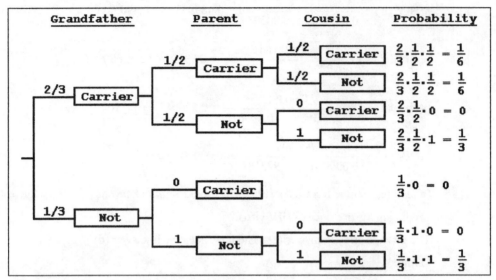

Since the husband ($H$) and wife ($W$) are related, the event that they are both carriers ($H \cap W$) is dependent. Thus, $p(H \cap W) = p(H \mid W) \cdot p(W)$. From the tree above, the probability that a cousin, the wife for example, is a carrier is 1/6. If the wife is a carrier, the grandfather must have been a carrier. The new tree for the husband follows:

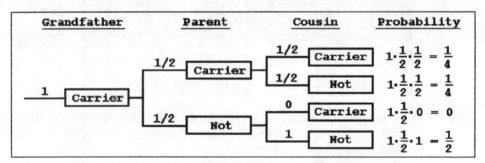

The probability the husband is a carrier given that the wife is a carrier is 1/4.

$$p(H \cap W) = p(H \mid W) \cdot p(W) = \frac{1}{4} \cdot \frac{1}{6} = \frac{1}{24}$$

29. a) Continued.

Therefore, the probability of both cousins being a carrier would be 1/24. And the probability of their child having cystic fibrosis (Type A) would be 1/96 from the following tree diagram.

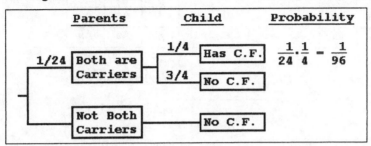

b)   Since husband and wife are unrelated, the probabilities of being carriers are independent. The husband and wife both have the same type of carrier in their background so $p(H) = p(W) = 1/6$.

$$p(H \cap W) = \frac{1}{6} \cdot \frac{1}{6} = \frac{1}{36}$$

And the probability of their child having cystic fibrosis would be 1/144 as seen in the following tree diagram.

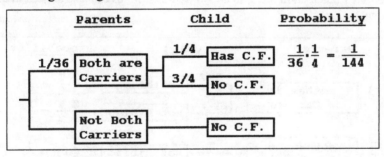

33.    Mr. Jones has genes aa. Ms. Jones could have genes Aa or AA with a $p(Aa) = 2/3$ and $p(AA) = 1/3$ since we know she is not aa. See Type A in Exercise 29. a).

Possible Punnett squares are:

|   | a | a |
|---|---|---|
| A | Aa | Aa |
| a | aa | aa |

← Mr. Jones

↑
Ms. Jones (carrier)

|   | a | a |
|---|---|---|
| A | Aa | Aa |
| A | Aa | Aa |

← Mr. Jones

↑
Ms. Jones (not a carrier)

$$p(Aa) = \frac{2}{4} = \frac{1}{2}$$

$$p(aa) = \frac{2}{4} = \frac{1}{2}$$

$$p(Aa) = \frac{4}{4} = 1$$

$$p(aa) = \frac{0}{4} = 0$$

$D$ = albino child (gene is aa)
$E$ = mother is a carrier (gene is Aa), $p(E) = 2/3$
$F$ = mother is not a carrier (gene is AA), $p(F) = 1/3$

$$p(D) = p(D \mid E) \cdot p(E) + p(D \mid F) \cdot p(F) = \frac{1}{2} \cdot \frac{2}{3} + 0 \cdot \frac{1}{3} = \frac{1}{3} + 0 = \frac{1}{3}$$

The probability of their child being an albino would be 1/3 as can also be seen in the following tree diagram.

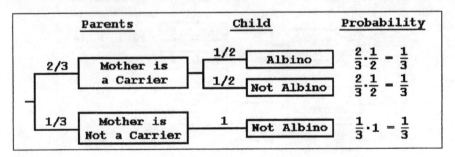

37. See Figure 3.33 for gene pairs.

   Mrs. York has black hair so she has genes $M^{Bk}M^{Bk}$ and $R^-R^-$. Mr. Wilson has dark red hair and genes $M^{Bd}M^{Bw}$ and $R^+R^+$.

   The Punnett squares are:

| Brownness | | | | Redness | | |
|---|---|---|---|---|---|---|
| | $M^{Bk}$ | $M^{Bk}$ | | | $R^-$ | $R^-$ |
| $M^{Bd}$ | $M^{Bd}M^{Bk}$ | $M^{Bd}M^{Bk}$ | | $R^+$ | $R^+R^-$ | $R^+R^-$ |
| $M^{Bw}$ | $M^{Bw}M^{Bk}$ | $M^{Bw}M^{Bk}$ | | $R^+$ | $R^+R^-$ | $R^+R^-$ |

   $p(M^{Bd}M^{Bk}) = \dfrac{2}{4} = \dfrac{1}{2}$, $p(M^{Bw}M^{Bk}) = \dfrac{2}{4} = \dfrac{1}{2}$, $p(R^+R^-) = \dfrac{4}{4} = 1$

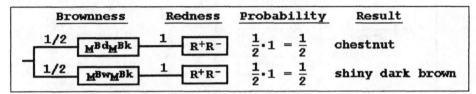

   The child has equal probability of receiving either chestnut hair or shiny dark brown hair.

41. See Figure 3.33 for gene pairs.

   Mrs. Landres has chestnut hair so she has genes $M^{Bd}M^{Bk}$ and $R^+R^-$. Mr. Landres has shiny dark brown hair and genes $M^{Bw}M^{Bk}$ and $R^+R^-$.

   The Punnett squares are:

| Brownness | | | | Redness | | |
|---|---|---|---|---|---|---|
| | $M^{Bd}$ | $M^{Bk}$ | | | $R^+$ | $R^-$ |
| $M^{Bw}$ | $M^{Bw}M^{Bd}$ | $M^{Bw}M^{Bk}$ | | $R^+$ | $R^+R^+$ | $R^+R^-$ |
| $M^{Bk}$ | $M^{Bk}M^{Bd}$ | $M^{Bk}M^{Bk}$ | | $R^-$ | $R^-R^+$ | $R^-R^-$ |

   $p(M^{Bw}M^{Bd}) = \dfrac{1}{4}$, $p(M^{Bw}M^{Bk}) = \dfrac{1}{4}$, $p(M^{Bk}M^{Bd}) = \dfrac{1}{4}$, $p(M^{Bk}M^{Bk}) = \dfrac{1}{4}$

   $p(R^+R^+) = \dfrac{1}{4}$, $p(R^+R^-) = \dfrac{2}{4} = \dfrac{1}{2}$, $p(R^-R^-) = \dfrac{1}{4}$

**41. Continued.**

The child's possible hair colors and the probability of each color are indicated on the tree diagram below.

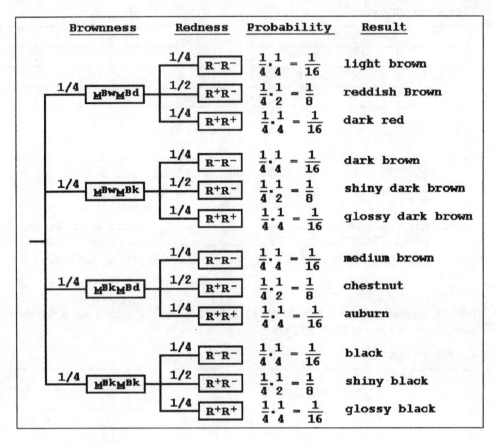

**45. Omitted.**

HHH
HHT
HTT
TTT

| **Chapter 3** | **Review** |

1.      Omitted.

5.      For parts (a), (b), and (c) use the following:

$D$ = rolling doubles = {(1,1), (2,2), (3,3), (4,4), (5,5), (6,6)}
$E$ = rolling a 7 = {(1,6), (2,5), (3,4), (4,3), (5,2), (6,1)}
$F$ = rolling an 11 = {(5,6), (6,5)}
$n(S) = 36, n(D) = 6, n(E) = 6, n(F) = 2$

a)      $p(E) = \dfrac{n(E)}{n(S)} = \dfrac{6}{36} = \dfrac{1}{6}$

b)      $p(F) = \dfrac{n(F)}{n(S)} = \dfrac{2}{36} = \dfrac{1}{18}$

c)   Using Rule 5 (Section 3.3) ($D$, $E$ and $F$ are mutually exclusive)

$p(D \cup E \cup F) = p(D) + p(E) + p(F)$

$= \dfrac{n(D)}{n(S)} + \dfrac{n(E)}{n(S)} + \dfrac{n(F)}{n(S)} = \dfrac{6}{36} + \dfrac{6}{36} + \dfrac{2}{36} = \dfrac{14}{36} = \dfrac{7}{18}$

For parts (d), (e), and (f) use Figure 3.6 to find the following:

$G$ = rolling an odd sum = {sum is 3, 5, 7, 9, or 11}

$H$ = rolling a sum greater than 8 = {sum is 9, 10, 11 or 12}

$G \cap H = \{9, 11\}$

$n(G) = 18, n(H) = 10, n(G \cap H) = 6$

d)      odd and greater than eight = rolling a 9 **or** an 11

$p(G \cap H) = \dfrac{n(G \cap H)}{n(S)} = \dfrac{6}{36} = \dfrac{1}{6}$

e)      odd or greater than eight = rolling a 3, 5, 7, 9 or 11, **or** a 9, 10, 11 or 12 (not mutually exclusive so use Rule 4 from Section 3.3)

$p(G \cup H) = p(G) + p(H) - p(G \cap H)$

$= \dfrac{n(G)}{n(S)} + \dfrac{n(H)}{n(S)} - \dfrac{n(G \cap H)}{n(S)} = \dfrac{18}{36} + \dfrac{10}{36} - \dfrac{6}{36} = \dfrac{22}{36} = \dfrac{11}{18}$

f)      neither odd nor greater than eight = rolling the complement of part e.

$p((G \cup H)') = 1 - p(G \cup H) = 1 - \dfrac{22}{36} = \dfrac{14}{36} = \dfrac{7}{18}$

9. The long-stemmed peas have one gene L and one gene s. The short-stemmed peas have two genes s. The Punnett square would look like the following:

| | L | s | |
|---|---|---|---|
| s | sL | ss | ← offspring |
| s | sL | ss | ← offspring |

↑ long-stemmed's genes (to the right of table: ← long-stemmed's genes)

↑
short-stemmed's genes

a) $A = \{sL\}$ (long-stemmed offspring)

$$p(A) = \frac{n(A)}{n(S)} = \frac{2}{4} = \frac{1}{2}$$

b) $B = \{ss\}$ (short-stemmed offspring)

$$p(B) = \frac{n(B)}{n(S)} = \frac{2}{4} = \frac{1}{2}$$

13. One parent is a carrier of Tay-Sachs disease and one parent is not. Let T denote the disease-free gene and t denote the recessive Tay-Sachs disease gene. The Punnett square would look like the following:

| | T | t | |
|---|---|---|---|
| T | TT | tT | ← offspring |
| T | TT | tT | ← offspring |

← carrier parent's genes

↑
non-carrier parent's genes

a) $D = \{tt\} = \varnothing$ (child has the disease)

$$p(D) = \frac{n(D)}{n(S)} = \frac{0}{4} = 0$$

b) $C = \{tT\}$ (child is a carrier)

$$p(C) = \frac{n(C)}{n(S)} = \frac{2}{4} = \frac{1}{2}$$

c) $H = \{TT\}$ (child does not have the disease and is not a carrier)

$$p(H) = \frac{n(H)}{n(S)} = \frac{2}{4} = \frac{1}{2}$$

17.  a)    Selecting all nine winning spots in nine-spot keno requires having 9 of a possible 20
            winning numbers ($n = 20$, $r = 9$) combined with none that are not winning ($n = 60$, $r = 0$).
            The sample space requires 9 numbers from a possible 80 ($n = 80$, $r = 9$).

            The order is **not** important so use **combinations**

$$_{20}C_9 = \frac{20!}{9! \cdot (20-9)!} = \frac{20!}{9! \cdot 11!} = 167{,}960$$

$$_{60}C_0 = \frac{60!}{0! \cdot (60-0)!} = \frac{60!}{0! \cdot 60!} = 1$$

$$_{80}C_9 = \frac{80!}{9! \cdot (80-9)!} = \frac{80!}{9! \cdot 71!} = 2.319 \times 10^{11}$$

$$p(\text{selecting all 9 winners}) = \frac{_{20}C_9 \cdot {}_{60}C_0}{_{80}C_9} \approx 0.0000007$$

    b)    Selecting eight winning spots in nine-spot keno requires having 8 of a possible 20 winning
            numbers ($n = 20$, $r = 8$) combined with 1 that is not winning ($n = 60$, $r = 1$).  The sample
            space requires 9 numbers from a possible 80 ($n = 80$, $r = 9$).

            The order is **not** important so use a **combinations**.

$$_{20}C_8 = \frac{20!}{8! \cdot (20-8)!} = \frac{20!}{8! \cdot 12!} = 125{,}970$$

$$_{60}C_1 = \frac{60!}{1! \cdot (60-1)!} = \frac{60!}{1! \cdot 59!} = 60$$

$$_{80}C_9 = \frac{80!}{9! \cdot (80-9)!} = \frac{80!}{9! \cdot 71!} = 2.319 \times 10^{11}$$

$$p(\text{selecting eight winners}) = \frac{_{20}C_8 \cdot {}_{60}C_1}{_{80}C_9} \approx 0.00003$$

21. The event is all possible five-card hands that include two tens and three jacks from a deck of 52 cards. There are 3 categories.

$T$ = selecting 2 tens from 4 possible suits
$J$ = selecting 3 jacks from 4 possible suits
$N$ = selecting no cards from the remaining 44 cards

The order is **not** important so use a **combinations**.

$$p(T) = {}_4C_2 = \frac{4!}{2! \cdot (4-2)!} = \frac{4!}{2! \cdot 2!} = 6$$

$$p(J) = {}_4C_3 = \frac{4!}{3! \cdot (4-3)!} = \frac{4!}{3! \cdot 1!} = 4$$

$$p(N) = {}_{44}C_0 = \frac{44!}{0! \cdot (44-0)!} = \frac{44!}{0! \cdot 44!} = 1$$

$$n(S) = {}_{52}C_5 = \frac{52!}{5! \cdot (52-5)!} = \frac{52!}{5! \cdot 47!} = 2,598,960$$

$$p(2 \text{ tens and 3 jacks}) = \frac{{}_4C_2 \cdot {}_4C_3 \cdot {}_{44}C_0}{{}_{52}C_5} = \frac{6 \cdot 4 \cdot 1}{2,598,960} \approx 0.0000092$$

# 4  Statistics

| Section 4.1 | Population, Sample, and Data |
|---|---|

1.   a)   The data set consists of a few distinct values, so utilize single-valued classes of data.

| Number of Visits | Tally | Frequency | Relative Frequency |
|---|---|---|---|
| 1 | ⁺⁺⁺⁺ IIII | 9 | 9/30 = 0.3000 = 30% |
| 2 | ⁺⁺⁺⁺ III | 8 | 8/30 = 0.2667 = 27% |
| 3 | II | 2 | 2/30 = 0.0667 = 7% |
| 4 | ⁺⁺⁺⁺ | 5 | 5/30 = 0.1667 = 17% |
| 5 | ⁺⁺⁺⁺ I | 6 | 6/30 = 0.2000 = 20% |
| | | $n = 30$ | total = 101% (rounding error) |

   b)   Multiply the relative frequency from part (a) by $360°$ to calculate the Central Angle and then using a protractor construct the pie chart.

| Number of Visits | Relative Frequency | Central Angle |
|---|---|---|
| 1 | 9/30 = 0.3000 | $9/30 \times 360° = 108°$ |
| 2 | 8/30 = 0.2667 | $8/30 \times 360° = 96°$ |
| 3 | 2/30 = 0.0667 | $2/30 \times 360° = 24°$ |
| 4 | 5/30 = 0.1667 | $5/30 \times 360° = 60°$ |
| 5 | 6/30 = 0.2000 | $6/30 \times 360° = 72°$ |
| | total = 1.0001 (rounding error) | total = $360°$ |

1. b) Continued.

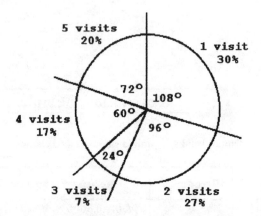

**Number of Visits to the Library
by a Student during the Past Week**

c) Use the data from part (a) to construct the following histogram.

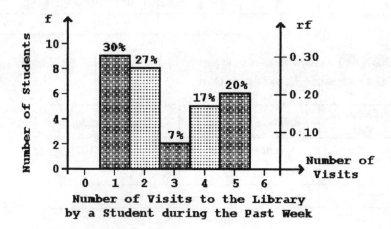

**Number of Visits to the Library
by a Student during the Past Week**

5.   a)   Range = largest − smallest = 80 − 51 = 29

$$\text{Interval width} = \frac{\text{range}}{\text{number of intervals}} = \frac{29}{6} = 4.833 \approx 5$$

| Speed | Tally | Frequency | Relative Frequency |
|-------|-------|-----------|--------------------|
| $51 \leq x < 56$ | \|\| | 2 | 2/40 = 0.05 =  5% |
| $56 \leq x < 61$ | \|\|\|\| | 4 | 4/40 = 0.10 = 10% |
| $61 \leq x < 66$ | ⋕⋕ \|\| | 7 | 7/40 = 0.175 = 17.5% |
| $66 \leq x < 71$ | ⋕⋕ ⋕⋕ | 10 | 10/40 = 0.25 = 25% |
| $71 \leq x < 76$ | ⋕⋕ \|\|\|\| | 9 | 9/40 = 0.225 = 22.5% |
| $76 \leq x < 81$ | ⋕⋕ \|\|\| | 8 | 8/40 = 0.20 = 20% |
|  |  | n = 40 | total = 100% |

b)

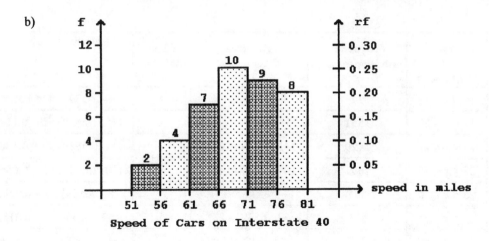

Speed of Cars on Interstate 40

9.

| Hourly Wage | Frequency | Relative Frequency |
|-------------|-----------|--------------------|
| $\$4.00 \leq x < 5.50$ | 21 | 21/152 = 0.138157 = 13.8% |
| $5.50 \leq x < 7.00$ | 35 | 35/152 = 0.230263 = 23.0% |
| $7.00 \leq x < 8.50$ | 42 | 42/152 = 0.276316 = 27.6% |
| $8.50 \leq x < 10.00$ | 27 | 27/152 = 0.177632 = 17.8% |
| $10.00 \leq x < 11.50$ | 18 | 18/152 = 0.118421 = 11.8% |
| $11.50 \leq x < 13.00$ | 9 | 9/152 = 0.059211 =  5.9% |
|  | n = 152 | total = 99.9% (due to rounding) |

9. Continued.

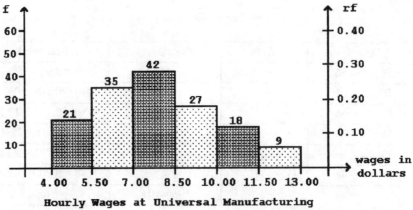

Hourly Wages at Universal Manufacturing

13.

| Age of Males | Frequency (in thousands) $f$ | Relative Frequency, $f/n$ | Class Width, $w$ | Relative Frequency Density, $rfd = (f/n) \div w$ |
|---|---|---|---|---|
| $14 \leq x < 18$ | 94 | $94/6539 \approx 1\%$ | 4 | $(94/6539) \div 4 = 0.0036$ |
| $18 \leq x < 20$ | 1551 | $1551/6539 \approx 24\%$ | 2 | $(1551/6539) \div 2 = 0.1186$ |
| $20 \leq x < 22$ | 1420 | $1420/6539 \approx 22\%$ | 2 | $(1420/6539) \div 2 = 0.1086$ |
| $22 \leq x < 25$ | 1091 | $1091/6539 \approx 17\%$ | 3 | $(1091/6539) \div 3 = 0.0556$ |
| $25 \leq x < 30$ | 865 | $865/6539 \approx 13\%$ | 5 | $(865/6539) \div 5 = 0.0265$ |
| $30 \leq x < 35$ | 521 | $521/6539 \approx 8\%$ | 5 | $(521/6539) \div 5 = 0.0159$ |
| $35 \leq x \leq 60$ | 997 | $997/6539 \approx 15\%$ | 25 | $(997/6539) \div 25 = 0.0061$ |
| | $n = 6539$ | total = 100% | | |

13. Continued.

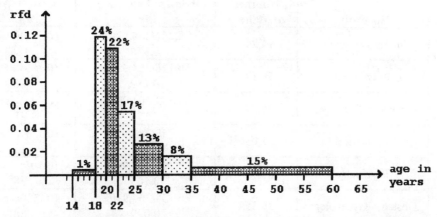

Age Composition of Male Students in Higher Education

17.

| Reason | Number of Patients (in Thousands, $f$ | Relative Frequency, $f/n$ | Central Angle $(f/n)(360°)$ |
|---|---|---|---|
| Stomach pain | 6610 | 6610/22,301 = 30% | (6610/22,301)(360°) = 107° |
| Chest pain | 5608 | 5608/22,301 = 25% | (5608/22,301)(360°) = 91° |
| Fever | 4678 | 4678/22,301 = 21% | (4678/22,301)(360°) = 76° |
| Headache | 2809 | 2809/22,301 = 13% | (2809/22,301)(360°) = 45° |
| Cough | 2596 | 2596/22,301 = 12% | (2596/22,301)(360°) = 42° |
| | $n = 22,301$ | total = 101% (rounding) | total = 361° (rounding) |

Use a protractor to mark off the appropriate central angles.

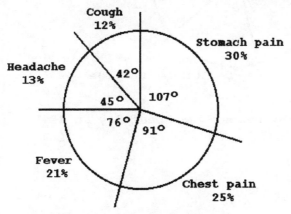

Reasons Given by Patients for
Emergency Room Vists, 1999

21. a)

| Specialty | Number of Males, $f$ | Relative Frequency, $f/n$ | Central Angle $(f/n)(360°)$ |
|---|---|---|---|
| Internal Medicine | 94,244 | 94,244/264,990 = 0.356 | (0.356)(360°) = 128° |
| Family Practice | 50,176 | 50,176/264,990 = 0.189 | (0.189)(360°) = 68° |
| General Surgery | 35,572 | 35,572/264,990 = 0.134 | (0.134)(360°) = 48° |
| Pediatrics | 31,050 | 31,050/264,990 = 0.117 | (0.117)(360°) = 42° |
| Psychiatry | 27,790 | 27,790/264,990 = 0.105 | (0.105)(360°) = 38° |
| Obstetric/Gynecology | 26,158 | 26,158/264,990 = 0.099 | (0.099)(360°) = 36° |
| | $n = 264,990$ | total = 1.000 | total = 360° |

Use a protractor to mark off the appropriate central angles.

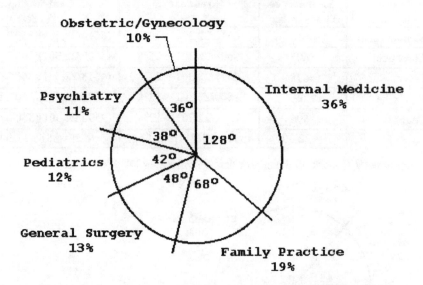

Male Physicians by Gender and Specialty in 1999

21. Continued.

b)

| Specialty | Number of Females, $f$ | Relative Frequency, $f/n$ | Central Angle $(f/n)(360°)$ |
|---|---|---|---|
| Internal Medicine | 34,465 | 34,465/110,061 = 0.313 | (0.313)(360°) = 113° |
| Family Practice | 18,887 | 18,887/110,061 = 0.172 | (0.172)(360°) = 62° |
| General Surgery | 3,739 | 3,739/110,061 = 0.034 | (0.034)(360°) = 12° |
| Pediatrics | 28,499 | 28,499/110,061 = 0.259 | (0.259)(360°) = 93° |
| Psychiatry | 11,266 | 11,266/110,061 = 0.102 | (0.102)(360°) = 37° |
| Obstetric/Gynecology | 13,205 | 13,205/110,061 = 0.120 | (0.120)(360°) = 43° |
| | $n = 110,061$ | total = 1.000 | total = 360° |

Use a protractor to mark off the appropriate central angles.

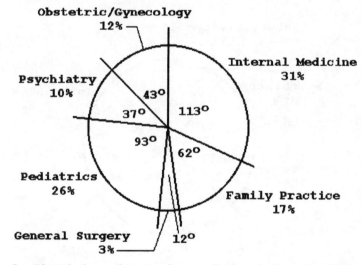

Female Physicians by Gender and Specialty in 1999

21. Continued.

c)    Add the males and females together by specialty for the total data.

| Specialty | Total Physicians, $f$ | Relative Frequency, $f/n$ | Central Angle $(f/n)(360°)$ |
|---|---|---|---|
| Internal Medicine | 128,709 | 128,709/375,051 = 0.343 | (0.343)(360°) = 124° |
| Family Practice | 69,063 | 69,063/375,051 = 0.184 | (0.184)(360°) = 66° |
| General Surgery | 39,311 | 39,311/375,051 = 0.105 | (0.105)(360°) = 38° |
| Pediatrics | 59,549 | 59,549/375,051 = 0.159 | (0.159)(360°) = 57° |
| Psychiatry | 39,056 | 39,056/375,051 = 0.104 | (0.104)(360°) = 37° |
| Obstetric/Gynecology | 39,363 | 39,363/375,051 = 0.105 | (0.105)(360°) = 38° |
| | $n = 375,051$ | total = 1.000 | total = 360° |

Use a protractor to mark off the appropriate central angles.

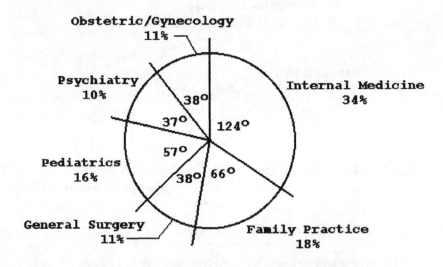

Total Physicians by Gender and Specialty in 1999

25.    Omitted.

29.   a)   A sample of a spreadsheet and chart created in Microsoft Excel 2000 would appear as follows:

| percent | bin numbers | Bin | Frequency | group boundaries | frequency |
|---------|-------------|------|-----------|------------------|-----------|
| 40 | 8 | 8 | 20 | 0-9 | 20 |
| 38 | 17 | 17 | 16 | 10-18 | 16 |
| 37 | 26 | 26 | 7 | 19-27 | 7 |
| 35 | 35 | 35 | 4 | 28-36 | 4 |
| 33 | 44 | 44 | 3 | 37-45 | 3 |
| 32 | | More | 0 | | |
| 31 | | | | | |
| 26 | | | | | |
| 25 | | | | | |
| 25 | | | | | |
| 24 | | | | | |
| 21 | | | | | |
| 18 | | | | | |
| 18 | | | | | |
| 17 | | | | | |
| 17 | | | | | |
| 15 | | | | | |
| 14 | | | | | |
| 14 | | | | | |
| 14 | | | | | |

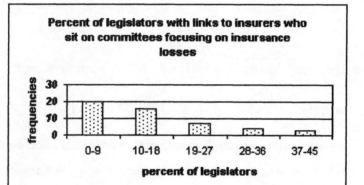

In general, the histogram should appear as follows:

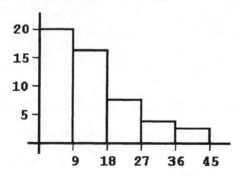

b)   Omitted.

| Section 4.2 | **Measures of Central Tendency** |

1.   mean: $\bar{x} = \dfrac{\sum x}{n} = \dfrac{9 + 12 + 8 + 10 + \dots + 10}{14} = \dfrac{175}{14} = 12.5$

median:  Arrange the numbers in order from low to high.  Because there are $n = 14$ data points, the

   location of the median is $L = \dfrac{14 + 1}{2} = 7.5$; add the 7th and 8th number and divide by 2.

   8, 9, 9, 9, 10, 10, 11, 12, 12, 14, 15, 15, 20, 21
                       ↑   ↑
   $\dfrac{11 + 12}{2} = 11.5$

mode:  The number with the highest frequency is 9.  It occurs three times.

5.   a)   mean: $\bar{x} = \dfrac{\sum x}{n} = \dfrac{9 + 9 + 10 + 11 + 12 + 15}{6} = \dfrac{66}{6} = 11$

median:  Arrange the numbers in order from low to high.  Because there are $n = 6$ data

   points, the location of the median is $L = \dfrac{6 + 1}{2} = 3.5$; add the 3rd and 4th number and

   divide by 2.

   9, 9, 10, 11, 12, 15
           ↑   ↑
   $\dfrac{10 + 11}{2} = 10.5$

mode:  The number with the highest frequency is 9.  It occurs two times.

b)   mean: $\bar{x} = \dfrac{\sum x}{n} = \dfrac{9 + 9 + 10 + 11 + 12 + 102}{6} = \dfrac{153}{6} = 25.5$

median:  Arrange the numbers in order from low to high.  Because there are $n = 6$ data

   points, the location of the median is $L = \dfrac{6 + 1}{2} = 3.5$; add the 3rd and 4th number and

   divide by 2.

   9, 9, 10, 11, 12, 102
           ↑   ↑
   $\dfrac{10 + 11}{2} = 10.5$

mode:  The number with the highest frequency is 9.  It occurs two times.

5. Continued.

    c)    The means in parts (a) and (b) are different because the mean in part (b) has been inflated by the extreme value of 102. The medians and the modes in parts (a) and (b) are equal.

9.    mean: $\bar{x} = \dfrac{\sum x}{n} = \dfrac{3+15+22+...+9}{15} = \dfrac{185}{17} = 10.8824... \approx 10.9$

    median:  Arrange the numbers in order from low to high. Because there are $n = 6$ data points, the location of the median is $L = \dfrac{17+1}{2} = 9$; select the $9^{th}$ number as the median.

$$1, 3, 5, 7, 8, 9, 9, 9, 10, 13, 13, 14, 15, 15, 15, 17, 22$$
$$\uparrow$$
$$\text{median value} = 10$$

    mode:    9 and 15 have the highest frequencies (3) so they are both modes.

13.

| Score, $x$ | Number of Students, $f$ | $f \cdot x$ |
|:---:|:---:|:---:|
| 10 | 3 | 30 |
| 9 | 10 | 90 |
| 8 | 9 | 72 |
| 7 | 8 | 56 |
| 6 | 10 | 60 |
| 5 | 2 | 10 |
| | $n = 42$ | $\sum(f \cdot x) = 318$ |

    mean: $\bar{x} = \dfrac{\sum x}{n} = \dfrac{318}{42} = 7.5714... \approx 7.6$

    median:  The location of the median is $L = \dfrac{42+1}{2} = 21.5$. If the numbers were arranged in order from low to high, there would be 2 fives, 10 sixes, and 8 sevens to reach 20 numbers. So the $21^{st}$ and $22^{nd}$ numbers would both be 8. The median would be $\dfrac{8+8}{2} = 8$

    mode:    both 6 and 9 are modes with a frequency of 10.

17.  If $y$ is the score on the 5$^{\text{th}}$ exam, then $\bar{x} = \dfrac{71+69+85+83+y}{5} = \dfrac{308+y}{5}$

   a)  Solve for $y$ when $\bar{x} \geq 70$.

   $$\frac{308+y}{5} \geq 70$$
   $$308+y \geq 350$$
   $$y \geq 42$$

   b)  Solve for $y$ when $\bar{x} \geq 80$.

   $$\frac{308+y}{5} \geq 80$$
   $$308+y \geq 400$$
   $$y \geq 92$$

   c)  Solve for $y$ when $\bar{x} \geq 90$.

   $$\frac{308+y}{5} \geq 90$$
   $$308+y \geq 450$$
   $$y \geq 142$$

   If there are 100 points possible on each exam, then $\bar{x} \geq 90$ would not be possible.

21.  Time to Milwaukee $= \dfrac{90 \text{ mi}}{60 \text{ mph}} = 1.5 \text{ hr}$, time returning $= \dfrac{90 \text{ mi}}{45 \text{ mph}} = 2 \text{ hr}$

   mean speed $= \dfrac{\text{total distance}}{\text{total time}} = \dfrac{90+90}{1.5+2} = \dfrac{180}{3.5} = 51.42857\ldots \approx 51.4 \text{ mph}$

25.  $n = 8$, $\bar{x} = \$40{,}000$, median $= \$42{,}000$

   a)  original $\bar{x} = \dfrac{\sum \text{original salaries}}{8} = 40{,}000$

   $\sum \text{original salaries} = 8(40{,}000) = 320{,}000$

   $\sum \text{new salaries} = 320{,}000 + 6{,}000 = 320{,}000 + 6{,}000 = 326{,}000$

   new $\bar{x} = \dfrac{\sum \text{new salaries}}{8} = \dfrac{326{,}000}{8} = \$40{,}750$

   b)  The new median salary would still be \$42,000, because only the highest salary was increased.

29.

| Age | Number of People (in thousands), $f$ | Midpoint, $x$ | $f \bullet x$ |
|---|---|---|---|
| $0 < x < 5$ | 19,176 | 2.5 | 47,940 |
| $5 \le x < 10$ | 20,550 | 7.5 | 154,125 |
| $10 \le x < 15$ | 20,528 | 12.5 | 256,600 |
| $15 \le x < 25$ | 39,184 | 20.0 | 783,680 |
| $25 \le x < 35$ | 39,892 | 30.0 | 1,196,760 |
| $35 \le x < 45$ | 44,149 | 40.0 | 1,765,960 |
| $45 \le x < 55$ | 37,678 | 50.0 | 1,883,900 |
| $55 \le x < 65$ | 24,275 | 60.0 | 1,456,500 |
| $65 \le x < 85$ | 30,752 | 75.0 | 2,306,400 |
| $85 \le x \le 100$ | 4,240 | 92.5 | 392,200 |
| Totals | 280,424 | | 10,244,065 |

a)    Subtract the "$85 \le x \le 100$" row from the totals

$$\text{mean age} = \bar{x} = \frac{\Sigma(f \bullet x)}{n} = \frac{10,244,065 - 392,200}{280,424 - 4,240} = \frac{9,851,865}{276,184} = 35.6713... \approx 35.7$$

b)    $$\text{mean age} = \bar{x} = \frac{\Sigma(f \bullet x)}{n} = \frac{10,244,065}{280,424} = 36.5306... \approx 36.5$$

## Section 4.3                                    Measures of Dispersion

1.    $n = 6$, mean $= \bar{x} = \dfrac{\Sigma x}{n} = \dfrac{42}{6} = 7$

| Data $x$ | Deviation $x - \bar{x}$ | Deviation Squared $(x - \bar{x})^2$ | $x^2$ |
|---|---|---|---|
| 3 | $3 - 7 = -4$ | 16 | 9 |
| 8 | $8 - 7 = 1$ | 1 | 64 |
| 5 | $5 - 7 = -2$ | 4 | 25 |
| 3 | $3 - 7 = -4$ | 16 | 9 |
| 10 | $10 - 7 = 3$ | 9 | 100 |
| 13 | $13 - 7 = 6$ | 36 | 169 |
| $\Sigma x = 42$ | | $\Sigma(x - \bar{x})^2 = 82$ | $\Sigma x^2 = 376$ |

1. Continued.

a)    variance: $s^2 = \dfrac{\Sigma(x-\bar{x})^2}{n-1} = \dfrac{82}{5} = 16.4$

standard deviation: $s = \sqrt{\text{variance}} = \sqrt{s^2} = \sqrt{16.4} = 4.0496... \approx 4.0$

b)    variance $= s^2 = \dfrac{1}{n-1}\left(\Sigma x^2 - \dfrac{(\Sigma x)^2}{n}\right) = \dfrac{1}{6-1}\left(376 - \dfrac{(42)^2}{6}\right) = \dfrac{1}{5}(376-294) = \dfrac{82}{5} = 16.4$

standard deviation $= s = \sqrt{\text{variance}} = \sqrt{s^2} = \sqrt{16.4} = 4.0496... \approx 4.0$

5.    a)                                                         b)

| $x$ | $x^2$ |
|---|---|
| 12 | 144 |
| 16 | 256 |
| 20 | 400 |
| 24 | 576 |
| 28 | 784 |
| 32 | 1024 |
| $\Sigma x = 132$ | $\Sigma x^2 = 3184$ |

| $x$ | $x^2$ |
|---|---|
| 600 | 360,000 |
| 800 | 640,000 |
| 1000 | 1,000,000 |
| 1200 | 1,440,000 |
| 1400 | 1,960,000 |
| 1600 | 2,560,000 |
| $\Sigma x = 6600$ | $\Sigma x^2 = 7,960,000$ |

a)    $n = 6$, mean $= \bar{x} = \dfrac{\Sigma x}{n} = \dfrac{132}{6} = 22$

variance $= s^2 = \dfrac{1}{n-1}\left(\Sigma x^2 - \dfrac{(\Sigma x)^2}{n}\right) = \dfrac{1}{6-1}\left(3184 - \dfrac{(132)^2}{6}\right) = \dfrac{1}{5}(3184-2904) = \dfrac{280}{5} = 56$

standard deviation $= s = \sqrt{\text{variance}} = \sqrt{s^2} = \sqrt{56} = 7.4833... \approx 7.5$

b)    $n = 6$, mean $= \bar{x} = \dfrac{\Sigma x}{n} = \dfrac{6600}{6} = 1100$

variance $= s^2 = \dfrac{1}{n-1}\left(\Sigma x^2 - \dfrac{(\Sigma x)^2}{n}\right) = \dfrac{1}{6-1}\left(7,960,000 - \dfrac{(6600)^2}{6}\right)$

$= \dfrac{1}{5}(7,960,000 - 7,260,000) = \dfrac{700,000}{5} = 140,000$

standard deviation $= s = \sqrt{\text{variance}} = \sqrt{s^2} = \sqrt{140,000} = 374.1657... \approx 374.2$

c)    The data in (b) represents the data in (a) multiplied by 50.

d)    The mean in (b) is the mean in (a) multiplied by 50. The standard deviation in (b) is 50 times the standard deviation in (a).

9.

| $x$ | $x^2$ |
|-----|-------|
| 16 | 256 |
| 25 | 625 |
| 24 | 576 |
| 19 | 361 |
| 33 | 1089 |
| 25 | 625 |
| 34 | 1156 |
| 46 | 2116 |
| 37 | 1369 |

| $x$ | $x^2$ |
|-----|-------|
| 33 | 1089 |
| 42 | 1764 |
| 40 | 1600 |
| 37 | 1369 |
| 34 | 1156 |
| 49 | 2401 |
| 73 | 5329 |
| 46 | 2116 |
| $\Sigma x = 613$ | $\Sigma x^2 = 24{,}997$ |

$$n = 613, \text{ mean } = \bar{x} = \frac{\Sigma x}{n} = \frac{613}{17} = 36.0588\ldots \approx 36.1$$

$$\text{variance } = s^2 = \frac{1}{n-1}\left(\Sigma x^2 - \frac{(\Sigma x)^2}{n}\right) = \frac{1}{17-1}\left(24{,}997 - \frac{(613)^2}{17}\right)$$

$$= \frac{1}{16}(24{,}997 - 22{,}104.0588\ldots) = \frac{2892.9412\ldots}{16} \approx 180.8088\ldots$$

$$\text{standard deviation } = s = \sqrt{\text{variance}} = \sqrt{s^2} = \sqrt{180.8088\ldots} = 13.4465\ldots \approx 13.4$$

13.    a)

| Rainfall, $x$ | $x^2$ |
|---------------|-------|
| 5.4 | 29.16 |
| 4.0 | 16.00 |
| 3.8 | 14.44 |
| 2.5 | 6.25 |
| 1.8 | 3.24 |
| 1.6 | 2.56 |
| 0.9 | 0.81 |
| 1.2 | 1.44 |
| 1.9 | 3.61 |
| 3.3 | 10.89 |
| 5.7 | 32.49 |
| 6.0 | 36.00 |
| $\Sigma x = 38.1$ | $\Sigma x^2 = 156.89$ |

13. a) Continued.

$$n = 12, \text{ mean} = \overline{x} = \frac{\sum x}{n} = \frac{38.1}{12} = 3.175$$

$$\text{variance } = s^2 = \frac{1}{n-1}\left(\sum x^2 - \frac{(\sum x)^2}{n}\right) = \frac{1}{12-1}\left(156.89 - \frac{(38.1)^2}{12}\right)$$

$$= \frac{1}{15}(156.89 - 120.9675) = \frac{35.9225}{15} \approx 2.394833...$$

$$\text{standard deviation} = s = \sqrt{\text{variance}} = \sqrt{s^2} = \sqrt{3.26568...} = 1.80711... \approx 1.807$$

b)    $[\overline{x} - s, \overline{x} + s] = [3.175 - 1.807, 3.175 + 1.807] = [1.368, 4.982]$

Arrange the data in order from smallest to largest to find one standard deviation from the mean.

0.9, 1.2, 1.6, 1.8, 1.9, 2.5, 3.3, 3.8, 4.0, 5.4, 5.7, 6.0

| $\overline{x} - s$ | $\overline{x}$ | $\overline{x} + s$ |
|---|---|---|
| 1.368 | 3.175 | 4.982 |

7 of the 12 observations lie between 1.368 and 4.982. Therefore, 7/12 = 0.5833... or 58% of the data lie within one standard deviation of the mean.

c)    $\left[\overline{x} - 2s, \overline{x} + 2s\right] = [3.175 - 2(1.807), 3.175 + 2(1.807)]$
$$= [3.175 - 3.614, 3.175 + 3614] = [-0.439, 6.789]$$

Arrange the data in order from smallest to largest to find one standard deviation from the mean.

0.9, 1.2, 1.6, 1.8, 1.9, 2.5, 3.3, 3.8, 4.0, 5.4, 5.7, 6.0

| $\overline{x} - 2s$ | $\overline{x}$ | $\overline{x} + 2s$ |
|---|---|---|
| −0.439 | 3.175 | 6.789 |

All 12 of the observations lie between −0.439 and 6.789. Therefore, 100% of the data lie within two standard deviations of the mean.

17.

| Weight (in ounces), $x$ | Mid-point, $x$ | Number of Boxes, $f$ | $f \bullet x$ | $f \bullet x^2 = (f \bullet x)x$ |
|---|---|---|---|---|
| $15.3 \leq x < 15.6$ | 15.45 | 13 | 200.85 | 3,103.1325 |
| $15.6 \leq x < 15.9$ | 15.75 | 24 | 378.00 | 5,953.5000 |
| $15.9 \leq x < 16.2$ | 16.05 | 84 | 1348.20 | 21,638.6100 |
| $16.2 \leq x < 16.5$ | 16.35 | 19 | 310.65 | 5,079.1275 |
| $16.5 \leq x < 16.8$ | 16.65 | 10 | 166.50 | 2,772.2250 |
| **Totals** | | $n = 150$ | $\Sigma(f \bullet x) =$ 2404.20 | $\Sigma(f \bullet x^2) =$ 38,546.5950 |

$$\text{variance } = s^2 = \frac{1}{n-1}\left(\Sigma f \bullet x^2 - \frac{(\Sigma f \bullet x)^2}{n}\right) = \frac{1}{150-1}\left(38,546.5950 - \frac{(2,404.2)^2}{150}\right)$$

$$= \frac{1}{149}(38,546.5950 - 38,534.518) = \frac{12.0774}{149} \approx 0.081056...$$

$$\text{standard deviation} = s = \sqrt{\text{variance}} = \sqrt{s^2} = \sqrt{0.081056...} = 0.28470... \approx 0.285$$

21.    Omitted.

25.    Enter the data following the directions for your computer or calculator. Don't forget to enter 11 states with 0%.

a)    $\bar{x} = 13.46$, $\Sigma x = 673$, $\Sigma x^2 = 15473$, $S_x = 11.441439...$, $\sigma_x = 11.3264...$, $n = 50$

The mean is $\bar{x} = 13.46 \approx 13.5$

The population standard deviation is $\sigma_x = 11.3264... \approx 11.3$

b)    Omitted.

| Section 4.4 | The Normal Distribution |

1.  a)  Use the body table (Appendix F) to look up 1.00.

$p(0 < z < 1) = 0.3413$

So, 34.13% of the $z$-distribution lies between $z = 0$ and $z = 1$.

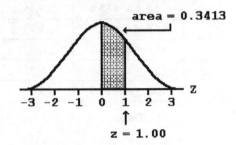

area = 0.3413

z = 1.00

b)  Use the body table to look up 1.00 using the symmetry of the distribution.

$p(-1 < z < 0) = p(0 < z < 1) = 0.3413$

So, 34.13% of the $z$-distribution lies between $z = -1$ and $z = 0$.

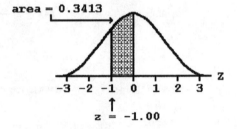

area = 0.3413

z = -1.00

c)  Use the body table to look up 1.00, then add the left body to the right body.

$p(-1 < z < 1)$
$= p(-1 < z < 0) + p(0 < z < 1)$
$= 0.3413 + 0.3413 = 0.6826$

So, 68.26% of the $z$-distribution lies between $z = -1$ and $z = 1$.

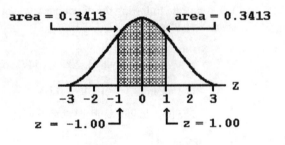

area = 0.3413    area = 0.3413

z = -1.00    z = 1.00

5.  $\mu = 24.7$, $\sigma = 2.3$

a)  One standard deviation of the mean:
$$\mu \pm 1\sigma = 24.7 \pm 1(2.3)$$
$$= 24.7 \pm 2.3$$
$$= [22.4, 27.0]$$
Two standard deviations of the mean:
$$\mu \pm 2\sigma = 24.7 \pm 2(2.3)$$
$$= 24.7 \pm 4.6$$
$$= [20.1, 29.3]$$
Three standard deviations of the mean:
$$\mu \pm 3\sigma = 24.7 \pm 3(2.3)$$
$$= 24.7 \pm 6.9$$
$$= [17.8, 31.6]$$

5. Continued.

b)      From Figure 4.87
              [22.4, 27.0] has 68.26% of the data
              [20.1, 29.3] has 95.44% of the data
              [17.8, 31.6] has 99.74% of the data

c)

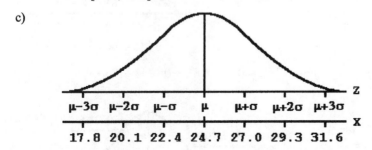

9.      a)      Find 0.1331 in the interior of the body table and
              read the edges to find $z$-number = 0.34.

              $p(0 < z < 0.34) = 0.1331$.

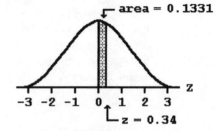

b)      Find 0.4812 in the interior of the body
              table and read the edges to find
              $z$-number = 2.08.  Since $c$ is on the left
              of zero, $c = -2.08$.

              $p(-2.08 < z < 0) = 0.4812$.

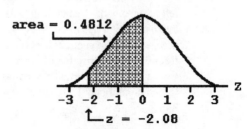

c)      Divide 0.4648 by 2 and find
              0.2324 in the interior of the body
              table.  Read the edges to find
              $z$-number = 0.62.  Looking at both
              sides of zero, $c = 0.62$ and
              $-c = -0.62$.

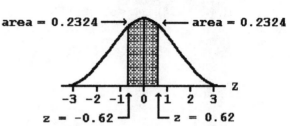

$p(-0.62 < z < 0.62) = p(-0.62 < z < 0) + p(0 < z < 0.62) = 0.2324 + 0.2324 = 0.4648$

9. Continued.

d)    $p(z > c) = 0.6064$ means $c$ is to the left of the mean.  Subtract 0.5 from 0.6064 and find 0.1064 in the interior of the body table.  Read the edges to find $z$-number $= 0.27$ and change to a negative.  So, $c = -0.27$.

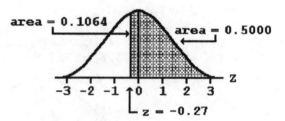

$$p(z > -0.27) = p(z > 0) + p(-0.27 < z < 0) = 0.5000 + 0.1064 = 0.6064$$

e)    $p(z > c) = 0.0505$, means $c$ is to the right of the mean.  Because 50% of the distribution lies to the right of zero, we need to subtract 0.0505 from 0.5 to find the area of the body from $z = 0$ to $z = c$.  Read the edges to find $z$-number $= 1.64$.  So, $c = 1.64$.

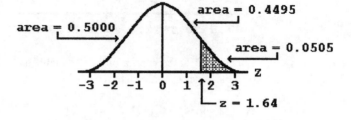

$$p(z > 1.64) = p(z > 0) - p(0 < z < 1.64) = 0.5000 - 0.4495 = 0.0505$$

f)    $p(z < c) = 0.1003$ means $c$ is to the left of the mean.  Because 50% of the distribution lies to the left of zero, we need to subtract 0.1003 from 0.5 to find the area of the body from $z = 0$ to $z = c$.  Read the edges to find $z$-number $= 1.28$ and change it to a negative.  So, $c = -1.28$.

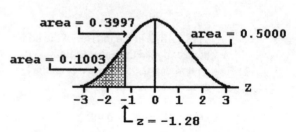

$$p(z < -1.28) = p(z < 0) - p(-1.28 < z < 0) = 0.5000 - 0.3997 = 0.1003.$$

13.     Convert a given $x$ to a $z$–number using $\mu = 36.8$ and $\sigma = 2.5$ in the formula $z = \dfrac{x - \mu}{\sigma}$. Use the body table to find the probability.

a)     If $x = 36.8$, $z = \dfrac{36.8 - 36.8}{2.5} = \dfrac{0}{2.5} = 0$

        If $x = 39.3$, $z = \dfrac{39.3 - 36.8}{2.5} = \dfrac{2.5}{2.5} = 1$

        $p(36.8 < x < 39.3) = p(0 < z < 1) = 0.3413$

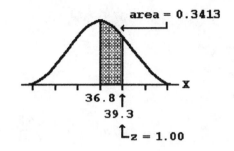

b)     If $x = 34.2$, $z = \dfrac{34.2 - 36.8}{2.5} = \dfrac{-2.6}{2.5} = -1.04$

        If $x = 38.7$, $z = \dfrac{38.7 - 36.8}{2.5} = \dfrac{1.9}{2.5} = 0.76$

        $p(34.2 < x < 38.7) = p(-1.04 < z < 0.76)$
        $= p(-1.04 < z < 0) + p(0 < z < 0.76)$
        $= 0.3508 + 0.2764 = 0.6272$

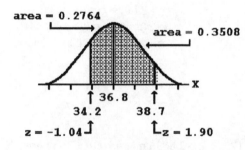

c)     If $x = 40.0$, $z = \dfrac{40.0 - 36.8}{2.5} = \dfrac{3.2}{2.5} = 1.28$

        $p(x < 40.0) = p(z < 1.28)$
        $= p(z < 0) + p(0 < z < 1.28)$
        $= 0.5000 + 0.3997 = 0.8997$

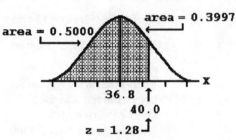

d)     If $x = 32.3$, $z = \dfrac{32.3 - 36.8}{2.5} = \dfrac{-4.5}{2.5} = -1.80$

        If $x = 41.3$, $z = \dfrac{41.3 - 36.8}{2.5} = \dfrac{4.5}{2.5} = 1.80$

        $p(32.3 < x < 41.3) = p(-1.8 < z < 1.8)$
        $= p(-1.8 < z < 0) + p(0 < z < 1.8)$
        $= 0.4641 + 0.4641 = 0.9282$

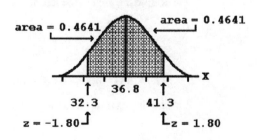

13. Continued.

    e)    If $x = 37.9$, there is no probability. (A line has no area.)

          $p(x = 37.9) = 0.0$

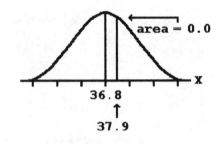

    f)    If $x = 37.9$, $z = \dfrac{37.9 - 36.8}{2.5} = \dfrac{1.1}{2.5} = 0.44$

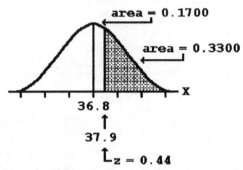

        $p(x > 37.9) = p(z > 0.44)$
            $= p(z > 0) - p(0 < z < 0.44)$
            $= 0.5000 - 0.1700 = 0.3300$

17.    $\mu = 2$ hr 36 min $= 2(60) + 36$ min $= 156$ min, $\sigma = 24$ min

    a)    2 hr 15 min $= 2(60) + 15$ min $= 135$ min

        Convert $x = 135$ to a $z$-number.  $z = \dfrac{x - \mu}{\sigma} = \dfrac{135 - 156}{24} = \dfrac{-21}{24} = -0.875 \approx -0.88$

        Find $-0.88$ on the edge of the body table and read the interior to find 0.3106. Subtract 0.3106 from 0.5000 to find the area to the left of $z = -0.88$.

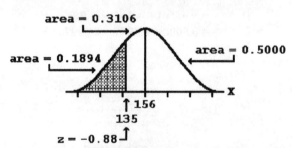

    $p(x < 135) = p(z < -0.88) = p(z < 0) - p(-0.88 < z < 0) = 0.5000 - 0.3106 = 0.1894$

    The drying time has a probability of 18.9% of being less that 2 hours and 15 minutes.

17. Continued.

    b)    2 hr = 2(60) min = 120 min, 3 hr = 3(60) min = 180 min

Convert $x = 120$ to a $z$-number. $z = \dfrac{x-\mu}{\sigma} = \dfrac{120-156}{24} = \dfrac{-36}{24} = -1.50$

Convert $x = 180$ to a $z$-number. $z = \dfrac{x-\mu}{\sigma} = \dfrac{180-156}{24} = \dfrac{24}{24} = 1.00$

Find 1.50 on the edge of the body table and read the interior to find 0.4332.

Find 1.00 on the edge of the body table and read the interior to find 0.3413.

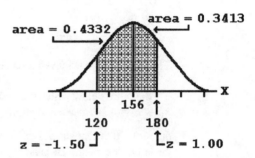

$p(120 < x < 180) = p(-1.50 < z < 1.00)$
$$= p(-1.50 < z < 0) + p(0 < z < 1.00) = 0.4332 + 0.3413 = 0.7745$$

The drying time has a probability of 77.5% of being between 2 and 3 hours.

21.    $\mu = 72$, $\sigma = 12$

    a)    $c$ = the score above which the top realtors will receive the certificate.
        $p(x > c) = 0.10$, $p(x < c) = 0.90$

$p(z > c) = 0.10$, means $c$ is to the right of the mean. Because 50% of the distribution lies to the right of zero, we need to subtract 0.10 from 0.50 to find the area of the body from $z = 0$ to $z = c$. The closest number to 0.40 is 0.3997 which corresponds to $z$-number = 1.28.

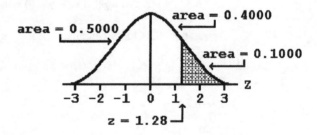

$p(z > 1.28) = p(z > 0) - p(0 < z < 1.28) = 0.5000 - 0.3997 = 0.1003 \approx 0.10$

21. a)  Continued.

Convert $z = 1.28$ to its $x$-number.

$$z = \frac{x - \mu}{\sigma}$$

$$1.28 = \frac{c - 72}{12}$$

$$12(1.28) = c - 72$$

$$15.36 = c - 72$$

$$c = 87.36 \approx 87$$

The realtors who achieved a score of at least 87 will receive a certificate.

b)    $c$ = the score below which the realtors will need to attend the workshop.
$p(x < c) = 0.20, \ p(x > c) = 0.80$

$p(z < c) = 0.20$, means $c$ is to the left of the mean.  Because 50% of the distribution lies to the left of zero, we need to subtract 0.20 from 0.50 to find the area of the body from $z = 0$ to $z = c$.  The closest number to 0.30 is 0.2995 which corresponds to $z$-number $= -0.84$.

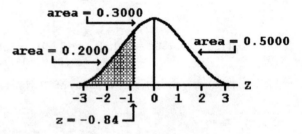

$p(z < 0.84) = p(z < 0) - p(-0.84 < z < 0) = 0.5000 - 0.2995 = 0.2005 \approx 0.20$

Convert $z = -0.84$ to its $x$-number.

$$z = \frac{x - \mu}{\sigma}$$

$$-0.84 = \frac{c - 72}{12}$$

$$12(-0.84) = c - 72$$

$$-10.08 = c - 72$$

$$c = 61.92 \approx 62$$

The realtors who achieved a score of less than 62 will need to attend the workshop.

## Section 4.5                                         Polls and Margin of Error

1.    Find each value in the interior of the body table and read the edges for the $z$-number.

   a)    $z_{0.2517} = 0.68$, that is $p(0 < z < 0.68) = 0.2517$.

   b)    $z_{0.1217} = 0.31$, that is $p(0 < z < 0.31) = 0.1217$.

   c)    $z_{0.4177} = 1.39$, that is $p(0 < z < 1.39) = 0.4177$.

   d)    $z_{0.4960} = 2.65$, that is $p(0 < z < 2.65) = 0.4960$.

5.    For a 92% level of confidence, $\alpha = 0.92$, so $\alpha/2 = 0.46$.

   Find 0.46 in the interior of the body table.  The closest value is 0.4599 which corresponds to $z$-number $= 1.75$.

   $$z_{\alpha/2} = z_{0.46} = 1.75$$

9.    $n = 1018$

   a)    For a 90% level of confidence, $\alpha = 0.90$, so $\alpha/2 = 0.45$.

   Find 0.45 in the interior of the body table.  The closest values are 0.4495 which corresponds to $z$-number $= 1.64$ and 0.4505 which corresponds to $z$-number $= 1.65$.  Use 1.645 which is halfway between the two $z$-numbers.  $z_{\alpha/2} = z_{0.45} = 1.645$

   $$\text{MOE} = \frac{z_{\alpha/2}}{2\sqrt{n}} = \frac{1.645}{2\sqrt{1018}} = 0.025779... \approx 0.026$$

   The margin of error associated with the sample proportion is plus or minus 2.6 percentage points at the 90% level of confidence.

   b)    For a 98% level of confidence, $\alpha = 0.98$, so $\alpha/2 = 0.49$.

   Find 0.49 in the interior of the body table.  The closest value is 0.4901 which corresponds to $z$-number $= 2.33$.  $z_{\alpha/2} = z_{0.49} = 2.33$

   $$\text{MOE} = \frac{z_{\alpha/2}}{2\sqrt{n}} = \frac{2.33}{2\sqrt{1018}} = 0.036513... \approx 0.037$$

   The margin of error associated with the sample proportion is plus or minus 3.7 percentage points at the 98% level of confidence.

13.    $n = 2710$ = number surveyed
yes = 2141 = number who have bought a lottery ticket
no = 569 = number who have not bought a lottery ticket

a)    $p = \dfrac{yes}{n} = \dfrac{2141}{2710} = 0.790037 \approx 0.790$

Sample proportion who bought tickets is 79.0%.

b)    $p = \dfrac{no}{n} = \dfrac{569}{2710} = 0.209963 \approx 0.210$

Sample proportion who have not purchased tickets is 21.0%.

c)    For a 90% level of confidence, $\alpha = 0.90$, so $\alpha/2 = 0.45$.

Find 0.45 in the interior of the body table. The closest values are 0.4495 which corresponds to z-number = 1.64 and 0.4505 which corresponds to z-number = 1.65. Use 1.645 which is halfway between the two z-numbers. $z_{\alpha/2} = z_{0.45} = 1.645$

$$\text{MOE} = \frac{z_{\alpha/2}}{2\sqrt{n}} = \frac{1.645}{2\sqrt{2710}} = 0.015800... \approx 0.016$$

The margin of error associated with the sample proportion is plus or minus 1.6 percentage points at the 90% level of confidence.

17.    $n = 2035$ = teenage Americans surveyed
yes = 1160 = number who said yes
no = 875 = number who said no

a)    $p = \dfrac{yes}{n} = \dfrac{1160}{2035} = 0.570025 \approx 0.570$

Sample proportion who know kids who carry weapons in school is 57.0%.

b)    $p = \dfrac{no}{n} = \dfrac{875}{2035} = 0.429975 \approx 0.430$

Sample proportion who do not know kids who carry weapons in school is 43.0%.

c)    For a 95% level of confidence, $\alpha = 0.95$, so $\alpha/2 = 0.475$.

Find 0.475 in the interior of the body table. The closest value is 0.4750 which corresponds to z-number = 1.96. $z_{\alpha/2} = z_{0.475} = 1.96$

$$\text{MOE} = \frac{z_{\alpha/2}}{2\sqrt{n}} = \frac{1.96}{2\sqrt{2035}} = 0.021724... \approx 0.022$$

The margin of error associated with the sample proportion is plus or minus 2.2 percentage points at the 95% level of confidence.

21.    For a 95% level of confidence, $\alpha = 0.95$, and $\alpha/2 = 0.475$.

Find 0.475 in the interior of the body table.  The closest value is 0.4750 which corresponds to $z$-number $= 1.96$.  $z_{\alpha/2} = z_{0.475} = 1.96$

a)    $n =$ number of males $= 430$

$$\text{MOE} = \frac{z_{\alpha/2}}{2\sqrt{n}} = \frac{1.96}{2\sqrt{430}} = 0.047260... \approx 0.047$$

The margin of error associated with the sample proportion is plus or minus 4.7 percentage points at the 95% level of confidence.

b)    $n =$ number of females $= 765$

$$\text{MOE} = \frac{z_{\alpha/2}}{2\sqrt{n}} = \frac{1.96}{2\sqrt{765}} = 0.035432... \approx 0.035$$

The margin of error associated with the sample proportion is plus or minus 3.5 percentage points at the 95% level of confidence.

c)    $n =$ total number $= 430 + 765 = 1195$

$$\text{MOE} = \frac{z_{\alpha/2}}{2\sqrt{n}} = \frac{1.96}{2\sqrt{1195}} = 0.028349... \approx 0.028$$

The margin of error associated with the sample proportion is plus or minus 2.8 percentage points at the 95% level of confidence.

25.    $n =$ total sample size $= 640 + 820 = 1460$
MOE $= 2.6\% = 0.026$

$$\text{MOE} = \frac{z_{\alpha/2}}{2\sqrt{n}}$$

$$0.026 = \frac{z_{\alpha/2}}{2\sqrt{1460}} = \frac{z_{\alpha/2}}{76.41989}$$

$z_{\alpha/2} = (0.026)(76.41989) = 1.9869 \approx 1.99$

Find $z$-number $= 1.99$ on the edges of the body table and locate $\alpha = 0.4767$ in the interior.

If $\alpha/2 = 0.4767$, then $\alpha = 2(0.4767) = 0.9534$

Based on a sample size of 1460, we are 95.3% confident that the margin of error is at most 2.6 percentage points.

| Section 4.6 | Linear Regression |
| --- | --- |

1.    a)    $n = 6, \Sigma x = 64, \Sigma x^2 = 814, \Sigma y = 85, \Sigma y^2 = 1351, \Sigma xy = 1039$

$$m = \frac{n(\Sigma xy) - (\Sigma x)(\Sigma y)}{n(\Sigma x^2) - (\Sigma x)^2} = \frac{6(1039) - 64(85)}{6(814) - (64)^2} = \frac{6234 - 5440}{4884 - 4096} = \frac{794}{788} = 1.00761421$$

$$b = \bar{y} - m\bar{x} = \frac{85}{6} - 1.00761421\left(\frac{64}{6}\right) = 14.16666667 - 10.74788494 = 3.41878173$$

$$\hat{y} = mx + b = 1.00761421x + 3.41878173$$

Rounding gives $\hat{y} = 1.0x + 3.4$, the line of best fit.

    b)    Substitute $x = 11$.

$$\hat{y} = 1.0x + 3.4 = 1.0(11) + 3.4 = 11 + 3.4 = 14.4$$

    c)    Substitute $y = 19$ for $\hat{y}$.

$$\hat{y} = 1.0x + 3.4$$

$$19 = 1.0x + 3.4$$

$$x = 19 - 3.4 = 15.6$$

    d)
$$r = \frac{n(\Sigma xy) - (\Sigma x)(\Sigma y)}{\sqrt{n(\Sigma x^2) - (\Sigma x)^2}\sqrt{n(\Sigma y^2) - (\Sigma y)^2}} = \frac{6(1039) - 64(85)}{\sqrt{6(814) - (64)^2}\sqrt{6(1351) - (85)^2}}$$

$$= \frac{794}{\sqrt{4884 - 4096}\sqrt{8106 - 7225}} = \frac{794}{\sqrt{788}\sqrt{881}} = 0.9529485$$

    e)    Yes, the predictions in parts (b) and (c) are reliable because the coefficient of correlation is close to 1.

5.   a)   Yes, the ordered pairs do exhibit a linear trend.

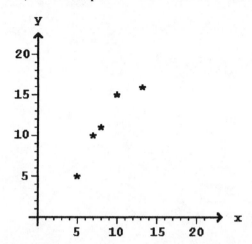

b)

| $(x, y)$ | $x$ | $x^2$ | $y$ | $y^2$ | $xy$ |
|---|---|---|---|---|---|
| (5, 5) | 5 | 25 | 5 | 25 | 25 |
| (7, 10) | 7 | 49 | 10 | 100 | 70 |
| (8, 11) | 8 | 64 | 11 | 121 | 88 |
| (10, 15) | 10 | 100 | 15 | 225 | 150 |
| (13, 16) | 13 | 169 | 16 | 256 | 208 |
| $n = 5$ | $\Sigma x = 43$ | $\Sigma x^2 = 407$ | $\Sigma y = 57$ | $\Sigma y^2 = 727$ | $\Sigma xy = 541$ |

$$m = \frac{n(\Sigma xy) - (\Sigma x)(\Sigma y)}{n(\Sigma x^2) - (\Sigma x)^2} = \frac{5(541) - (43)(57)}{5(407) - (43)^2} = \frac{2705 - 2451}{2035 - 1849} = \frac{254}{186} = 1.36559140$$

$$b = \overline{y} - m\overline{x} = \frac{57}{5} - 1.36559140\left(\frac{43}{5}\right) = 11.4 - 11.744408602 = -0.34408602$$

$\hat{y} = mx + b = 1.36559140x - 0.34408602$

Rounding gives $\hat{y} = 1.4x - 0.3$, the line of best fit.

c)   Substitute $x = 9$.

$\hat{y} = 1.4x - 0.3 = 1.4(9) - 0.3 = 12.6 - 0.3 = 12.3$

5. Continued.

d)

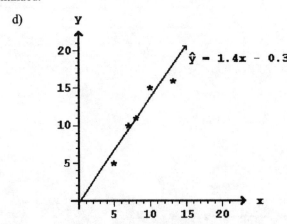

$$\hat{y} = 1.4x - 0.3$$

e)

$$r = \frac{n(\sum xy) - (\sum x)(\sum y)}{\sqrt{n(\sum x^2) - (\sum x)^2} \sqrt{n(\sum y^2) - (\sum y)^2}} = \frac{5(541) - (43)(57)}{\sqrt{5(407) - (43)^2} \sqrt{5(727) - (57)^2}}$$

$$= \frac{2705 - 2451}{\sqrt{2035 - 1849} \sqrt{3635 - 3249}} = \frac{254}{\sqrt{186} \sqrt{386}} = 0.9479459$$

f)    Yes, the prediction in part (c) is reliable because the coefficient of correlation is close to 1.

9.    a)    Yes, the data does exhibit a linear trend.

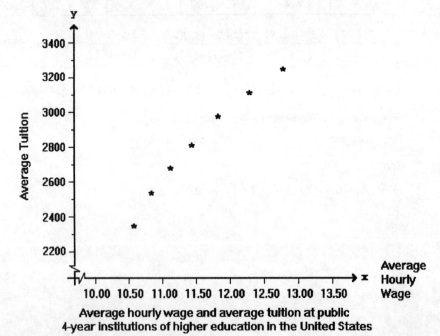

Average hourly wage and average tuition at public
4-year institutions of higher education in the United States

9. Continued.

b)

| Year | Wage, $x$ | $x^2$ | Tuition, $y$ | $y^2$ | $xy$ |
|------|-----------|-------|--------------|-------|------|
| 1992 | 10.57 | 111.7249 | 2349 | 5,517,801 | 24,828.93 |
| 1993 | 10.83 | 117.2889 | 2537 | 6,436,369 | 27,475.71 |
| 1994 | 11.12 | 123.6544 | 2681 | 7,187,761 | 29,812.72 |
| 1995 | 11.43 | 130.6449 | 2811 | 7,901,721 | 32,129.73 |
| 1996 | 11.82 | 139.7124 | 2975 | 8,850,625 | 35,164.50 |
| 1997 | 12.28 | 150.7984 | 3111 | 9,678,321 | 38,203.08 |
| 1998 | 12.77 | 163.0729 | 3247 | 10,543,009 | 41,464.19 |
| $n=7$ | $\Sigma x =$ 80.82 | $\Sigma x^2 =$ 936.8968 | $\Sigma y =$ 19,711 | $\Sigma y^2 =$ 56,115,607 | $\Sigma xy =$ 229,078.86 |

$$m = \frac{n(\Sigma xy)-(\Sigma x)(\Sigma y)}{n(\Sigma x^2)-(\Sigma x)^2} = \frac{7(229,078.86)-(80.82)(19,711)}{7(936.8968)-(80.82)^2}$$

$$= \frac{1,603,522.02-1,593,043.02}{6558.2776-6531.8724} = \frac{10,509}{26.4052} = 397.9898$$

$$b = \bar{y} - m\bar{x} = \frac{19,711}{7} - 397.9898\left(\frac{80.82}{7}\right) = 2815.857 - 4595.076 = -1779.219$$

$$\hat{y} = mx + b = 397.9898x - 1779.219$$

Rounding gives $\hat{y} = 398.0x - 1779.2$, the line of best fit.

c)   Substitute $x = 12.50$ to predict the average tuition when the average hourly wage is $12.50.

$$\hat{y} = 398.0(12.50) - 1779.2 = 4975.0 - 1779.2 = \$3195.80$$

d)   Substitute $\hat{y} = 3000$ to predict the average hourly wage when the average tuition is $3,000.

$$\hat{y} = 398.0x - 1779.22$$

$$3000 = 398.0x - 1779.22$$

$$398.0x = 3000 + 1779.2 = 4779.2$$

$$x = \frac{4779.2}{398.0} = 12.0080 \approx \$12.01$$

**9. Continued.**

e)
$$r = \frac{n(\Sigma xy) - (\Sigma x)(\Sigma y)}{\sqrt{n(\Sigma x^2) - (\Sigma x)^2}\sqrt{n(\Sigma y^2) - (\Sigma y)^2}}$$

$$= \frac{7(229,078.86) - (80.82)(19,711)}{\sqrt{7(936.8968) - (80.82)^2}\sqrt{7(56,115,607) - (19,711)^2}}$$

$$= \frac{1,603,522.02 - 1,593,043.02}{\sqrt{6558.2776 - 6531.8724}\sqrt{392,809,249 - 388,523,521}}$$

$$= \frac{10,509}{\sqrt{26.4052}\sqrt{4,285,728}} = \frac{10,509}{(5.1386)(2070.20)}$$

$$= 0.987880 \approx 0.988$$

f)      Yes, the predictions in parts (c) and (d) are reliable because the coefficient of correlation is close to 1.

---

**Chapter 4**                                                                                     **Review**

---

1.

| $x$ | $x^2$ |     | $x$ | $x^2$ |
|-----|-------|-----|-----|-------|
| 5   | 25    |     | 10  | 100   |
| 8   | 64    |     | 6   | 36    |
| 10  | 100   |     | 8   | 64    |
| 4   | 16    |     | 7   | 49    |
| 8   | 64    |     | 5   | 25    |
|     |       |     | $\Sigma x = 71$ | $\Sigma x^2 = 543$ |

a)      $n = 10$, mean $= \bar{x} = \dfrac{\Sigma x}{n} = \dfrac{71}{10} = 7.1$

b)      median:   Arrange the numbers in order from low to high.  Because there are $n = 10$ data points, the location of the median is $L = \dfrac{10+1}{2} = 5.5$; add the 5th and 6th number and divide by 2.

    4, 5, 5, 6, 7, 8, 8, 8, 10, 10
              ↑ ↑
          $\dfrac{7+8}{2} = 7.5$

c)      mode:  The number with the highest frequency is 8.  It occurs three times.

1. Continued.

    d)    variance:

$$s^2 = \frac{1}{n-1}\left(\sum x^2 - \frac{(\sum x)^2}{n}\right) = \frac{1}{10-1}\left(543 - \frac{(71)^2}{10}\right) = \frac{1}{9}(543 - 504.1) = \frac{38.9}{9} = 4.322...$$

    standard deviation $= s = \sqrt{\text{variance}} = \sqrt{s^2} = \sqrt{4.322...} = 2.07899... \approx 2.1$

5.    $x$ = the last test score

$$\bar{x} = \frac{74 + 65 + 85 + 76 + x}{5} \geq 80$$

$$\frac{300 + x}{5} \geq 80$$

$$300 + x \geq 80(5)$$

$$300 + x \geq 400$$

$$x \geq 400 - 300 = 100$$

You must score at least 100 on the next exam to average at least 80.

9.    a)    "Weights of motorcycles" are **continuous**, because they would not necessarily be a whole number.

    b)    "Colors of motorcycles" are **neither** because they do not have a number value associated with them.

    c)    "Number of motorcycles" is **discrete** because you count whole numbers of motorcycles.

    d)    "Ethnic background of students" is **neither** because numbers are not associated with backgrounds.

    e)    "Number of students" is **discrete** because you count whole numbers of students.

    f)    "Amounts of time spent studying" is **continuous** because you can measure time in parts of minutes or seconds.

13.    $\mu = 420$, $\sigma = 45$

$p(z < c) = 0.3400$, means $c$ is to the left of the mean. Because 50% of the distribution lies to the left of zero, we need to subtract 0.34 from 0.5 to find the area of the body from $z = 0$ to $z = c$. The closest number to 0.16 is 0.1591 which corresponds to $z$-number $= 0.41$. Since $c$ is left of the mean, $z$-number $= -0.41$.

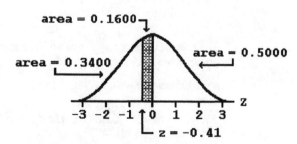

$p(z < -0.41) = p(z < 0) - p(-0.41 < z < 0) = 0.5000 - 0.1600 = 0.3400$.

13. Continued.

Convert $z = -0.41$ to its $x$-number.

$$z = \frac{x - \mu}{\sigma}$$

$$-0.41 = \frac{x - 420}{45}$$

$$45(-0.41) = x - 420$$

$$-18.45 = x - 420$$

$$x = 401.55$$

$$p(z < -0.41) = p(x < 401.55) \approx 0.3400$$

Students who score 401 or lower will have to take the review course. (Do not round to 402 in this situation because a student with that score would not need to take the review course.)

17. For a 95% level of confidence, $\alpha = 0.95$, and $\alpha/2 = 0.475$.

Find 0.475 in the interior of the body table. The closest value is 0.4750 which corresponds to $z$-number = 1.96.

$$z_{\alpha/2} = z_{0.475} = 1.96$$

a) $n$ = number of males = 580

$$\text{MOE} = \frac{z_{\alpha/2}}{2\sqrt{n}} = \frac{1.96}{2\sqrt{580}} = 0.040692 \approx 0.041$$

The margin of error associated with the sample proportion is plus or minus 4.1 percentage points at the 95% level of confidence.

b) $n$ = number of females = 970

$$\text{MOE} = \frac{z_{\alpha/2}}{2\sqrt{n}} = \frac{1.96}{2\sqrt{970}} = 0.031466 \approx 0.031$$

The margin of error associated with the sample proportion is plus or minus 3.1 percentage points at the 95% level of confidence.

c) $n$ = total number = 580 + 970 = 1550

$$\text{MOE} = \frac{z_{\alpha/2}}{2\sqrt{n}} = \frac{1.96}{2\sqrt{1550}} = 0.024892 \approx 0.025$$

The margin of error associated with the sample proportion is plus or minus 2.5 percentage points at the 95% level of confidence.

21.  a)    Yes, the data does exhibit a linear trend.

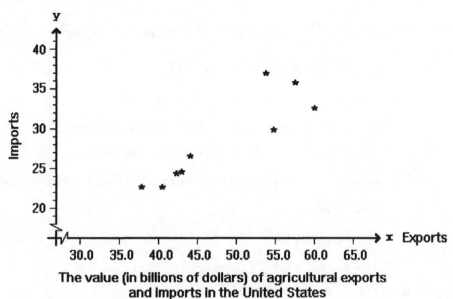

The value (in billions of dollars) of agricultural exports
and imports in the United States

b)    The data is given in thousands of dollars.  Data in the table has been rounded to three decimal places, but all digits were used in the computations.

| Year | Exports, $x$ | $x^2$ | Imports, $y$ | $y^2$ | $xy$ |
|------|------|------|------|------|------|
| 1990 | 40.4 | 1632.16 | 22.7 | 515.29 | 917.08 |
| 1991 | 37.8 | 1428.84 | 22.7 | 515.29 | 858.06 |
| 1992 | 42.6 | 1814.76 | 24.5 | 600.25 | 1043.70 |
| 1993 | 42.9 | 1840.41 | 24.6 | 605.16 | 1055.34 |
| 1994 | 44.0 | 1936.00 | 26.6 | 707.56 | 1170.40 |
| 1995 | 54.7 | 2992.09 | 29.9 | 894.01 | 1635.53 |
| 1996 | 59.9 | 3588.01 | 32.6 | 1062.76 | 1952.74 |
| 1997 | 57.4 | 3294.76 | 35.8 | 1281.64 | 2054.92 |
| 1998 | 53.7 | 2883.69 | 37.0 | 1369.00 | 1986.90 |
| $n = 9$ | $\sum x =$ 433.4 | $\sum x^2 =$ 21,410.72 | $\sum y =$ 256.4 | $\sum y^2 =$ 7550.96 | $\sum xy =$ 12,674.67 |

$$m = \frac{n(\sum xy) - (\sum x)(\sum y)}{n(\sum x^2) - (\sum x)^2} = \frac{9(12,674.67) - (433.4)(256.4)}{9(21,410.72) - (433.4)^2}$$

$$= \frac{114,072.03 - 111,123.76}{192,696.48 - 187,835.56} = \frac{2948.27}{4860.92} = 0.606525$$

21. b) Continued.

$$b = \bar{y} - m\bar{x} = \frac{256.4}{9} - 0.606525\left(\frac{433.4}{9}\right) = 28.488889 - 29.207553 = -0.718664$$

$$\hat{y} = mx + b = 0.606525x - 0.718664$$

Rounding gives $\hat{y} = 0.607x - 0.719$, the line of best fit.

c)    If the value of agricultural exports is 50 billion dollars, substitute $x = 50$.

$$\hat{y} = 0.607(50) - 0.719 = 30.35 - 0.719 = 29.631 \text{ billion dollars}$$

d)    Substitute $\hat{y} = 35.0$ to predict the value of agricultural exports when the value of imports is 35.0 billion dollars.

$$\hat{y} = 0.607x - 0.719$$

$$35.0 = 0.607x - 0.719$$

$$0.607x = 35.0 + 0.719 = 35.719$$

$$x = \frac{35.719}{0.607} = 58.845140 \approx 58.845 \text{ billion dollars}$$

e)
$$r = \frac{n(\Sigma xy) - (\Sigma x)(\Sigma y)}{\sqrt{n(\Sigma x^2) - (\Sigma x)^2}\sqrt{n(\Sigma y^2) - (\Sigma y)^2}}$$

$$= \frac{9(12,674.67) - (433.4)(256.4)}{\sqrt{9(21,410.72) - (433.4)^2}\sqrt{9(7550.96) - (256.4)^2}}$$

$$= \frac{114,072.03 - 111,123.76}{\sqrt{192,696.48 - 187,835.56}\sqrt{67,958.64 - 65,740.96}}$$

$$= \frac{2948.27}{\sqrt{4860.92}\sqrt{2217.68}} = \frac{2948.27}{(69.7203)(47.0922)}$$

$$= 0.897963 \approx 0.898$$

f)    Yes, the predictions in parts (c) and (d) are reliable because the coefficient of correlation is close to 1.

# 5  Finance

It is recommended that the method of using the calculator that is given in the textbook be followed in solving the exercises from this chapter, however, the intermediate calculations have been supplied to assist in checking your work. All decimal digits were used in calculating the final answers even though intermediate steps may be shown as rounded. Final answers and intermediate results may differ slightly depending upon the accuracy of your calculator.

1.  a)  $P = \$2000, r = 8\% = 0.08, t = 3$ years.  Calculate $I$.

   $I = Prt = (2000)(0.08)(3) = \$480.00$

   b)  The lender will receive $2000 plus $480 interest at the end of 3 years.

5.  a)  $P = \$1410, r = 12\frac{1}{4}\% = 0.1225, t = 325$ days $= \dfrac{325}{365}$ years.  Calculate $I$.

   $I = Prt = (1410)(0.1225)\left(\dfrac{325}{365}\right) = 153.79623 \approx \$153.80$

   b)  The lender will receive $1410 plus $153.80 interest at the end of 325 days.

9.  a)  $P = \$12,430, r = 5\frac{7}{8}\% = 0.05875, t = 2$ years 3 months $= (24 + 3)$ months $= \dfrac{27}{12}$ years

   $FV = P(1 + rt)$ \qquad\qquad Calculate $FV$.

   $FV = 12,430\left(1 + 0.05875\left(\dfrac{27}{12}\right)\right) = 12,430(1 + 0.1321875)$

   $= 12,430(1.1321875) = 14,073.0906 \approx \$14,073.09$

   b)  The lender will receive $14,073.09 at the end of 2 years 3 months.

13. a)  $P = \$5900, r = 14\frac{1}{2}\% = 0.145, t = 112$ days $= \dfrac{112}{365}$ years

   $FV = P(1 + rt)$ \qquad\qquad Calculate $FV$.

   $FV = 5900\left(1 + 0.145\left(\dfrac{112}{365}\right)\right) = 6162.5096 \approx \$6162.51$

   b)  The lender will receive $6162.51 at the end of 112 days.

17.  a)  $FV = \$8600, r = 9\frac{1}{2}\% = 0.095, t = 3$ years

$FV = P(1 + rt)$        Calculate $P$.

$8600 = P(1 + (0.095)(3))$

$8600 = P(1.285)$        Divide by 1.285.

$P = \dfrac{8600}{1.285} = 6692.6070 \approx \$6692.61$

b)  A principal of \$6692.61 would generate a future value of \$8600 in 3 years.

21.  a)  $FV = \$1311, r = 6\frac{1}{2}\% = 0.065, t = 317$ days $= \dfrac{317}{365}$ years

$FV = P(1 + rt)$        Calculate $P$.

$1311 = P\left(1 + 0.065\left(\dfrac{317}{365}\right)\right)$

$1311 = P(1.056452)$        Divide by 1.056452.

$P = \dfrac{1311}{1.056452} = 1240.9460 \approx \$1240.95$

b)  A principal of \$1240.95 would generate a future value of \$1311 in 317 days.

25.  $P = \$1312.82, FV = \$1615.00, r = 6\frac{7}{8}\% = 0.06875$

$FV = P(1 + rt)$        Calculate $t$.

$1615.00 = 1312.82(1 + 0.06875t)$        Divide by 1312.82.

$\dfrac{1615.00}{1312.82} = 1 + 0.06875t$

$1.2301763 = 1 + 0.06875t$        Subtract 1.

$0.2301763 = 0.06875t$        Divide by 0.06875.

$t = \dfrac{0.2301763}{0.06875} = 3.348 \approx 3\frac{1}{3}$ years

29.  $P = $ Loan Amount $= \$3700 - \$500 = \$3200, r = 9.8\% = 0.098, t = 36$ months $= 3$ years

$FV = P(1 + rt)$    Calculate $FV$ and then divide by 36

$FV = 3200(1 + (0.098)(3)) = \$4140.80$

Monthly Payment $= \dfrac{4140.80}{36} = 115.0222 \approx \$115.02$

33.   Billing period = March 1 to March 31,
      Previous Balance = $157.14

| Time Interval | Days | Daily Balance |
|---|---|---|
| March 1 - March 4 | 4 | 157.14 |
| March 5 - March 16 | 12 | $157.14 - 25.00 = 132.14$ |
| March 17 - March 31 | 15 | $132.14 + 36.12 = 168.26$ |

$$\text{Average Daily Balance} = \frac{4(157.14) + 12(132.14) + 15(168.26)}{4 + 12 + 15} = \frac{4738.14}{31} = 152.8432 \approx \$152.84$$

$P = \$152.8432, r = 21\% = 0.21, t = 31 \text{ days} = \dfrac{31}{365} \text{ years}$

$I = Prt = (152.8432)(0.21)\left(\dfrac{31}{365}\right) = 2.726053 \approx \$2.73$

37.   a)    Down Payment = 10% of purchase price = 389,400(0.10) = \$38,940

      b)    Borrowed from bank = 80% of purchase price = 389,400(0.80) = \$311,520

      c)    Borrowed from seller = 10% of purchase price = 389,400(0.10) = \$38,940

      d)    $P$ = 10% of purchase price = 389,400(0.10) = \$38,940,

            $r = 11\% = 0.11, t = 1 \text{ month} = \dfrac{1}{12} \text{ year.}$

            $I = Prt = (38,940)(0.11)\left(\dfrac{1}{12}\right) = \$356.95$

      e)    Total income = 10% of purchase price + 10% Promissory Note
                          + monthly interest payments for 48 months

            $= (0.10)(389,400) + (0.10)(389,400) + 48(356.95)$

            $= 38,940 + 38,940 + 17,133.60 = \$95,013.60$

      f)    Income from bank = 80% of purchase price − 6% commission

            $= (0.80)(389,400) - (0.06)(389,400)$

            $= 311,520 - 23,364 = \$288,156$

      g)    Total income = income from bank + down payment income
            $= 288,156 + 95,013.60 = \$383,169.60$

---

| **Section 5.2** | **Compound Interest** |
|---|---|

1.   Compound rate $= 12\% = 0.12$,   $i = $ periodic rate $= \dfrac{\text{compound rate}}{\text{number of periods per year}}$

   a) quarterly $= 4$ times per year, so $i = \dfrac{0.12}{4} = 0.03$

   b) monthly $= 12$ times per year, so $i = \dfrac{0.12}{12} = 0.01$

   c) daily $= 365$ times per year, so $i = \dfrac{0.12}{365} = 0.000328767$

   d) biweekly $= 26$ times per year, so $i = \dfrac{0.12}{26} = 0.0046153846$

   e) semimonthly $= 24$ times per year, so $i = \dfrac{0.12}{24} = 0.005$

5.   Compound Rate $= 9.7\% = 0.097$.  See problem 1 for formula and times per year for each section.

   a) quarterly:  $i = \dfrac{0.097}{4} = 0.02425$

   b) monthly:  $i = \dfrac{0.097}{12} = 0.008083333$

   c) daily:  $i = \dfrac{0.097}{365} = 0.000265753$

   d) biweekly:  $i = \dfrac{0.097}{26} = 0.003730769$

   e) semimonthly:  $i = \dfrac{0.097}{24} = 0.0040416667$

9.   $t = $ number of years $= 30$, $n = $ number of periods $= $ (number of years)$\dfrac{\text{number of periods}}{1 \text{ year}}$

   a) quarter:  $n = (30 \text{ years})\left(\dfrac{4 \text{ quarters}}{1 \text{ year}}\right) = 120$ quarters

   b) month:  $n = (30 \text{ years})\left(\dfrac{12 \text{ months}}{1 \text{ year}}\right) = 360$ months

   c) day:  $n = (30 \text{ years})\left(\dfrac{365 \text{ days}}{1 \text{ year}}\right) = 10{,}950$ days

13.   a)   $P = \$5200, i = \dfrac{6.75\%}{4} = \dfrac{0.0675}{4}, n = \left(8\dfrac{1}{2} \text{ years}\right)\left(\dfrac{4 \text{ quarters}}{1 \text{ year}}\right) = 34 \text{ quarters}$

$$FV = P(1+i)^n = 5200\left(1+\dfrac{0.0675}{4}\right)^{34}$$

$$= 5200(1.016875)^{34} = 5200(1.7664339) = \$9185.4563 \approx \$9185.46$$

   b)   This means the value of $5200 earning $6\dfrac{3}{4}\%$ interest compounded quarterly after $8\dfrac{1}{2}$ years would be $9185.46.

17.   a)   $i = \dfrac{8\%}{12} = \dfrac{0.08}{12}, n = \dfrac{12 \text{ months}}{1 \text{ year}} = 12 \text{ months}, t = 1 \text{ year}$

   $FV \text{ (compound interest)} = FV \text{ (simple interest)}$

$$P(1+i)^n = P(1+rt) \qquad \text{Calculate } r.$$

$$(1+i)^n = (1+rt) \qquad \text{Divide by } P.$$

$$\left(1+\dfrac{0.08}{12}\right)^{12} = 1 + r(1)$$

$$(1.0066667)^{12} = 1 + r$$

$$1.0829995 = 1 + r$$

$$r = 0.0829995 \approx 8.30\%$$

   b)   This means that in 1 year's time 8% interest compounded monthly would have the same effect as 8.30% simple interest.

21.   a)   $i = \dfrac{10\%}{4} = \dfrac{0.10}{4}, n = \dfrac{4 \text{ quarters}}{1 \text{ year}} = 4 \text{ quarters}, t = 1 \text{ year}$

$$(1+i)^n = 1 + rt \qquad \text{Calculate } r.$$

$$\left(1+\dfrac{0.10}{4}\right)^4 = 1 + r(1)$$

$$(1.025)^4 = 1 + r$$

$$1.1038129 = 1 + r$$

$$r = 0.1038129 \approx 10.38\%$$

   This means that in 1 year's time 10% interest compounded quarterly would have the same effect as 10.38% simple interest.

21. Continued.

b) $i = \dfrac{10\%}{12} = \dfrac{0.10}{12}$, $n = \dfrac{12 \text{ months}}{1 \text{ year}} = 12 \text{ months}$, $t = 1 \text{ year}$

$(1 + i)^n = 1 + rt$ \qquad Calculate $r$.

$\left(1 + \dfrac{0.10}{12}\right)^{12} = 1 + r(1)$

$(1.00833333)^{12} = 1 + r$

$1.1047131 = 1 + r$

$r = 0.1047131 \approx 10.47\%$

This means that in 1 year's time 10% interest compounded monthly would have the same effect as 10.47% simple interest.

c) $i = \dfrac{10\%}{365} = \dfrac{0.10}{365}$, $n = \dfrac{365 \text{ days}}{1 \text{ year}} = 365 \text{ days}$, $t = 1 \text{ year}$

$(1 + i) = 1 + rt$ \qquad Calculate $r$.

$\left(1 + \dfrac{0.10}{365}\right)^{365} = 1 + r(1)$

$(1.00027397)^{365} = 1 + r$

$1.1051558 = 1 + r$

$r = 0.1051558 \approx 10.52\%$

This means that in 1 year's time 10% interest compounded daily would have the same effect as 10.52% simple interest.

25. a) $FV = \$3,758$, $i = \dfrac{11\frac{7}{8}\%}{12} = \dfrac{0.11875}{12}$

$n = 17 \text{ years} \left(\dfrac{12 \text{ months}}{1 \text{ year}}\right) + 7 \text{ months} = 17(12) + 7 = 211 \text{ months}$

$FV = P(1 + i)^n$ \qquad Calculate $P$.

25. a) Continued.

$$3{,}758 = P\left(1 + \frac{0.11875}{12}\right)^{211}$$

$$3{,}758 = P(1.00989583)^{211}$$

$$3{,}758 = 7.9865360P$$

$$P = \frac{3{,}758}{7.9865360} = 470.54192 \approx \$470.54$$

b)   This means \$470.54 must be invested now at $11\frac{7}{8}$ % interest compounded monthly for 17 years 7 months to obtain \$3758.

29.   Interest compounded daily: $i = \dfrac{6.5\%}{365} = \dfrac{0.065}{365}$, $n = 365$ days, $t = 1$ year (Calculating annual yield)

$FV$(compound interest) = $FV$(simple interest)

$$P(1 + i)^{n} = P(1 + rt) \qquad\qquad \text{Calculate } r.$$

$$(1 + i)^{n} = (1 + rt)$$

$$\left(1 + \frac{0.065}{365}\right)^{365} = 1 + r(1)$$

$$(1.00017808)^{365} = 1 + r$$

$$1.0671529 = 1 + r$$

$$r = 0.0671529 \approx 6.72\% \qquad \text{which verifies the advertised yield}$$

33.   $P = \$3000$, $r = 6\dfrac{1}{2}\% = 0.065\%$, $i = \dfrac{0.065}{365}$

a)    $n = 18 \text{ years}\left(\dfrac{365 \text{ days}}{1 \text{ year}}\right) = 6570 \text{ days}$

$$FV = P(1 + i)^{n} \qquad\qquad \text{Calculate } FV.$$

$$FV = 3000\left(1 + \frac{0.065}{365}\right)^{6570}$$

$$= 3000(1.000178082)^{6570}$$

$$= 9664.9653 \approx \$9664.97$$

33. Continued.

    b)    $P = \$9664.97$, $i = \dfrac{0.065}{365}$, $n = 30$ days

$$FV = P(1+ i)^{\,n} \quad \text{Calculate } FV$$

$$FV = 9664.97\left(1 + \dfrac{0.065}{365}\right)^{30}$$

$$= 9664.97(1.000178082)^{30}$$

$$= 9716.7383 \approx \$9716.74$$

Interest earned = Future Value – Principal = 9716.74 – 9664.97 = \$51.77

37.    $FV = \$100,000$ at age 65

    a)    $i = \dfrac{8\frac{3}{8}\%}{365} = \dfrac{0.08375}{365}$, $n = (65 - 35 \text{ years})(365 \text{ days}) = 30(365) = 10,950$ days

$$FV = P(1 + i)^{\,n} \qquad\qquad \text{Calculate } P.$$

$$100,000 = P\left(1 + \dfrac{0.08375}{365}\right)^{10,950}$$

$$100,000 = P(1.00022945)^{\,10,950}$$

$$100,000 = P(12.332176)$$

$$P = \dfrac{100,000}{12.332176} = 8108.8738 \approx \$8108.87$$

    b)    $P = \$100,000$, $i = \dfrac{0.08375}{365}$, $n = 30$ days

$$FV = P(1 + i)^{\,n} \qquad\qquad \text{Calculate } FV.$$

$$FV = 100,000\left(1 + \dfrac{0.08375}{365}\right)^{30}$$

$$= 100,000(1.000229452)^{30}$$

$$= 100,690.65 \approx \$100,690.65$$

Interest earned = Future Value – Principal = 100,690.65 – 100,000 = \$690.65

41. Find $r$ when $t = 1$ year.

$$P(1 + i)^{n} = P(1 + rt)$$

$P(1 + i)^{n} = P(1 + r(1))$      Substitute $t = 1$.

$(1 + i)^{n} = 1 + r$      Divide by $P$.

$(1 + i)^{n} - 1 = r$      Subtract 1.

$r = (1 + i)^{n} - 1$      Rewriting.

45. $r = 5\frac{5}{8}\% = 0.05625$. Use the formula, $r = (1 + i)^{n} - 1$, derived in Exercise 41

   a) semiannually: $n = 2$, $i = \dfrac{0.05625}{2}$

   $$r = \left(1 + \frac{0.05625}{2}\right)^{2} - 1 = (1.028125)^{2} - 1 = (1.0570410) - 1 = 0.0570410 \approx 5.70\%$$

   b) quarterly: $n = 4$, $i = \dfrac{0.05625}{4}$

   $$r = \left(1 + \frac{0.05625}{4}\right)^{4} - 1 = (1.0140625)^{4} - 1 = (1.0574477) - 1 = 0.0574477 \approx 5.74\%$$

   c) monthly: $n = 12$, $i = \dfrac{0.05625}{12}$

   $$r = \left(1 + \frac{0.05625}{12}\right)^{12} - 1 = (1.0046875)^{12} - 1 = (1.0577231) - 1 = 0.0577231 \approx 5.77\%$$

   d) daily: $n = 365$, $i = \dfrac{0.05625}{365}$

   $$r = \left(1 + \frac{0.05625}{365}\right)^{365} - 1 = (1.0001541)^{365} - 1 = (1.0578575) - 1 = 0.0578575 \approx 5.79\%$$

   e) biweekly: $n = 26$, $i = \dfrac{0.05625}{26}$

   $$r = \left(1 + \frac{0.05625}{26}\right)^{26} - 1 = (1.0021635)^{26} - 1 = (1.0577978) - 1 = 0.0577978 \approx 5.78\%$$

   f) semimonthly: $n = 24$, $i = \dfrac{0.05625}{24}$

   $$r = \left(1 + \frac{0.05625}{24}\right)^{24} - 1 = (1.00234375)^{24} - 1 = (1.0577925) - 1 = 0.0577925 \approx 5.78\%$$

57. The discrepancy would be due to rounding in the computations.

61.    <u>First part</u>

$$FV = \$15,000,\ P = \$10,000,\ i = \frac{8.125\%}{365} = \frac{0.08125}{365}$$

$$FV = P(1+i)^n \qquad\qquad \text{Calculate } n.$$

$$15,000 = 10,000\left(1+\frac{0.08125}{365}\right)^n$$

$$1.5 = \left(1+\frac{0.08125}{365}\right)^n$$

Equations: $y_1 = 1.5$ and $y_2 = \left(1+\frac{0.08125}{365}\right)^n$

Simple Interest Approximation:

$$15,000 = 10,000(1 + 0.08125t)$$
$$1.5 = 1 + 0.08125t$$
$$0.5 = 0.08125t$$
$$t = \frac{0.5}{0.8125} = 6.15384\ldots \approx 6 \text{ years}$$

$$x\text{max} = 6(365) = 2190 \approx 3000$$

$y\text{max} = 2$ for graphing purposes because $y_1 = 1.5$.

$x\text{scl} = 500$ and $y\text{scl} = 0.5$

Graph and find the intersection: $x = 1821.6767,\ y = 1.5$

Number of periods $= 1822$ days  or  $\dfrac{1822}{365} = 4.99$ years

<u>Second part</u>

$$FV = \$100,000,\ P = \$10,000,\ i = \frac{8.125\%}{365} = \frac{0.08125}{365}$$

$$FV = P(1+i)^n \qquad\qquad \text{Calculate } n.$$

$$100,000 = 10,000\left(1+\frac{0.08125}{365}\right)^n$$

$$10 = \left(1+\frac{0.08125}{365}\right)^n$$

Equations: $y_1 = 10$ and $y_2 = \left(1+\frac{0.08125}{365}\right)^n$

61. Continued.

Simple Interest Approximation:

$$10,000 = 10,000(1 + 0.08125t)$$
$$10 = 1 + 0.08125t$$
$$9 = 0.08125t$$
$$t = \frac{9}{0.8125} = 110.77... \approx 110 \text{ years}$$

$xmax = 110(365) = 40,150 \approx 40,000$

$ymax = 20$ for graphing purposes because $y_1 = 10$.

$xscl = 10,000$ and $yscl = 1$

Graph and find the intersection: $x = 10,345.07..., y = 10$

Number of periods = 10,346 days  or  $\dfrac{10,346}{365} = 28.3452... \approx 28.35$ years

---

## Section 5.3 | Annuities

1.  a)  pymt = \$120 monthly, $i = \dfrac{5\frac{3}{4}\%}{12} = \dfrac{0.0575}{12}$, $n = 1$ year = 12 months

$$FV = \text{pymt} \, \frac{(1+i)^n - 1}{i} \qquad\qquad \text{Calculate ordinary } FV.$$

$$FV = 120 \, \frac{\left(1 + \dfrac{0.0575}{12}\right)^{12} - 1}{\dfrac{0.0575}{12}}$$

$$= 120 \, \frac{(1.0047917)^{12} - 1}{0.0047917} = 120 \left( \frac{1.0590398 - 1}{0.0047917} \right) = 120 \left( \frac{0.0590398}{0.0047917} \right)$$

$$= 120(12.321356) = 1478.5627 \approx \$1478.56$$

b)  A monthly payment of \$120 for 1 year compounded at $5\frac{3}{4}\%$ would yield a future value of \$1478.56.

5.    pymt = \$75 for February to November, $i = \dfrac{7\%}{12} = \dfrac{0.07}{12}$, $n = 10$ months

   a)    $FV = \text{pymt}\,\dfrac{(1+i)^n - 1}{i}$            Calculate ordinary $FV$.

   $$FV = 75\,\dfrac{\left(1+\dfrac{0.07}{12}\right)^{10} - 1}{\dfrac{0.07}{12}} = 75(10.266625) = 769.9969 \approx \$770.00$$

   b)    total contribution = \$75(10 months) = \$750.00

   c)    total interest = future value – contributions = 770.00 – 750.00 = \$20.00

9.    pymt = \$175 monthly, $i = \dfrac{10\frac{1}{2}\%}{12} = \dfrac{0.105}{12}$, $n = (65 - 39 \text{ years})(12 \text{ months}) = 26(12) = 312$ months

   a)    $FV = \text{pymt}\,\dfrac{(1+i)^n - 1}{i}$            Calculate ordinary $FV$.

   $$FV = 175\,\dfrac{\left(1+\dfrac{0.105}{12}\right)^{312} - 1}{\dfrac{0.105}{12}} = 175(1617.359188) = 283{,}037.8579 \approx \$283{,}037.86$$

   b)    total contribution = \$175(312 months) = \$54,600.00

   c)    total interest = future value – contributions = 283,037.86 – 54,600.00 = \$228,437.86

13.   a)    pymt = \$120 monthly, $i = \dfrac{5\frac{3}{4}\%}{12} = \dfrac{0.0575}{12}$, $FV(\text{ordinary}) = \$1478.5627$,

   $n = 1$ year = 12 months

   $FV(\text{Ordinary}) = FV(\text{Lump Sum})$

   $$FV = P(1 + i)^n$$            Calculate $P$.

   $$1478.5627 = P\left(1+\dfrac{0.0575}{12}\right)^{12}$$

   $$1478.5627 = P(1.0047917)^{12}$$

   $$1478.5627 = P(1.0590398)$$

   $$P = \dfrac{1478.5627}{1.0590398} = 1396.1351 \approx \$1396.14$$

13. Continued.

b)    This means that $1396.14 would need to be deposited at $5\frac{3}{4}$ % compounded monthly at the beginning of the year to yield the same amount of money ($1478.56) at the end of the year as $120 deposited monthly.

17.    a)    pymt = $175 monthly, $i = \dfrac{10\frac{1}{2}\%}{12} = \dfrac{0.105}{12}$, $FV = \$283{,}037.8579$

$n = (65 - 39 \text{ years})(12 \text{ months}) = 26(12) = 312$ months

$$FV = P(1 + i)^{n} \qquad \text{Calculate } P.$$

$$FV = P\left(1 + \frac{0.105}{12}\right)^{312}$$

$$283{,}037.8579 = P(1.00875)^{312}$$

$$283{,}037.8579 = P(15.1518929)$$

$$P = \frac{283{,}037.8579}{15.1518929} = 18680.0329 \approx \$18{,}680.03$$

b)    This means that $18,680.03 would need to be deposited at $10\frac{1}{2}$ % compounded monthly at age 39 to yield the same amount of money ($283,037.86) at age 65 as $175 deposited monthly.

21.    $FV = \$250{,}000$, $i = \dfrac{10\frac{1}{2}\%}{12} = \dfrac{0.105}{12}$, $n = (40 \text{ years})(12 \text{ months}) = 40(12) = 480$ months

$$FV = \text{pymt}\,\frac{(1+i)^{n} - 1}{i} \qquad \text{Calculate ordinary pymt.}$$

$$250{,}000 = \text{pymt}\,\frac{\left(1 + \dfrac{0.105}{12}\right)^{480} - 1}{\dfrac{0.105}{12}}$$

$$250{,}000 = \text{pymt}\,\frac{(1.00875)^{480} - 1}{0.00875}$$

$$250{,}000 = \text{pymt}\left(\frac{64.4791315}{0.00875}\right)$$

$$250{,}000 = \text{pymt}(7369.0436)$$

$$\text{pymt} = \frac{250{,}000}{7369.0436} = 33.925705 \approx \$33.93$$

25. pymt = $100 biweekly, $i = \dfrac{8\frac{1}{8}\%}{26} = \dfrac{0.08125}{26}$

$n = 35\dfrac{1}{2}$ years = (35.5 years)(26 periods) = 923 periods

a) $FV = \text{pymt}\ \dfrac{(1+i)^n - 1}{i}$          Calculate ordinary $FV$.

$$FV = 100\ \dfrac{\left(1 + \dfrac{0.08125}{26}\right)^{923} - 1}{\dfrac{0.08125}{26}}$$

$$= 100\ \dfrac{(1.003125)^{923} - 1}{0.003125}$$

$$= 100\left(\dfrac{16.8120916}{0.003125}\right)$$

$$= 100(5379.8693) = 537{,}986.9319 \approx \$537{,}986.93$$

b) $i = \dfrac{6.1\%}{12} = \dfrac{0.061}{12}$ , $n = 1$ month

Interest $= FV - P = P(1 + i)^n - P = P((1 + i)^n - 1)$

$$= P\left(\left(1 + \dfrac{0.061}{12}\right)^1 - 1\right) = P(0.005083333)$$

<u>Calculations for Month 1 (other months will be similar)</u>

Beginning Account Balance = $537,986.93

Interest for the Month = P(0.005083333)
                          = 537,986.93(0.005083333)
                          = 2734.7669 ≈ $2734.77

Withdrawal = $650.00

Ending Account Balance = Beginning Balance + Interest − Withdrawal
                             = 537,986.93 + 2734.77 − 650.00 = $540,071.70

25. b) Continued.

| Month Number | Account Balance (Beginning of Month) | Interest for the Month | Withdrawal | Account Balance (End of Month) |
|---|---|---|---|---|
| 1 | 537,986.93 | 2734.77 | 650.00 | 540,071.70 |
| 2 | 540,071.70 | 2745.36 | 650.00 | 542,167.06 |
| 3 | 542,167.06 | 2756.02 | 650.00 | 544,273.08 |
| 4 | 544,273.08 | 2766.72 | 650.00 | 546,389.80 |
| 5 | 546,389.80 | 2777.48 | 650.00 | 548,517.28 |

29. pymt = $1000 + $1000 = $2000 each year, $i = \dfrac{8.5\%}{1} = \dfrac{0.085}{1}$, $n = 65 - 14 = 51$ years

a) $FV = \text{pymt}\,\dfrac{(1+i)^n - 1}{i}$          Calculate ordinary $FV$.

$FV = 2000\,\dfrac{(1+0.085)^{51} - 1}{0.085} = 2000(742.4547333) = 1,484,909.46666 \approx \$1,484,909.47$

b) total contribution = $2000(51 \text{ years}) = \$102,000.00$

c) total interest = future value − contributions $= 1,484,909.47 - 102,000.00 = \$1,382,909.47$

d) $n = 65 - 19 = 46$ years

$FV = 2000\,\dfrac{(1+0.085)^{46} - 1}{0.085} = 2000(489.825480) = \$979,650.96$

e) $n = 65 - 24 = 41$ years

$FV = 2000\,\dfrac{(1+0.085)^{41} - 1}{0.085} = 2000(321.815552) = \$643,631.10$

33. $FV = \$1200$, $i = \dfrac{9\%}{12} = \dfrac{0.09}{12}$, $n = (2 \text{ years})(12 \text{ months}) = 24$ months

$FV = \text{pymt}\,\dfrac{(1+i)^n - 1}{i}$          Calculate ordinary pymt.

$1200 = \text{pymt}\,\dfrac{\left(1+\dfrac{0.09}{12}\right)^{24} - 1}{\dfrac{0.09}{12}}$

$1200 = \text{pymt}\,\dfrac{(1.0075)^{24} - 1}{0.0075}$

33. Continued.

$$1200 = \text{pymt}\left(\frac{0.19641353}{0.0075}\right)$$

$$1200 = \text{pymt}(26.1884706)$$

$$\text{pymt} = \frac{1200}{26.1884706} = 45.8216907 \approx \$45.82$$

37.        $FV$(Lump Sum) = $FV$(Ordinary Annuity)

$$P(1 + i)^{n} = \text{pymt}\,\frac{(1+i)^{n}-1}{i} \qquad\qquad \text{Calculate } P.$$

$$P(1 + i)^{n}\left(\frac{1}{(1+i)^{n}}\right) = \text{pymt}\left(\frac{(1+i)^{n}-1}{i}\right)\left(\frac{1}{(1+i)^{n}}\right) \qquad \text{Multiply by } \frac{1}{(1+i)^{n}} = (1+i)^{-n}.$$

$$P = \text{pymt}\,\frac{(1+i)^{n}-1}{i(1+i)^{n}} \qquad\qquad \text{Simplify.}$$

$$P = \text{pymt}\left(\frac{(1+i)^{n}}{(1+i)^{n}}\right)\left(\frac{1-(1+i)^{-n}}{i}\right) \qquad \text{Factor.}$$

$$P = \text{pymt}\,\frac{1-(1+i)^{-n}}{i} \qquad\qquad \text{Simplify.}$$

41.    pymt = \$75 for 10 months, $i = \dfrac{7\%}{12} = \dfrac{0.07}{12}$ , $n = 10$ months

$$P = \text{pymt}\,\frac{1-(1+i)^{-n}}{i} \qquad\qquad \text{Calculate } P.$$

$$P = 75\,\frac{1-\left(1+\dfrac{0.07}{12}\right)^{-10}}{\dfrac{0.07}{12}} = 75\left(\frac{1-(1.0058333)^{-10}}{0.0058333}\right) = 75\left(\frac{0.056504659}{0.0058333}\right)$$

$$= 726.4885 \approx \$726.49$$

45.    Omitted.

49.    $FV = \$500,000$, pymt $= \$100$ twice-monthly, $i = \dfrac{5\%}{24} = \dfrac{0.05}{24}$

$$FV = \text{pymt}\,\frac{(1+i)^n - 1}{i} \qquad \text{Calculate } n.$$

$$500{,}000 = 100\,\frac{\left(1+\dfrac{0.05}{24}\right)^n - 1}{\dfrac{0.05}{24}}$$

$$5000\left(\frac{0.05}{24}\right) = \left(1+\frac{0.05}{24}\right)^n - 1 \qquad \text{divide by 100 and multiply by denominator}$$

$$5000\left(\frac{0.05}{24}\right) + 1 = \left(1+\frac{0.05}{24}\right)^n$$

$$y_1 = 5000\left(\frac{0.05}{24}\right) + 1 = 11.4166\ldots,\ y_2 = \left(1+\frac{0.05}{24}\right)^n$$

Simple interest approximation:
$$y_1 = 1 + rt$$
$$11.4166667 = 1 + 0.05t$$
$$10.4166667 = 0.05t$$
$$t = \frac{10.41666667}{0.05} = 208 \text{ years}$$

$x\text{max} = 208(12) = 2496$
$y\text{max} = 15$ for graphing purposes because $y_1 = 11.4166667$.
$x\text{scl} = 500$ and $y\text{scl} = 5$

Graph and find the intersection: $x = 1170.0527\ldots, y = 11.416667$

Number of months $= 1171$ or $\dfrac{1171}{24} = 48.79$ years or 48 years $9\dfrac{1}{2}$ months

Comparing the total time to the 48 years and 10 months in Exercise 47, the loan would be paid off one-half month sooner, with a savings of $100.

---

| **Section 5.4** | **Amortized Loans** |
|---|---|

---

1.  $P = \$5000$, $i = \dfrac{9\frac{1}{2}\%}{12} = \dfrac{0.095}{12}$, n = (4 years)(12 months) = 48 months

   a)   <u>monthly payment</u>

$$\text{pymt} \frac{(1+i)^n - 1}{i} = P(1+i)^n \qquad \text{Calculate pymt.}$$

$$\text{pymt} \frac{\left(1 + \dfrac{0.095}{12}\right)^{48} - 1}{\dfrac{0.095}{12}} = 5000\left(1 + \dfrac{0.095}{12}\right)^{48}$$

$$\text{pymt} \frac{(1.007916667)^{48} - 1}{0.007916667} = 5000(1.007916667)^{48}$$

$$\text{pymt} \left(\frac{0.46009824}{0.007916667}\right) = 7300.4912$$

$$\text{pymt}(58.1176726) = 7300.4912$$

$$\text{pymt} = \frac{7300.4912}{58.1176726} = 125.6157 \approx \$125.62$$

   b)   $\text{Total Interest} = \begin{pmatrix}\text{Number of} \\ \text{Payments}\end{pmatrix}(\text{Payment}) - \text{Principal}$

   Total Interest = 48(125.62) − 5000 = 6,029.76 − 5000.00 = $1,029.76

5.  $P = \$155,000$, $i = \dfrac{9\frac{1}{2}\%}{12} = \dfrac{0.095}{12}$, $n = $ (30 years)(12 months) = 360 months

   a)   <u>monthly payment</u>

$$\text{pymt} \frac{(1+i)^n - 1}{i} = P(1+i)^n \qquad \text{Calculate pymt.}$$

$$\text{pymt} \frac{\left(1 + \dfrac{0.095}{12}\right)^{360} - 1}{\dfrac{0.095}{12}} = 155,000\left(1 + \dfrac{0.095}{12}\right)^{360}$$

$$\text{pymt} \frac{(1.007916667)^{360} - 1}{0.007916667} = 155,000(1.007916667)^{360}$$

5. a) Continued.

$$\text{pymt}\left(\frac{16.0948614}{0.007916667}\right) = 2,649,703.52$$

$$\text{pymt}(2033.03512) = 2,649,703.52$$

$$\text{pymt} = \frac{2,649,703.52}{2033.03512} = 1303.3240 \approx \$1303.32$$

b)    $\text{Interest} = \left(\begin{array}{c}\text{Number of}\\\text{Payments}\end{array}\right)(\text{Payment}) - \text{Principal}$

   $= 360(1303.32) - 155,000 = 469,195.20 - 155,000.00 = \$314,195.20$

9.    $P = \$212,500 - \text{Down Payment} = 212,500 - 212,500(20\%) = 212,500 - 42,500 = \$170,000$

   $i = \dfrac{10\frac{7}{8}\%}{12} = \dfrac{0.10875}{12}, n = 30 \text{ years}(12 \text{ months}) = 360 \text{ months}$

a)    <u>monthly payment</u>

$$\text{pymt}\frac{(1+i)^n - 1}{i} = P(1+i)^n \qquad \text{Calculate pymt.}$$

$$\text{pymt}\frac{\left(1+\dfrac{0.10875}{12}\right)^{360} - 1}{\dfrac{0.10875}{12}} = 170,000\left(1+\frac{0.10875}{12}\right)^{360}$$

$$\text{pymt}\frac{(1.0090625)^{360} - 1}{0.0090625} = 170,000(1.0090625)^{360}$$

$$\text{pymt}\left(\frac{24.7338053}{0.0090625}\right) = 4,374,746.90$$

$$\text{pymt}(2729.24748) = 4,374,746.90$$

$$\text{pymt} = \frac{4,374,746.90}{2729.24748} = 1602.91323 \approx \$1602.91$$

b)    $\text{Interest} = \left(\begin{array}{c}\text{Number of}\\\text{Payments}\end{array}\right)(\text{Payment}) - \text{Principal}$

   $= (360)(1602.91) - 170,000 = 577,047.60 - 170,000.00 = \$407,047.60$

9. Continued.

    c)    <u>Amortization Schedule</u>

$$r = 10\frac{7}{8}\% = 0.10875, \; t = 1 \text{ month} = \frac{1}{12} \text{ year}$$

<u>Payment #1</u>

$$P = \$170,000.00$$

$$I = Prt = 170,000(0.10875)\left(\frac{1}{12}\right) = 1540.625 \approx 1540.63$$

Principal Portion = Payment − Interest Portion = 1602.91 − 1540.63 = 62.28

Amount Due After Payment = Previous Principal − Principal Portion
$$= 170,000 - 62.28 = 169,937.72$$

<u>Payment #2</u> (similar to #1)

$$P = \$169,937.72$$

$$I = Prt = 169,937.72(0.10875)\left(\frac{1}{12}\right) = 1540.061 \approx 1540.06$$

Principal Portion = Payment − Interest Portion = 1602.91 − 1540.06 = 62.85

Amount Due After Payment = Previous Principal − Principal Portion
$$= 169,937.72 - 62.85 = 169,874.87$$

| Payment Number | Principal Portion | Interest Portion | Total Payment | Amount Due After Payment |
|:---:|:---:|:---:|:---:|:---:|
| 0 | --- | --- | --- | 170,000.00 |
| 1 | 62.28 | 1540.63 | 1602.91 | 169,937.72 |
| 2 | 62.85 | 1540.06 | 1602.91 | 169,874.87 |

    d)    38% of monthly income must be greater than monthly payment. (Assume that there are only home loan payments.)

$$0.38(\text{Income}) > 1602.91$$

$$\text{Income} > \frac{1602.91}{0.38} = \$4218.18$$

13.    Original Cost = \$15,829.32

<u>Loan through the Car Dealer</u>  4 Year Add-On Interest

$$P = \$15,829.32 - \text{Down Payment} = 15,829.32 - 1000 = 14,829.32, \; r = 7\frac{3}{4}\% = 0.0775, \; t = 4 \text{ years}$$

13. Continued.

a) <u>monthly payment</u>  (Calculate $FV$ and divide by 48)

$$FV = P(1 + rt)$$

$$= (14{,}829.32)(1 + 0.0775(4))$$

$$= 14{,}829.32(1.31) = 19{,}426.4092$$

$$\text{Payment} = \frac{19{,}426.4092}{48} = 404.7169 \approx \$404.72$$

b) $\text{Interest} = \begin{pmatrix} \text{Number of} \\ \text{Payments} \end{pmatrix}(\text{Payment}) - \text{Principal}$

$$= 48(404.72) - 14{,}829.32$$

$$= 19{,}426.56 - 14{,}829.32 = \$4597.24$$

<u>Loan through Bank</u>   4 Year Simple Interest Amortized

$P = \$15{,}829.32 - \text{Down Payment}$
  $= 15{,}829.32 - 15{,}829.32(10\%) = 15{,}829.32 - 1582.93 = \$14{,}246.39$

$i = \dfrac{8\frac{7}{8}\%}{12} = \dfrac{0.08875}{12}$ , $n = 4 \text{ years}(12 \text{ months}) = 48 \text{ months}$

a) <u>monthly payment</u>

$$\text{pymt}\frac{(1+i)^n - 1}{i} = P(1 + i)^n \qquad \text{Calculate pymt}$$

$$\text{pymt}\frac{\left(1+\dfrac{0.08875}{12}\right)^{48} - 1}{\dfrac{0.08875}{12}} = (14{,}246.39)\left(1+\dfrac{0.08875}{12}\right)^{48}$$

$$\text{pymt}\frac{(1.00739583)^{48} - 1}{0.00739583} = 14{,}246.39(1.00739583)^{48}$$

$$\text{pymt}\left(\frac{0.42431882}{0.00739583}\right) = 20{,}291.4013$$

$$\text{pymt}(57.3726851) = 20{,}291.4013$$

$$\text{pymt} = \frac{20{,}291.4013}{57.3726851} = 353.6770 \approx \$353.68$$

13. Continued.

b) $\quad \text{Interest} = \begin{pmatrix} \text{Number of} \\ \text{Payments} \end{pmatrix} (\text{Payment}) - \text{Principal}$

$\quad = (48)(353.68) - 14{,}246.39 = 16{,}976.64 - 14{,}246.39 = \$2730.25$

c) $\quad$ Dennis should choose the simple interest loan from the bank. Although it has a higher down payment, the monthly payments are less, and the total amount of interest paid is less.

17. $\quad P = \$100{,}000, \; n = (30 \text{ years})(12 \text{ months}) = 360 \text{ months}$

a) $\quad i = \dfrac{6\%}{12} = \dfrac{0.06}{12}$

$$\text{pymt} \frac{(1+i)^n - 1}{i} = P(1+i)^n \qquad \text{Calculate pymt.}$$

$$\text{pymt} \frac{\left(1 + \dfrac{0.06}{12}\right)^{360} - 1}{\dfrac{0.06}{12}} = 100{,}000\left(1 + \dfrac{0.06}{12}\right)^{360}$$

$$\text{pymt} \frac{(1.005)^{360} - 1}{0.005} = 100{,}000(1.005)^{360}$$

$$\text{pymt} \left(\frac{5.02257521}{0.005}\right) = 602{,}257.5212$$

$$\text{pymt}(1004.515042) = 602{,}25.5212$$

$$\text{pymt} = \frac{602{,}257.5212}{1004.515042} = 599.5505 \approx \$599.55$$

$$\text{Interest} = \begin{pmatrix} \text{Number of} \\ \text{Payments} \end{pmatrix} (\text{Payment}) - \text{Principal}$$

$$= 360(599.55) - 100{,}000 = 215{,}838.00 - 100{,}000.00 = \$115{,}838.00$$

17. Continued.

b)    $i = \dfrac{7\%}{12} = \dfrac{0.07}{12}$

$$\text{pymt}\,\frac{(1+i)^n - 1}{i} = P(1+i)^n \qquad \text{Calculate pymt.}$$

$$\text{pymt}\,\frac{\left(1+\dfrac{0.07}{12}\right)^{360} - 1}{\dfrac{0.07}{12}} = 100{,}000\left(1+\dfrac{0.07}{12}\right)^{360}$$

$$\text{pymt}\,\frac{(1.00583333)^{360} - 1}{0.00583333} = 100{,}000(1.00583333)^{360}$$

$$\text{pymt}\left(\frac{7.11649747}{0.00583333}\right) = 811{,}649.747$$

$$\text{pymt}(1219.970996) = 811{,}649.747$$

$$\text{pymt} = \frac{811{,}649.747}{1219.970996} = 665.3025 \approx \$665.30$$

$$\text{Interest} = \binom{\text{Number of}}{\text{Payments}}(\text{Payment}) - \text{Principal}$$

$$= 360(665.30) - 100{,}000 = 239{,}508.00 - 100{,}000.00 = \$139{,}508.00$$

c)    $i = \dfrac{8\%}{12} = \dfrac{0.08}{12}$

$$\text{pymt}\,\frac{(1+i)^n - 1}{i} = P(1+i)^n \qquad \text{Calculate pymt.}$$

$$\text{pymt}\,\frac{\left(1+\dfrac{0.08}{12}\right)^{360} - 1}{\dfrac{0.08}{12}} = 100{,}000\left(1+\dfrac{0.08}{12}\right)^{360} 1+$$

$$\text{pymt}\,\frac{(1.00666667)^{360} - 1}{0.00666667} = 100{,}000(1.00666667)^{360}$$

$$\text{pymt}\left(\frac{9.9357294}{0.00666667}\right) = 1{,}093{,}572.94$$

17. c) Continued.

$$\text{pymt}(1,490.35941) = 1,093,572.94$$

$$\text{pymt} = \frac{1,093,572.94}{1490.35941} = 733.7646 \approx \$733.76$$

$$\text{Interest} = \left(\begin{array}{c}\text{Number of} \\ \text{Payments}\end{array}\right)(\text{Payment}) - \text{Principal}$$

$$= 360(733.76) - 100,000 = 264,153.60 - 100,000.00 = \$164,153.60$$

d) $\quad i = \dfrac{9\%}{12} = \dfrac{0.09}{12}$

$$\text{pymt}\frac{(1+i)^n - 1}{i} = P(1+i)^n \qquad\qquad \text{Calculate pymt.}$$

$$\text{pymt}\frac{\left(1+\dfrac{0.09}{12}\right)^{360} - 1}{\dfrac{0.09}{12}} = 100,000\left(1+\frac{0.09}{12}\right)^{360}$$

$$\text{pymt}\frac{(1.0075)^{360} - 1}{0.0075} = 100,000(1.0075)^{360}$$

$$\text{pymt}\left(\frac{13.73057612}{0.0075}\right) = 1,473,057.612$$

$$\text{pymt}(1830.74348) = 1,473,057.612$$

$$\text{pymt} = \frac{1,473,057.612}{1830.74348} = 804.6226 \approx \$804.62$$

$$\text{Interest} = \left(\begin{array}{c}\text{Number of} \\ \text{Payments}\end{array}\right)(\text{Payment}) - \text{Principal}$$

$$= 360(804.62) - 100,000 = 289,663.20 - 100,000.00 = \$189,663.20$$

17. Continued.

e)    $i = \dfrac{10\%}{12} = \dfrac{0.10}{12}$

$$\text{pymt}\dfrac{(1+i)^n - 1}{i} = P(1+i)^n \qquad \text{Calculate pymt.}$$

$$\text{pymt}\dfrac{\left(1 + \dfrac{0.10}{12}\right)^{360} - 1}{\dfrac{0.10}{12}} = 100{,}000\left(1 + \dfrac{0.10}{12}\right)^{360}$$

$$\text{pymt}\dfrac{(1.00833333)^{360} - 1}{0.00833333} = 100{,}000(1.00833333)^{360}$$

$$\text{pymt}\left(\dfrac{18.8373994}{0.00833333}\right) = 1{,}983{,}739.937$$

$$\text{pymt}(2260.48792) = 1{,}983{,}739.937$$

$$\text{pymt} = \dfrac{1{,}983{,}739.937}{2260.48792} = 877.5715 \approx \$877.57$$

$$\text{Interest} = \left(\begin{array}{c}\text{Number of}\\\text{Payments}\end{array}\right)(\text{Payment}) - \text{Principal}$$

$$= 360(877.57) - 100{,}000 = 315{,}925.20 - 100{,}000.00 = \$215{,}925.20$$

f)    $i = \dfrac{11\%}{12} = \dfrac{0.11}{12}$

$$\text{pymt}\dfrac{(1+i)^n - 1}{i} = P(1+i)^n \qquad \text{Calculate pymt.}$$

$$\text{pymt}\dfrac{\left(1 + \dfrac{0.11}{12}\right)^{360} - 1}{\dfrac{0.11}{12}} = 100{,}000\left(1 + \dfrac{0.11}{12}\right)^{360}$$

$$\text{pymt}\dfrac{(1.009166667)^{360} - 1}{0.009166667} = 100{,}000(1.009166667)^{360}$$

$$\text{pymt}\left(\dfrac{25.7080970}{0.009166667}\right) = 2{,}670{,}809.70$$

17. f)  Continued.

$$\text{pymt}(2804.51967) = 2{,}670{,}809.70$$

$$\text{pymt} = \frac{2{,}670{,}809.70}{2804.51967} = 952.32340 \approx \$952.32$$

$$\text{Interest} = \begin{pmatrix} \text{Number of} \\ \text{Payments} \end{pmatrix}(\text{Payment}) - \text{Principal}$$

$$= 360(952.32) - 100{,}000 = 342{,}835.20 - 100{,}000.00 = \$242{,}835.20$$

Summary of Exercise 17.

|     | Interest Rate | Monthly Payment | Total Interest |
|-----|---------------|-----------------|----------------|
| a)  | 6.0%          | $599.55         | $115,838.00    |
| b)  | 7.0%          | 665.30          | 139,508.00     |
| c)  | 8.0%          | 733.76          | 164,153.60     |
| d)  | 9.0%          | 804.62          | 189,663.20     |
| e)  | 10.0%         | 877.57          | 215,925.20     |
| f)  | 11.0%         | 952.32          | 242,835.20     |

21.  $P = \$100{,}000, \ r = 10\% = 0.10$

   a)    Monthly Payments:

$$n = (30 \text{ years})(12 \text{ months}) = 360 \text{ months} , \ i = \frac{10\%}{12} = \frac{0.10}{12}$$

$$\text{pymt}\frac{(1+i)^n -1}{i} = P(1 + i)^n \qquad \text{Calculate pymt.}$$

$$\text{pymt}\frac{\left(1+\dfrac{0.10}{12}\right)^{360} -1}{\dfrac{0.10}{12}} = 100{,}000\left(1+\dfrac{0.10}{12}\right)^{360}$$

$$\text{pymt}\frac{(1.00833333)^{360} -1}{0.00833333} = 100{,}000(1.00833333)^{360}$$

$$\text{pymt}\left(\frac{18.8373994}{0.00833333}\right) = 1{,}983{,}739.937$$

$$\text{pymt}(2260.48792) = 1{,}983{,}739.937$$

$$\text{pymt} = \frac{1{,}983{,}739.937}{2260.48792} = 877.5715 \approx \$877.57$$

21. a) Continued.

$$\text{Amount paid in 1 year} = 12(877.57) = \$10,530.84$$

$$\text{Interest} = \begin{pmatrix} \text{Number of} \\ \text{Payments} \end{pmatrix}(\text{Payment}) - \text{Principal}$$

$$= 360(877.57) - 100{,}000 = 315{,}925.20 - 100{,}000.00 = \$215{,}925.20$$

b)   <u>Biweekly Payments:</u>

$$n = (30 \text{ years})(26 \text{ periods}) = 780 \text{ periods}, \ i = \frac{10\%}{26} = \frac{0.10}{26}$$

$$\text{pymt}\frac{(1+i)^n - 1}{i} = P(1+i)^{\,n} \qquad \text{Calculate pymt.}$$

$$\text{pymt}\frac{\left(1+\dfrac{0.10}{26}\right)^{780} - 1}{\dfrac{0.10}{26}} = 100{,}000\left(1+\dfrac{0.10}{26}\right)^{780}$$

$$\text{pymt}\frac{(1.003846154)^{780} - 1}{0.003846154} = 100{,}000(1.003846154)^{780}$$

$$\text{pymt}\left(\frac{18.9702863}{0.003846154}\right) = 1{,}997{,}028.63$$

$$\text{pymt}(4932.27444) = 1{,}997{,}028.63$$

$$\text{pymt} = \frac{1{,}997{,}028.63}{4932.27444} = 404.8900 \approx \$404.89$$

$$\text{Amount paid in 1 year} = 26(404.89) = \$10,527.14$$

$$\text{Interest} = \begin{pmatrix} \text{Number of} \\ \text{Payments} \end{pmatrix}(\text{Payment}) - \text{Principal}$$

$$= 780(404.89) - 100{,}000 = 315{,}814.20 - 100{,}000.00 = \$215{,}814.20$$

Biweekly payments are $3.70 per year less than monthly payments for a total interest savings of $111.00. This is not a significant savings for the home owner but it might be more convenient to make biweekly payments if the owner receives a biweekly salary.

25.   $P = 48,000, i = \dfrac{9\frac{1}{4}\%}{12} = \dfrac{0.0925}{12}, n = 4$ months

a) <u>monthly payment</u>

$$\text{pymt}\dfrac{(1+i)^n - 1}{i} = P(1+i)^n \qquad \text{Calculate pymt.}$$

$$\text{pymt}\dfrac{\left(1 + \dfrac{0.0925}{12}\right)^4 - 1}{\dfrac{0.0925}{12}} = 48,000\left(1 + \dfrac{0.0925}{12}\right)^4$$

$$\text{pymt}\dfrac{(1.007708333)^4 - 1}{0.007708333} = 48,000(1.007708333)^4$$

$$\text{pymt}\left(\dfrac{0.03119168}{0.007708333}\right) = 49,497.2006$$

$$\text{pymt}(4.04648811) = 49,497.2006$$

$$\text{pymt} = \dfrac{49,497.2006}{4.04648811} = 12,232.1379 \approx \$12,232.14$$

b)   <u>Calculations for Month 1 (months 2 and 3 are similar)</u>

$P = \$48,000, r = 9\frac{1}{4}\% = 0.0925, t = 1$ month $= \dfrac{1}{12}$ year

$I = Prt = (48,000)(0.0925)\left(\dfrac{1}{12}\right) = 370.00$

Principal portion = payment − interest portion = 12,232.14 − 370.00 = 11,862.14

$\begin{pmatrix} \text{Amount due} \\ \text{after Payment} \end{pmatrix} = \begin{pmatrix} \text{Previous} \\ \text{principal} \end{pmatrix} - \begin{pmatrix} \text{Principal} \\ \text{portion} \end{pmatrix} = 48,000.00 - 11,862.14 = 36,137.86$

<u>Calculations for Month 4 (the LAST month)</u>

$P = 12,138.56$

$I = Prt = (12,138.56)(0.0925)\left(\dfrac{1}{12}\right) = 93.57$

Total payment = Principal portion + Interest portion = 12,138.56 + 93.57 = 12,232.13

25. b) Continued.

| Payment Number | Principal Portion | Interest Portion | Total Payment | Amount Due After Payment |
|---|---|---|---|---|
| 0 | --- | --- | --- | 48,000.00 |
| 1 | 11,862.14 | 370.00 | 12,232.14 | 36,137.86 |
| 2 | 11,953.58 | 278.56 | 12,232.14 | 24,184.28 |
| 3 | 12,045.72 | 186.42 | 12,232.14 | 12,138.56 |
| 4 | 12,138.56 | 93.57 | 12,232.13 | 0.00 |

29.    From Exercise 13:  Car Dealer Add-On Interest Loan

$P = \$14,829.32, r = 7\frac{3}{4}\% = 0.0775, t = \frac{1}{12}$ year, pymt = $404.72

Monthly interest $= Prt = 14,829.32(0.0775)\left(\frac{1}{12}\right) = \$95.77$

Monthly principal $=$ payment $-$ monthly interest $= 404.72 - 95.77 = \$308.95$

Number 1 Amount Due After Payment $=$ Previous Amount Due $-$ monthly principal
$= 14,829.32 - 308.95 = \$14,520.37$

Number 2 Amount Due After Payment $=$ Previous Amount Due $-$ monthly principal
$= 14,520.37 - 308.95 = \$14,211.42$

Number 3 Amount Due After Payment $=$ Previous Amount Due $-$ monthly principal
$= 14,211.42 - 308.95 = \$13,902.47$

| Payment Number | Principal Portion | Interest Portion | Total Payment | Amount Due After Payment |
|---|---|---|---|---|
| 0 | --- | --- | --- | 14,829.32 |
| 1 | 308.95 | 95.77 | 404.72 | 14,520.37 |
| 2 | 308.95 | 95.77 | 404.72 | 14,211.42 |
| 3 | 308.95 | 95.77 | 404.72 | 13,902.47 |

From Exercise 13:  Bank Simple Interest Amortized Loan

$P = 14,246.39, r = 8\frac{7}{8}\% = 0.08875, t = \left(\frac{1}{12}\right)$ year, pymt = $353.68

Payment Number 1 Calculations:  (Calculations for payments 2 and 3 are similar.)

interest $= Prt = 14,246.39(0.08875)\left(\frac{1}{12}\right) = \$105.36$

principal $=$ payment $-$ first month interest $= 353.68 - 105.36 = \$248.32$

Amount Due After Payment $=$ Previous Amount Due $-$ principal
$= 14,246.39 - 248.32 = 13,998.07$

29. Continued.

| Payment Number | Principal Portion | Interest Portion | Total Payment | Amount Due After Payment |
|---|---|---|---|---|
| 0 | --- | --- | --- | 14,246.39 |
| 1 | 248.32 | 105.36 | 353.68 | 13.998.07 |
| 2 | 250.15 | 103.53 | 353.68 | 13,747.92 |
| 3 | 252.00 | 101.68 | 353.68 | 13,495.92 |

The advantages of the simple interest amortized loan are the payments are less and interest is paid only on the remaining principal.

33.　down payment = 20% of \$212,500 = \$42,500

$P$ = Cost − down payment = 212,500 − 42,500 = \$170,000

$n$ = (30 years)(12 months) = 360 months, $i = \dfrac{10\frac{7}{8}\%}{12} = \dfrac{0.10875}{12}$

Calculate monthly payment

$$\text{pymt}\frac{(1+i)^n - 1}{i} = P(1+i)^n \qquad \text{Calculate pymt.}$$

$$\text{pymt}\frac{\left(1+\dfrac{0.10875}{12}\right)^{360} - 1}{\dfrac{0.10875}{12}} = 170,000\left(1+\dfrac{0.10875}{12}\right)^{360}$$

$$\text{pymt}\frac{(1.0090625)^{360} - 1}{0.0090625} = 170,000(1.0090625)^{360}$$

$$\text{pymt}\left(\frac{24.7338053}{0.0090625}\right) = 4,374,746.90$$

$$\text{pymt}(2729.24748) = 4,374,746.90$$

$$\text{pymt} = \frac{4,374,746.90}{2729.24748} = 1602.91323 \approx \$1602.91$$

33. Continued.

Calculate unpaid balance

$n = 8$ years, 2 months $= (8$ years$)(12$ months$) + 2$ months $= 98$ months, $i = \dfrac{10\frac{7}{8}\%}{12} = \dfrac{0.10875}{12}$

$$\text{Unpaid Balance} = P(1+i)^n - \text{pymt}\,\frac{(1+i)^n - 1}{i}$$

$$= 170{,}000\left(1+\frac{0.10875}{12}\right)^{98} - 1602.91\,\frac{\left(1+\dfrac{0.10875}{12}\right)^{98} - 1}{\dfrac{0.10875}{12}}$$

$$= 170{,}000(1.00990625)^{98} - 1602.91\,\frac{(1.0090625)^{98} - 1}{0.0090625}$$

$$= 411{,}547.02 - 1602.91\left(\frac{1.4208648}{0.0090625}\right)$$

$$= 411{,}547.02 - 251{,}312.38 = 160{,}234.64$$

37.  a)    Monthly payment on existing loan

$P = \$187{,}900$, $n = (30$ years$)(12$ months$) = 360$ months, $i = \dfrac{10\frac{1}{2}\%}{12} = \dfrac{0.105}{12}$

$$\text{pymt}\,\frac{(1+i)^n - 1}{i} = P(1+i)^n \qquad \text{Calculate pymt.}$$

$$\text{pymt}\,\frac{\left(1+\dfrac{0.105}{12}\right)^{360} - 1}{\dfrac{0.105}{12}} = 187{,}900\left(1+\frac{0.105}{12}\right)^{360}$$

$$\text{pymt}\,\frac{(1.00875)^{360} - 1}{0.00875} = 187{,}900(1.00875)^{360}$$

$$\text{pymt}\left(\frac{22.018509}{0.00875}\right) = 4{,}325{,}177.80$$

$$\text{pymt}(2516.401) = 4{,}325{,}177.80$$

$$\text{pymt} = \frac{4{,}325{,}177.80}{2516.401} = 1718.7951 \approx 1718.80$$

37. Continued.

b) <u>Find unpaid balance at end of 10 years on existing loan.</u>

$$n = (10 \text{ years})(12 \text{ months}) = 120 \text{ months}, \ i = \frac{10\frac{1}{2}\%}{12} = \frac{0.105}{12}$$

$$\text{Unpaid Balance} = P(1+i)^n - \text{pymt}\frac{(1+i)^n - 1}{i}$$

$$= 187,900\left(1+\frac{0.105}{12}\right)^{120} - 1718.80\frac{\left(1+\frac{0.105}{12}\right)^{120} - 1}{\frac{0.105}{12}}$$

$$= 187,900(1.00875)^{120} - 1718.80\frac{(1.00875)^{120} - 1}{0.00875}$$

$$= 534,505.9052 - 1718.80\left(\frac{1.8446296}{0.00875}\right)$$

$$= 534,505.9052 - 362,348.5015 = 172,157.4037 \approx 172,157.40$$

c) <u>Interest saved by prepayment (Use current principal and remaining number of payments.)</u>

$P = \$172,157.40, \ n = (20 \text{ years})(12 \text{ months}) = 240 \text{ months}$

$$\text{Interest} = \binom{\text{Number of}}{\text{Payments}}(\text{Payment}) - \text{Principal}$$

$$= 240(1718.80) - 172,157.40 = 412,512.00 - 172,157.40 = \$240,354.60$$

d) <u>Find future value of depositing half of payment in an ordinary annuity</u>

$$\text{pymt} = \frac{1718.80}{2} = 859.40 \text{ monthly}, \ i = \frac{9\%}{12} = \frac{0.09}{12}, \ n = 20 \text{ years} = 240 \text{ months}$$

$$FV = \text{pymt}\frac{(1+i)^n - 1}{i} \qquad\qquad \text{Calculate ordinary } FV.$$

$$FV = 859.40\frac{\left(1+\frac{0.09}{12}\right)^{240} - 1}{\frac{0.09}{12}}$$

$$= 859.40\frac{(1.0075)^{240} - 1}{0.0075} = 859.40\left(\frac{5.0091515}{0.0075}\right)$$

$$= 859.40(667.8869) = 573,981.9760 \approx \$573,981.98$$

37. Continued.

e)   <u>Find the future value of depositing the unpaid balance in an account.</u>

$P = \$172,157.40$, $n = $ (20 years)(12 months) $= 240$ months, $i = \dfrac{9\frac{3}{4}\%}{12} = \dfrac{0.0975}{12}$

$FV = P(1 + i)^n$

$= 172,157.40\left(1 + \dfrac{0.0975}{12}\right)^{240} = 172,157.40(1.008125)^{240} = 172,157.40(6.9735245)$

$= 1,200,543.8603 \approx \$1,200,543.86$

f)   The decision is between saving more for retirement or increasing their standard of living now.

41.   Omitted.

45.   Omitted.

49.   From Exercise 7:  $P = 14,502.44$, $r = 11.5\% = 0.115$, $n = 48$ months, pymt $= 378.35$

a)   Follow the instructions in the textbook to construct the following table:

| Payment Number | Principal Portion | Interest Portion | Total Payment | Balance |
|---|---|---|---|---|
| 0 | | | | 14,502.44 |
| 1 | 239.37 | 138.98 | 378.35 | 14,263.07 |
| 2 | 241.66 | 136.69 | 378.35 | 14,021.41 |
| 3 | 243.98 | 134.37 | 378.35 | 13,777.43 |
| 4 | 246.32 | 132.03 | 378.35 | 13,531.12 |
| 5 | 248.68 | 129.67 | 378.35 | 13,282.44 |
| 6 | 251.06 | 127.29 | 378.35 | 13,031.38 |
| 7 | 253.47 | 124.88 | 378.35 | 12,777.91 |
| 8 | 255.90 | 122.45 | 378.35 | 12,522.02 |
| 9 | 258.35 | 120.00 | 378.35 | 12,263.67 |
| 10 | 260.82 | 117.53 | 378.35 | 12,002.85 |
| 11 | 263.32 | 115.03 | 378.35 | 11,739.52 |
| 12 | 265.85 | 112.50 | 378.35 | 11,473.68 |

First Year Totals            1,511.44

b)   First year's interest $= \$1511.44$.

49. Continued.

c)   $n = 48 - 12$ months $= 36$ months

Unpaid Balance $= P(1 + i)^n - \text{pymt} \dfrac{(1+i)^n - 1}{i}$

$$= 14{,}502.44\left(1 + \frac{0.115}{12}\right)^{36} - 378.35 \frac{\left(1 + \dfrac{0.115}{12}\right)^{36} - 1}{\dfrac{0.115}{12}}$$

$$= 14{,}502.44(1.009583333)^{36} - 378.35 \frac{(1.009583333)^{36} - 1}{0.009583333}$$

$$= 20{,}443.6899 - 378.35\left(\frac{0.40967243}{0.009583333}\right)$$

$$= 20{,}443.6899 - 16{,}173.8677 = 4269.8222 \approx \$4269.82$$

d)   Follow the instructions for the last period to complete the following table.

| Payment Number | Principal Portion | Interest Portion | Total Payment | Balance |
|---|---|---|---|---|
| 36 | | | | 4,269.82 |
| 37 | 337.43 | 40.92 | 378.35 | 3,932.39 |
| 38 | 340.66 | 37.69 | 378.35 | 3,591.72 |
| 39 | 343.93 | 34.42 | 378.35 | 3,247.80 |
| 40 | 347.23 | 31.12 | 378.35 | 2,900.57 |
| 41 | 350.55 | 27.80 | 378.35 | 2,550.02 |
| 42 | 353.91 | 24.44 | 378.35 | 2,196.10 |
| 43 | 357.30 | 21.05 | 378.35 | 1,838.80 |
| 44 | 360.73 | 17.62 | 378.35 | 1,478.07 |
| 45 | 364.19 | 14.16 | 378.35 | 1,113.89 |
| 46 | 367.68 | 10.67 | 378.35 | 746.21 |
| 47 | 371.20 | 7.15 | 378.35 | 375.01 |
| 48 | 375.01 | 3.59 | 378.61 | 0.00 |

Last Year
Totals                     270.64

e)   Last year's interest $= \$270.64$

53.    a)    $P = \$350,000$, $n = 30$ years(12 months) = 360 months, $i = \dfrac{14\frac{1}{2}\%}{12} = \dfrac{0.145}{12}$

$$\text{pymt}\frac{(1+i)^n -1}{i} = P(1+i)^n \qquad \text{Calculate pymt.}$$

$$\text{pymt}\frac{\left(1+\dfrac{0.145}{12}\right)^{360} -1}{\dfrac{0.145}{12}} = 350{,}000\left(1+\dfrac{0.145}{12}\right)^{360}$$

$$\text{pymt}\frac{(1.01208333)^{360} -1}{0.01208333} = 350{,}000(1.01208333)^{360}$$

$$\text{pymt}\left(\frac{74.48459229}{0.01208333}\right) = 26{,}419{,}607.3014$$

$$\text{pymt}(6164.2421) = 26{,}419{,}607.3014$$

$$\text{pymt} = \frac{26{,}419{,}607.3014}{6164.2421} = 4285.9457 \approx 4285.95$$

a)    Follow the instructions in the textbook to construct the following table:

| Payment Number | Principal Portion | Interest Portion | Total Payment | Balance |
|---|---|---|---|---|
| 0 | | | | 350,000.00 |
| 1 | 56.78 | 4,229.17 | 4,285.95 | 349,943.22 |
| 2 | 57.47 | 4,228.48 | 4,285.95 | 349,885.75 |
| 3 | 58.16 | 4,227.79 | 4,285.95 | 349,827.58 |
| 4 | 58.87 | 4,227.08 | 4,285.95 | 349,768.72 |
| 5 | 59.58 | 4,226.37 | 4,285.95 | 349,709.14 |
| 6 | 60.30 | 4,225.65 | 4,285.95 | 349,648.84 |
| 7 | 61.03 | 4,224.92 | 4,285.95 | 349,587.81 |
| 8 | 61.76 | 4,224.19 | 4,285.95 | 349,526.05 |
| 9 | 62.51 | 4,223.44 | 4,285.95 | 349,463.54 |
| 10 | 63.27 | 4,222.68 | 4,285.95 | 349,400.27 |
| 11 | 64.03 | 4,221.92 | 4,285.95 | 349,336.24 |
| 12 | 64.80 | 4,221.15 | 4,285.95 | 349,271.44 |

First Year Totals          50,702.84

b)    First Year's Interest = $50,702.84

53. Continued.

   c)    $n = (30 - 1)$ years $= 29$ years$(12$ months$) = 348$ months

$$\text{Unpaid Balance} = P(1 + i)^{n} - \text{pymt}\,\frac{(1+i)^{n} - 1}{i}$$

$$= 350{,}000\left(1+\frac{0.145}{12}\right)^{348} - 4285.95\,\frac{\left(1+\dfrac{0.145}{12}\right)^{348} - 1}{\dfrac{0.145}{12}}$$

$$= 350{,}000(1.01208333)^{348} - 4285.95\,\frac{(1.01208333)^{348} - 1}{0.01208333}$$

$$= 22{,}873{,}418.7549 - 4285.95\left(\frac{64.3526250}{0.01208333}\right)$$

$$= 22{,}873{,}418.7549 - 22{,}825{,}831.7114 = 47{,}587.0435 \approx \$47{,}587.04$$

   d)    Follow the instructions for the last period to complete the following table.

| Payment Number | Principal Portion | Interest Portion | Total Payment | Balance |
|---|---|---|---|---|
| 348 | | | | 47,587.04 |
| 349 | 3,710.94 | 575.01 | 4,285.95 | 43,876.10 |
| 350 | 3,755.78 | 530.17 | 4,285.95 | 40,120.32 |
| 351 | 3,801.16 | 484.79 | 4,285.95 | 36,319.16 |
| 352 | 3,847.09 | 438.86 | 4,285.95 | 32,472.06 |
| 353 | 3,893.58 | 392.37 | 4,285.95 | 28,578.48 |
| 354 | 3,940.63 | 345.32 | 4,285.95 | 24,637.86 |
| 355 | 3,988.24 | 297.71 | 4,285.95 | 20,649.61 |
| 356 | 4,036.43 | 249.52 | 4,285.95 | 16,613.18 |
| 357 | 4,085.21 | 200.74 | 4,285.95 | 12,527.97 |
| 358 | 4,134.57 | 151.38 | 4,285.95 | 8,393.40 |
| 359 | 4,184.53 | 101.42 | 4,285.95 | 4,208.87 |
| 360 | 4,208.87 | 50.86 | 4,285.95 | 0.00 |

Last Year
Totals                  3,818.14

   e)    Last year's interest = $3818.14

57.    $P = 32{,}440$, $t = 1$ year $= 12$ months, $i = \dfrac{5.25\%}{12} = \dfrac{0.0525}{12}$

$$\text{pymt} \frac{(1+i)^n - 1}{i} = P(1 + i)^n \qquad \text{Calculate pymt.}$$

$$\text{pymt} \frac{\left(1 + \dfrac{0.0525}{12}\right)^{12} - 1}{\dfrac{0.0525}{12}} = 32{,}440\left(1 + \dfrac{0.0525}{12}\right)^{12}$$

$$\text{pymt} \frac{(1.004375)^{12} - 1}{0.004375} = 32{,}440(1.004375)^{12}$$

$$\text{pymt}\left(\frac{0.05378188}{0.004375}\right) = 34{,}184.6844$$

$$\text{pymt}(12.293002) = 34{,}184.6844$$

$$\text{pymt} = \frac{34{,}184.6844}{12.293002} = 2780.8246 \approx \$2780.82$$

Follow the instructions in the textbook to construct the following table:

| Payment Number | Principal Portion | Interest Portion | Total Payment | Balance |
|---|---|---|---|---|
| 0 | | | | 32,440.00 |
| 1 | 2,638.90 | 141.93 | 2,780.82 | 29,801.11 |
| 2 | 2,650.44 | 130.38 | 2,780.82 | 27,150.66 |
| 3 | 2,662.04 | 118.78 | 2,780.82 | 24,488.63 |
| 4 | 2,673.68 | 107.14 | 2,780.82 | 21,814.95 |
| 5 | 2,685.38 | 95.44 | 2,780.82 | 19,129.57 |
| 6 | 2,697.13 | 83.69 | 2,780.82 | 16,432.44 |
| 7 | 2,708.93 | 71.89 | 2,780.82 | 13,723.51 |
| 8 | 2,720.78 | 60.04 | 2,780.82 | 11,002.73 |
| 9 | 2,732.68 | 48.14 | 2,780.82 | 8,270.05 |
| 10 | 2,744.64 | 36.18 | 2,780.82 | 5,525.41 |
| 11 | 2,756.65 | 24.17 | 2,780.82 | 2,768.76 |
| 12 | 2,768.76 | 12.11 | 2,780.88 | 0.00 |

| First Year Totals | 32,440.00 | 929.90 | 33,369.90 | |
|---|---|---|---|---|

---

| Section 5.5 | Annual Percentage Rate on a Graphing Calculator |
|---|---|

1. down payment = 10% of $16,113.82 = $1611.38

$P$ = Cost – down payment = 16,113.82 – 1611.38 = $14,502.44

$i = \dfrac{11\frac{1}{2}\%}{12} = \dfrac{0.115}{12}$ , $n$ = (4 years)(12 months) = 48 months

$$\text{pymt}\frac{(1+i)^n - 1}{i} = P(1+i)^n \qquad\qquad \text{Calculate pymt}$$

$$\text{pymt}\frac{\left(1+\dfrac{0.115}{12}\right)^{48} - 1}{\dfrac{0.115}{12}} = 14{,}502.44\left(1+\frac{0.115}{12}\right)^{48}$$

$$\text{pymt}\frac{(1.009583333)^{48} - 1}{0.009583333} = 14{,}502.44(1.009583333)^{48}$$

$$\text{pymt}\left(\frac{0.58060837}{0.009583333}\right) = 22{,}922.68$$

$$\text{pymt}(60.5852206) = 22{,}922.68$$

$$\text{pymt} = \frac{22{,}922.68}{60.5852206} = 378.3543 \approx \$378.35$$

A.P.R. Financed Amount = Loan Amount – Loan Fees = 14,502.44 – 814.14 = **$13,688.30**

Substitute in the following formula to find the two equations to graph.

$$\text{pymt}\frac{(1+i)^n - 1}{i} = P(1+i)^n$$

$$378.35\frac{(1+i)^{48} - 1}{i} = 13{,}688.30(1+i)^{48}$$

$$y_1 = 378.35\frac{(1+x)^{48} - 1}{x}, \quad y_2 = 13{,}688.30(1+x)^{48}$$

Estimate the range from the calculation for the payment above.

$x\text{max} = i = 0.009583 \approx 0.02$,

$y$ should approximate the right side of the payment equation. $y\text{min} = 20{,}000$, $y\text{max} = 30{,}000$

Graph and find the intersection.

$x = 0.01218645271$, $y = 24482.428273$

1. Continued.

Convert $x$ to the annual rate.

annual rate $= 12(0.01218645271) = 0.146237 \approx 14.6\%$

This means that the 11.5% loan requires the same monthly payment as a 14.6% loan with no points or fees.

5.     $P = \text{Cost} = \$4600, n = (4 \text{ years})(12 \text{ months}) = 48 \text{ months}$

a)     Monthly Payment:  (Add-on Interest)  $r = 8\% = 0.08$

$FV = P(1 + rt) = 4600(1 + (0.08)(4)) = 4600(1.32) = 6072.00$

Monthly Payment $= \dfrac{6072.00}{48} = \$126.50$

b)     Verify the A.P.R.

Substitute in the following formula to find the two equations to graph.

$$\text{pymt}\frac{(1+i)^n - 1}{i} = P(1+i)^n$$

$$126.50\frac{(1+i)^{48} - 1}{i} = 4600(1+i)^{48}$$

$$y_1 = 126.50\frac{(1+x)^{48} - 1}{x}, \quad y_2 = 4600(1+x)^{48}$$

Estimate the range from the calculation for the payment above.

$$x\text{max} = i = \frac{0.08}{12} = 0.0066667 \approx 0.02,$$

$y$ should approximate the future value in the payment calculation.
$$y\text{min} = 0, y\text{max} = 10,000$$

Graph and find the intersection.
$$x = 0.01195440844, y = 8137.3551821$$

Convert $x$ to the annual rate.

annual rate $= 12(0.01195440844) = 0.143453 \approx 14.3\%$

This means that the 8% add-on loan requires the same monthly payment as a 14.3% simple interest amortized loan would have.  The A.P.R. is legally correct since 14.3453% − 14.25% = 0.095% which is less than 0.125%, the legal tolerance.

9.     It would seem that the savings and loan would be the less expensive loan, however, since the A.P.R. is not given it is not known what the effect of the finance charges would be and therefore it cannot be determined which loan would be least expensive.

13.    $P = 80\%$ of $119,000 = 95,200$, $n = (30 \text{ years})(12 \text{ months}) = 360$, $i = \dfrac{8\frac{1}{4}\%}{12} = \dfrac{0.0825}{12}$

Find the monthly payment:

$$\text{pymt}\,\frac{(1+i)^n - 1}{i} = P(1 + i)^n \qquad\qquad \text{Calculate pymt.}$$

$$\text{pymt}\,\frac{\left(1+\dfrac{0.0825}{12}\right)^{360} - 1}{\dfrac{0.0825}{12}} = 95,200\left(1+\dfrac{0.0825}{12}\right)^{360} 1 +$$

$$\text{pymt}\,\frac{(1.006875)^{360} - 1}{0.006875} = 95,200(1.006875)^{360}$$

$$\text{pymt}\left(\frac{10.78150562}{0.006875}\right) = 1,121,599.335$$

$$\text{pymt}(1568.218999) = 1,121,599.335$$

$$\text{pymt} = \frac{1,121,599.335}{1568.218999} = 715.2058 \approx \$715.21$$

Approximate the fees included in the finance charge:

$$i = \frac{\text{A.P.R.}}{12} = \frac{9.23\%}{12} = \frac{0.0923}{12}$$

$$\text{pymt}\,\frac{(1+i)^n - 1}{i} = P(1 + i)^n \qquad\qquad \text{Calculate } P.$$

$$715.21\,\frac{\left(1+\dfrac{0.0923}{12}\right)^{360} - 1}{\dfrac{0.0923}{12}} = P\left(1+\dfrac{0.0923}{12}\right)^{360}$$

$$715.21\,\frac{(1.007691667)^{360} - 1}{0.007691667} = P(1.007691667)^{360}$$

$$715.21\left(\frac{14.77466514}{0.007691667}\right) = P(15.77466514)$$

$$715.21(1920.86654) = 15.77466514P$$

$$1,373,822.958 = 15.77466514P$$

$$P = \frac{1,373,822.958}{15.77466514} = 87,090.4673 \approx \$87,090.47$$

Estimated points and fees $= 95,200.00 - 87,090.47 = \$8109.53$

## Section 5.6        Payout Annuities

1.    pymt $= 1200$, $n = (20 \text{ years})(12 \text{ months}) = 240$, $i = \dfrac{8\%}{12} = \dfrac{0.08}{12}$

      a)          $P(1+i)^n = \text{pymt} \dfrac{(1+i)^n - 1}{i}$     Calculate $P$.

$$P\left(1 + \frac{0.08}{12}\right)^{240} = 1200 \frac{\left(1 + \dfrac{0.08}{12}\right)^{240} - 1}{\dfrac{0.08}{12}}$$

$$P(1.006666667)^{240} = 1200 \frac{(1.006666667)^{240} - 1}{0.006666667}$$

$$P(4.92680277) = 1200\left(\frac{3.92680277}{0.006666667}\right)$$

$$4.92680277P = 706{,}824.499$$

$$P = \frac{706{,}824.499}{4.92680277} = \$143{,}465.15$$

      b)     Total payout $= (20 \text{ years})(12 \text{ months})(\$1200) = \$288{,}000$

5.    $FV = \$143{,}465.15$, $i = \dfrac{8\%}{12} = \dfrac{0.08}{12}$, $n = (30 \text{ years})(12 \text{ months}) = 360$

      a)          $\text{pymt} \dfrac{(1+i)^n - 1}{i} = \text{Future Value}$         Calculate pymt.

$$\text{pymt} \frac{\left(1 + \dfrac{0.08}{12}\right)^{360} - 1}{\dfrac{0.08}{12}} = 143{,}465.15$$

$$\text{pymt} \frac{(1.006666667)^{360} - 1}{0.006666667} = 143{,}465.15$$

$$\text{pymt}\left(\frac{9.9357297}{0.006666667}\right) = 143{,}465.15$$

5. a) Continued.

$$\text{pymt}(1490.35945) = 143,465.15$$

$$\text{pymt} = \frac{143,465.15}{1490.35945} = 96.2621 \approx \$96.26$$

b) Total paid in = (30 years)(12 months)($96.26) = $34,653.60

She receives $288,000.00 − $34,653.60 = $253,346.40 more than she paid into the annuity.

9. $t = 20$ years, pymt = $14,000, $c = 4\% = 0.04$

a) $r = 8\% = 0.08$

$$P = \text{pymt}\,\frac{1 - \left(\dfrac{1+c}{1+r}\right)^t}{r - c} \qquad \text{Calculate } P.$$

$$P = 14,000\,\frac{1 - \left(\dfrac{1+0.04}{1+0.08}\right)^{20}}{0.08 - 0.04} = 14,000\,\frac{1 - \left(\dfrac{1.04}{1.08}\right)^{20}}{0.04}$$

$$= 14,000\,\frac{1 - (0.96296296)^{20}}{0.04} = 14,000\,\frac{1 - 0.47010154}{0.04} = 14,000\,\frac{0.52989846}{0.04}$$

$$= 14,000(13.247461) = 185,464.4601 \approx \$185,464.46$$

b) First annual payout = $14,000 (there is no C.O.L.A. adjustment)

c) Second annual payout $= P(1 + c)^{n-1}$
$$= 14,000(1 + 0.04)^{2-1} = 14,000(1.04)^1 = \$14,560.00$$

d) Last annual payout $= P(1 + c)^{n-1}$
$$= 14,000(1 + 0.04)^{20-1} = 14,000(1.04)^{19}$$
$$= 14,000(2.10684918) = \$29,495.8885 \approx \$29,495.89$$

13. pymt = $1000 monthly for 25 years, $c = 3\% = 0.03$, $r = 10\%$ compounded monthly

a) $n = (1 \text{ year})(12 \text{ months}) = 12$

$$FV = \text{pymt}\,\frac{(1+i)^n - 1}{i} \qquad \text{Calculate ordinary } FV.$$

$$FV = 1000\,\frac{\left(1 + \dfrac{0.10}{12}\right)^{12} - 1}{\dfrac{0.10}{12}} = 1000(12.565568) = 12,565.5681 \approx \$12,565.57$$

13.  Continued.

b)    $r = \left(1 + \dfrac{0.10}{12}\right)^{12} - 1 = 1.104713067 - 1 = 0.104713067 \approx 10.4713067\%$

c)    $P = \text{pymt}\,\dfrac{1 - \left(\dfrac{1+c}{1+r}\right)^{t}}{r - c}$           Calculate $P$.

$$= 12{,}565.57\,\frac{1 - \left(\dfrac{1+0.03}{1+0.104713067}\right)^{25}}{0.104713067 - 0.03} = 12{,}565.57\,\frac{1 - \left(\dfrac{1.03}{1.104713067}\right)^{25}}{0.074713067}$$

$$= 12{,}565.57\,\frac{1 - 0.17365742}{0.074713067} = 12{,}565.57\,\frac{0.82634258}{0.074713067}$$

$$= 12565.57(11.0602149) = 138{,}977.905 \approx \$138{,}977.91$$

d)    $FV = \$143{,}465.15,\ i = \dfrac{10\%}{12} = \dfrac{0.10}{12},\ n = (30\text{ years})(12\text{ months}) = 360$

$$\text{pymt}\,\frac{(1+i)^{n} - 1}{i} = \text{Future Value}\qquad\text{Calculate pymt.}$$

$$\text{pymt}\,\frac{\left(1 + \dfrac{0.10}{12}\right)^{360} - 1}{\dfrac{0.10}{12}} = 138{,}977.90$$

$$\text{pymt}\,\frac{(1.00833333)^{360} - 1}{0.00833333} = 138{,}977.90$$

$$\text{pymt}\left(\frac{18.8373994}{0.00833333}\right) = 138{,}977.90$$

$$\text{pymt}(2260.48793) = 138{,}977.90$$

$$\text{pymt} = \frac{138{,}977.90}{2260.48793} = 61.4814 \approx \$61.48$$

17.  pymt = 50,000, $n$ = 20 years, $i$ = 8% = 0.08

$$P(1+i)^n = \text{pymt}\frac{(1+i)^n - 1}{i} \qquad \text{Calculate } P.$$

$$P(1+0.08)^{20} = 50{,}000\frac{(1+0.08)^{20} - 1}{0.08}$$

$$P(1.08)^{20} = 50{,}000\frac{(1.08)^{20} - 1}{0.08}$$

$$P(4.66095714) = 50{,}000\frac{3.66095714}{0.08}$$

$$4.66095714P = 2{,}288{,}098.215$$

$$P = \frac{2{,}288{,}098.215}{4.66095714} = 490{,}907.3704 \approx \$490{,}907.37$$

---

## Chapter 5           Review

1.  $P = \$8140, r = 9\frac{3}{4}\% = 0.0975, t = 11$ years

$$I = Prt = (8140)(0.0975)(11) = \$8730.15$$

5.  $P = \$7000, FV = $ Amount needed $= \$8000, r = 7\frac{1}{2}\% = 0.075$

$$FV = P(1 + rt) \qquad\qquad \text{Calculate } t.$$

$$8000 = 7000(1 + (0.075)t)$$

$$\frac{8000}{7000} = 1 + 0.075t$$

$$1.14285714 = 1 + 0.075t$$

$$0.14285714 = 0.075t$$

$$t = \frac{0.14285714}{0.075} = 1.904762 \approx 1.9 \text{ years}$$

9.    $FV = \$250,000$ at age 65

a)    $i = \dfrac{10\frac{1}{8}\%}{4} = \dfrac{0.10125}{4}$ , $n = (65 - 25 \text{ years})(4 \text{ quarters}) = 40(4) = 160 \text{ quarters}$

$$FV = P(1 + i)^{\,n} \qquad \text{Calculate } P.$$

$$250,000 = P\left(1 + \dfrac{0.10125}{4}\right)^{160}$$

$$250,000 = P(1.0253125)^{160}$$

$$250,000 = P(54.5758277)$$

$$P = \dfrac{250,000}{54.5758277} = 4580.7826 \approx \$4580.78$$

b)    $P = \$250,000$, $r = 10\dfrac{1}{8}\% = 0.10125$, $t = 1 \text{ month} = \dfrac{1}{12} \text{ year}$

$$I = Prt = 250,000(0.10125)\left(\dfrac{1}{12}\right) = 2109.375 \approx \$2109.38$$

13.    a)    $I = \$1300$, $r = 8\dfrac{1}{4}\% = 0.0825$, $t = 1 \text{ month} = \dfrac{1}{12} \text{ year}$

$I = Prt \qquad$ Calculate $P$.

$$1300 = P(0.0825)\left(\dfrac{1}{12}\right)$$

$$1300 = P(0.006875)$$

$$P = \dfrac{1300}{0.006875} = 189,090.9091 \approx \$189,090.91$$

b)    $FV = \$189,090.91$, $i = \dfrac{9\frac{3}{4}\%}{12} = \dfrac{0.0975}{12}$ , $n = (30 \text{ years})(12 \text{ months}) = 30(12) = 360 \text{ months}$

$$FV = \text{pymt}\dfrac{(1+i)^{n} - 1}{i} \qquad \text{Calculate ordinary pymt.}$$

$$189,090.91 = \text{pymt}\dfrac{\left(1 + \dfrac{0.0975}{12}\right)^{360} - 1}{\dfrac{0.0975}{12}}$$

$$189,090.91 = \text{pymt}\dfrac{(1.008125)^{360} - 1}{0.008125}$$

$$189,090.91 = \text{pymt}\left(\dfrac{17.4152875}{0.008125}\right)$$

**13. b)** Continued.

$$189,090.91 = \text{pymt}(2143.42000)$$

$$\text{pymt} = \frac{189,090.91}{2143.42000} = 88.219252 \approx \$88.22$$

**17.** pymt = 1700, $n = (25 \text{ years})(12 \text{ months}) = 300$

**a)** $i = \dfrac{6.1\%}{12} = \dfrac{0.061}{12}$

$$P(1+i)^{n} = \text{pymt}\frac{(1+i)^{n}-1}{i} \qquad \text{Calculate } P.$$

$$P\left(1+\frac{0.061}{12}\right)^{300} = 1700\frac{\left(1+\dfrac{0.061}{12}\right)^{300}-1}{\dfrac{0.061}{12}}$$

$$P(1.005083333)^{300} = 1700\frac{(1.005083333)^{300}-1}{0.005083333}$$

$$P(4.57742697) = 1700\left(\frac{3.57742697}{0.005083333}\right)$$

$$4.57742697P = 1,196,385.414$$

$$P = \frac{1,196,385.414}{4.57742697} = 261,366.357 \approx \$261,366.36$$

**b)** $FV = \$261,366.36$, $i = \dfrac{6.1\%}{12} = \dfrac{0.061}{12}$, $n = (30 \text{ years})(12 \text{ months}) = 360$

$$\text{pymt}\frac{(1+i)^{n}-1}{i} = \text{Future Value} \qquad \text{Calculate pymt.}$$

$$\text{pymt}\frac{\left(1+\dfrac{0.061}{12}\right)^{360}-1}{\dfrac{0.061}{12}} = 261,366.36$$

$$\text{pymt}\frac{(1.005083333)^{360}-1}{0.005083333} = 261,366.36$$

17. b)  Continued.

$$\text{pymt}\left(\frac{5.2050561}{0.005083333}\right) = 261{,}366.36$$

$$\text{pymt}(1023.94545) = 261{,}366.36$$

$$\text{pymt} = \frac{261{,}366.36}{1023.94545} = 255.2542 \approx \$255.25$$

c)     Total paid in = (30 years)(12 months)($255.25) = $91,890.00

Total received = (25 years)(12 months)($1700) = $510,000.00

# 6 Geometry

| Section 6.1 | Perimeter and Area |
|---|---|

1. Triangle: $b = 9.2$ cm, $h = 3.5$ cm

   $A = \dfrac{1}{2} bh = \dfrac{1}{2}(9.2 \text{ cm})(3.5 \text{cm}) = 16.1 \text{ cm}^2$

5. Parallelogram: $b = 6.2$ ft, $h = 3.5$ ft

   $A = bh = (6.2 \text{ ft})(3.5 \text{ ft}) = 21.7 \text{ ft}^2$

9. Circle: $d = 6.5$ in., $r = \dfrac{d}{2} = \dfrac{6.5 \text{ in.}}{2} = 3.25$ in.

   a) $\quad A = \pi r^2 = \pi (3.25 \text{ in.})^2 = \pi (10.5625 \text{ in.}^2) = 33.1831 \text{ in.}^2 \approx 33.2 \text{ in.}^2$

   b) $\quad C = \pi d = \pi (6.5 \text{ in.}) = 20.4204 \text{ in.} \approx 20.4 \text{ in.}$

13. Triangle: $a = 8$ m, $b = 8$ m, $c = 8$ m

    a) Use Heron's Formula

    $s = \dfrac{1}{2}(a + b + c) = \dfrac{1}{2}(8 \text{ m} + 8 \text{ m} + 8 \text{ m}) = \dfrac{1}{2}(24 \text{ m}) = 12 \text{ m}$

    $\begin{aligned} A &= \sqrt{s(s-a)(s-b)(s-c)} \\ &= \sqrt{12(12-8)(12-8)(12-8)} \\ &= \sqrt{12 \text{ m}(4 \text{ m})(4 \text{ m})(4 \text{ m})} = \sqrt{768 \text{ m}^4} = 27.7128 \text{ m}^2 \approx 27.7 \text{ m}^2 \end{aligned}$

    b) $\quad P = a + b + c = 8 \text{ m} + 8 \text{ m} + 8 \text{ m} = 24 \text{ m}$

17. Semicircle: $d = 100$ yd, $r = \dfrac{d}{2} = \dfrac{100 \text{ yd}}{2} = 50$ yd

    a) $\quad A = \dfrac{1}{2}(\pi r^2) = \dfrac{1}{2} \pi (50 \text{ yd})^2 = \dfrac{2500}{2} \pi \text{ yd}^2 = 3926.99082 \text{ yd}^2 \approx 3927.0 \text{ yd}^2$

17. Continued.

   b)   $C = \dfrac{1}{2}\pi d + d = \dfrac{1}{2}\pi(100 \text{ yd}) + 100 \text{ yd}$

   $= (50\pi + 100) \text{ yd} = (157.0796 + 100.00) \text{ yd} = 257.0796 \text{ yd} \approx 257.1 \text{ yd}$

21.   Norman window: $b = d = 5$ ft, $h = 8$ ft, $r = \dfrac{d}{2} = \dfrac{5 \text{ ft}}{2} = 2.5$ ft

   a)   $A_{\text{total}} = A_{\text{rectangle}} + A_{\text{semicircle}}$

   $= bh + \dfrac{1}{2}(\pi r^2)$

   $= (5 \text{ ft})(8 \text{ ft}) + \dfrac{1}{2}\pi(2.5 \text{ ft})^2 = 40 \text{ ft}^2 + \dfrac{6.25}{2}\pi \text{ ft}^2 = (40 + 3.125\,\pi) \text{ ft}^2$

   $= (40 + 9.8175) \text{ ft}^2 = 49.8175 \text{ ft2} \approx 49.8 \text{ ft}^2$

   b)   $P = \dfrac{1}{2}\text{Circumference} + 1 \text{ base} + 2 \text{ heights}$

   $= \dfrac{1}{2}(\pi d) + b + 2h$

   $= \dfrac{1}{2}\pi(5 \text{ ft}) + 5 \text{ ft} + 2(8 \text{ ft}) = (7.8540 + 5 + 16) \text{ ft} = 28.8540 \text{ ft} \approx 28.9 \text{ ft}$

25.   Square: $P = 18$ ft 8 in.  (Need to find $s$ = length of sides)

   a)   Area in square inches

   $P = 18 \text{ ft}\left(\dfrac{12 \text{ in.}}{1 \text{ ft}}\right) + 8 \text{ in.} = (216 + 8) \text{ in.} = 224 \text{ in.}$

   $s = \dfrac{P}{4} = \dfrac{224 \text{ in.}}{4} = 56 \text{ in.}$

   $A = s^2 = (56 \text{ in.})^2 = 3{,}136 \text{ in.}^2$

   b)   Area in square feet

   $P = 18 \text{ ft} + 8 \text{ in.}\left(\dfrac{1 \text{ ft}}{12 \text{ in.}}\right) = (18 + \dfrac{8}{12}) \text{ ft} = 18.6667 \text{ ft}$

   $s = \dfrac{18.6667}{4} = 4.66667 \text{ ft}$

   $A = s^2 = (4.6667 \text{ ft})^2 = 21.7778 \text{ ft}^2 \approx 21.8 \text{ ft}^2$

29.    Triangle:  Find the hypotenuse using the Pythagorean Theorem.

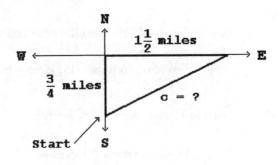

$a = \dfrac{3}{4}$ mi, $b = 1\dfrac{1}{2}$ mi

$c^2 = a^2 + b^2$

$= \left(\dfrac{3}{4}\right) \text{mi}^2 + \left(1\dfrac{1}{2}\right) \text{mi}^2$

$= \left(\dfrac{9}{16} + \dfrac{9}{4}\right) \text{mi}^2 = \dfrac{45}{16} \text{mi}^2 = 2.8125 \text{ mi}^2$

$c = \sqrt{2.8125} \ \text{mi}^2 = 1.67705 \text{ mi} \approx 1.68 \text{ mi}$

Miles jogged $= P = a + b + c$

$\qquad = \dfrac{3}{4} + 1\dfrac{1}{2} + 1.677 \text{ mi} = (0.75 + 1.50 + 1.677) \text{ mi} = 3.927 \text{ mi} \approx 3.9 \text{ mi}$

33.    Oval field:  $b = 100$ yd, $h = d = 40$ yd, $r = \dfrac{d}{2} = \dfrac{40 \text{ yd}}{2} = 20$ yd

   a)    $A_{\text{total}} = A_{\text{rectangle}} + 2 \bullet A_{\text{semicircle}}$

$\qquad = bh + 2\left(\dfrac{1}{2}\right)(\pi r^2)$

$\qquad = (100 \text{ yd})(40 \text{ yd}) + \pi(20 \text{ yd})^2$

$\qquad = 4000 \text{ yd}^2 + 400\pi \text{ yd}^2 = (4000 + 1256.6371) \text{ yd}^2 = 5256.6371 \text{ yd}^2 \approx 5256.6 \text{ yd}^2$

   b)    $P = 2(\text{Circumference of semicircle}) + 2 \text{ bases}$

$\qquad = 2\left(\dfrac{1}{2}\right)(\pi d) + 2b$

$\qquad = \pi (40 \text{ yd}) + 2(100 \text{ yd}) = (125.6637 + 200) \text{ yd} = 325.6637 \text{ yd} \approx 325.7 \text{ yd}$

37.    <u>Small</u> (13 in.) for \$11.75:  $d = 13$ in., $r = \dfrac{d}{2} = \dfrac{13 \text{ in.}}{2} = 6.5$ in.

$\qquad A = \pi r^2 = \pi (6.5 \text{ in.})^2 = 132.7323 \text{ in.}^2$

$\qquad \text{price per area} = \dfrac{\$11.75}{132.7323 \text{ in.}^2} = \$0.0885 \text{ per in.}^2$

37. Continued.

Large (16 in.) for \$14.75:  $d = 16$ in., $r = \dfrac{d}{2} = \dfrac{16 \text{ in.}}{2} = 8$ in.

$$A = \pi r^2 = \pi (8 \text{ in.})^2 = 201.0619 \text{ in.}^2$$

price per area $= \dfrac{\$14.75}{201.0619 \text{ in.}^2} = \$0.0734$ per in.$^2$

Super (19 in.) for \$22.75:  $d = 19$ in., $r = \dfrac{d}{2} = \dfrac{19 \text{ in.}}{2} = 9.5$ in.

$$A = \pi r^2 = \pi (9.5 \text{ in.})^2 = 283.5287 \text{ in.}^2$$

price per area $= \dfrac{\$22.75}{283.5287 \text{ in.}^2} = \$0.0802$ per in.$^2$

The best deal is the large pizza for \$14.75 since it costs only \$0.0734 per square inch.

---

## Section 6.2                     Volume and Surface Area

1.    Rectangular box:  $L = 5.2$ m, $w = 2.1$ m, $h = 3.5$ m

a)    $V = L \bullet w \bullet h = (5.2 \text{ m})(2.1 \text{ m})(3.5 \text{ m}) = 38.22 \text{ m}^3$

b)    $A_{\text{box}} = 2 \bullet A_{\text{base}} + 2 \bullet A_{\text{front}} + 2 \bullet A_{\text{side}}$

$= 2 \bullet L \bullet w + 2 \bullet L \bullet h + 2 \bullet w \bullet h$

$= [2(5.2)(2.1) + 2(5.2)(3.5) + 2(2.1)(3.5)] \text{ m}^2$

$= (21.84 + 36.40 + 14.70) \text{ m}^2$

$= 72.94 \text{ m}^2$

5.    Sphere:  $d = 1\dfrac{3}{4}$ in. $= 1.75$ in., $r = \dfrac{d}{2} = \dfrac{1.75 \text{ in.}}{2} = 0.875$ in.

a)    $V = \dfrac{4}{3}\pi r^3 = \dfrac{4}{3}\pi (0.875 \text{ in.})^3 = \dfrac{4}{3}\pi (0.6699 \text{ in.}^3) = 2.8062 \text{ in.}^3 \approx 2.81 \text{ in.}^3$

b)    $A = 4\pi r^2 = 4\pi (0.875 \text{ in.})^2 = 4\pi (0.7656 \text{ in.}^2) = 9.6211 \text{ in.}^2 \approx 9.62 \text{ in.}^2$

9. Pyramid with square base: $s = 4$ ft, $h = 4$ ft

$$V = \frac{1}{3} A_{\text{base}} \bullet h = \frac{1}{3} s^2 h = \frac{1}{3} (4 \text{ ft})^2 (4 \text{ ft}) = \frac{64}{3} \text{ ft}^3 = 21.3333 \text{ ft}^3 \approx 21.33 \text{ ft}^3$$

13. Two cylinders: $h = 5$ ft, $r_{\text{outer}} = 2$ ft, $r_{\text{inner}} = 1$ ft

$$V_{\text{total}} = V_{\text{outer}} - V_{\text{inner}} = A_{\text{outer base}} \bullet h - A_{\text{inner base}} \bullet h$$

$$= \pi (r_{\text{outer}})^2 \bullet h - \pi (r_{\text{inner}})^2 \bullet h$$

$$= \pi (2 \text{ ft})^2 (5 \text{ ft}) - \pi (1 \text{ ft})^2 (5 \text{ ft})$$

$$= 20\pi \text{ ft}^3 - 5\pi \text{ ft}^3 = (20\pi - 5\pi) \text{ ft}^3 = 15\pi \text{ ft}^3 = 47.12389 \text{ ft}^3 \approx 47.12 \text{ ft}^3$$

17. Silo: $h = 50$ ft, $d = 20$ ft, $r = \dfrac{20 \text{ ft}}{2} = 10$ ft

$$V_{\text{silo}} = V_{\text{cylinder}} + V_{\text{hemisphere}}$$

$$= A_{\text{base}} \bullet h + \frac{1}{2} V_{\text{sphere}}$$

$$= (\pi r^2) h + \frac{1}{2} \left( \frac{4}{3} \pi r^3 \right)$$

$$= \pi (10 \text{ ft})^2 (50 \text{ ft}) + \frac{2}{3} \pi (10 \text{ ft})^3 = 5000\pi \text{ ft}^3 + 666.67\pi \text{ ft}^3 = 5666.67\pi \text{ ft}^3$$

$$= 17{,}802.358 \text{ ft}^3 \approx 17{,}802.36 \text{ ft}^3$$

21. Earth: $d = 7920$ mi, $r = \dfrac{7920 \text{ mi}}{2} = 3960$ mi

$$V_{\text{earth}} = \frac{4}{3} \pi r^3 = \frac{4}{3} \pi (3960 \text{ mi})^3 = 2.601203 \times 10^{11} \text{ mi}^3$$

Moon: $d = 2160$ mi, $r = \dfrac{2160 \text{ mi}}{2}$ mi $= 1080$ mi

$$V_{\text{moon}} = \frac{4}{3} \pi r^3 = \frac{4}{3} \pi (1080 \text{ mi})^3 = 5.276669 \times 10^9 \text{ mi}^3$$

$$\frac{V_{\text{earth}}}{V_{\text{moon}}} = \frac{2.601203 \times 10^{11} \text{ mi}^3}{5.276669 \times 10^9 \text{ mi}^3} = 49.296$$

21. Continued.

Alternate Method:

$$\frac{V_{earth}}{V_{moon}} = \frac{\frac{4}{3}\pi(3960 \text{ mi})^3}{\frac{4}{3}\pi(1080 \text{ mi})^3} = \frac{(3960 \text{ mi})^3}{(1080 \text{ mi})^3} = 49.296$$

Approximately 49 moons could fit inside the earth.

25.    Volume of one cord of wood = 128 ft$^3$, price of one cord = $190
purchased:  $L = 10$ ft, $w = 4$ ft, $h = 2$ ft

$V_{purchased} = L \bullet w \bullet h = (10 \text{ ft})(4 \text{ ft})(2 \text{ ft}) = 80$ ft$^3$

You did not get a cord of wood so you did not get an honest deal.  You received only 80/128 = 0.625 of a cord so you should have paid (0.625)($190) = $118.75.

29.    Cement driveway:  $L = 36$ ft$\left(\dfrac{1 \text{ yd}}{3 \text{ ft}}\right) = 12$ yd, $w = 9$ ft$\left(\dfrac{1 \text{ yd}}{3 \text{ ft}}\right) = 3$ yd, $h = 6$ in.$\left(\dfrac{1 \text{ yd}}{36 \text{ in.}}\right) = \dfrac{1}{6}$ yd

$V_{driveway} = L \bullet w \bullet h = (12 \text{ yd})(3 \text{ yd})\left(\dfrac{1}{6} \text{ yd}\right) = 6$ yd$^3$

4 sacks of dry cement mix = 1 cubic yard of cement
6 cubic yards of cement require (6)(4) = 24 sacks of mix
24 sacks at $7.30 = 24($7.30) = $175.20

It will cost Ron $175.20 to buy the cement for his driveway.

33.    Water tank in shape of cone:  $d = 12$ ft, $r = \dfrac{12 \text{ ft}}{2} = 6$ ft

Use the Pythagorean Theorem to find the height of the cone.  $(a = 6 \text{ ft}, c = 10 \text{ ft}, b = h)$

$$c^2 = a^2 + b^2$$

$$10^2 = 6^2 + h^2$$

$$100 = 36 + h^2$$

$h^2 = 100 - 36 = 64,$  so $h = \sqrt{64} = 8$ ft

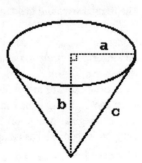

$V_{tank} = \dfrac{1}{3} A_{top} \bullet h = \dfrac{1}{3}(\pi r^2)h = \dfrac{1}{3}\pi(6 \text{ ft})^2(8 \text{ ft}) = 96\pi$ ft$^3$

$= 301.59289$ ft$^3 \approx 301.59$ ft$^3$

## Section 6.3                                              Egyptian Geometry

1.  Length of segments for the sides are 6, 8 and 10 units. The largest value $= c = 10$.

    $$a^2 + b^2 = 6^2 + 8^2 = 36 + 64 = 100 = 10^2 = c^2$$

    The segment lengths satisfy the Pythagorean Theorem, therefore the triangle is a right triangle with the right angle opposite the side of length 10.

5.  a)  Truncated pyramid: $h = 12$ cubits, $a = 2$ cubits, $b = 15$ cubits

    $$V = \frac{h}{3}(a^2 + ab + b^2)$$

    $$V = \frac{12 \text{ cubits}}{3}[(2 \text{ cubits})^2 + (2 \text{ cubits})(15 \text{ cubits}) + (15 \text{ cubits})^2]$$

    $$= (4 \text{ cubits})(4 \text{ cubits}^2 + 30 \text{ cubits}^2 + 225 \text{ cubits}^2) = (4 \text{ cubits})(259 \text{ cubits}^2)$$

    $$= 1036 \text{ cubits}^3\left(\frac{\frac{3}{2} \text{ khar}}{1 \text{ cubits}^3}\right) = 1554 \text{ khar}$$

    b)  Regular pyramid: $h = 18$ cubits, $b = 15$ cubits

    $$V = \frac{h}{3}(b^2) = \left(\frac{18 \text{ cubits}}{3}\right)(15 \text{ cubits})^2 = (6 \text{ cubits})(225 \text{ cubits}^2)$$

    $$= 1350 \text{ cubits}^3\left(\frac{\frac{3}{2} \text{ khar}}{1 \text{ cubits}^3}\right) = 2025 \text{ khar}$$

    The regular pyramid has the larger storage capacity.

9.  Construct a square whose sides are 24 units each, and inscribe a circle in the square. Thus, the diameter of the circle is 24 units. Now divide the sides of the square into three equal segments of 8 units each. Remove the triangular corners as shown and an irregular octagon is formed. The area of this octagon is then used as an approximation of the area of the circle: $A_{\text{circle}} \approx A_{\text{octagon}}$

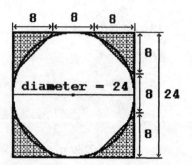

**9. Continued.**

The area of the octagon is equal to the area of the large square minus the area of the four triangular corners. Rearranging these triangles, we can see that the area of two triangular corners is the same as the area of one square of side 8 units.

$$A_{\text{circle}} \approx A_{\text{octagon}} = A_{\text{square of side 24}} - 2A_{\text{square of side 8}}$$
$$= (24)^2 - 2(8)^2 = 576 - 128 = 448 \text{ square units}$$

$\sqrt{448} = 21.1660105 \approx 21$ units. Therefore, we can say

$$A_{\text{circle of diameter 24}} \approx 448 \approx 441 = A_{\text{square of side 21}}$$

**13.**    Circle: $r = 3$ palms

a)    Egyptian approximation of pi $\left( \pi = \dfrac{256}{81} \right)$

$$A = \pi r^2 = \frac{256}{81}(3 \text{ palms})^2 = \frac{256}{81}(9 \text{ palms}^2) = \frac{256}{9} \text{ palms}^2$$

$$= 28.44444444 \text{ palms}^2 \approx 28.44 \text{ palms}^2$$

b)    Calculator value of pi.

$$A = \pi r^2 = \pi (3 \text{ palms})^2 = 9\pi \text{ palms}^2 = 28.27433388 \text{ palms}^2 \approx 28.27 \text{ palms}^2$$

c)    Error of Egyptian calculation relative to the calculator value.

$$A_{\text{Egyptian}} - A_{\text{Calculator}} = 28.44444444 - 28.27433388 = 0.17011056 \text{ palms}^2$$

$$\text{relative error} = \frac{0.17011056}{28.27433388} = 0.0060164 \approx 0.6\%$$

**17.**    $a = 2$ cubits, 5 palms, 3 fingers, $b = 3$ cubits, 3 palms, 2 fingers, $c = 4$ cubits, 4 palms, 3 fingers

$$P = a + b + c = \quad 2 \text{ cubits, } 5 \text{ palms, 3 fingers}$$
$$+ 3 \text{ cubits, } 3 \text{ palms, 2 fingers}$$
$$\underline{+ 4 \text{ cubits, } 4 \text{ palms, 3 fingers}}$$
$$9 \text{ cubits, 12 palms, 8 fingers}$$

Convert fingers to palms (1 palm = 4 fingers):
$$9 \text{ cubits, 12 palms, } 8 \text{ fingers}$$
$$\underline{+ 2 \text{ palms, } -8 \text{ fingers}}$$
$$9 \text{ cubits, 14 palms}$$

17. Continued.

Convert palms to cubits (1 cubit = 7 palms):

9 cubits,  14 palms

+ 2 cubits, −14 palms

11 cubits

The perimeter of the triangle is 11 cubits.

21. Triangle: $a = 150$ cubits, $b = 200$ cubits, $c = 250$ cubits

Use Heron's Formula

$$s = \frac{1}{2}(a+b+c) = \frac{1}{2}(150 + 200 + 250) = \frac{1}{2}(600 \text{ cubits}) = 300 \text{ cubits}$$

$$A = \sqrt{s(s-a)(s-b)(s-c)}$$

$$= \sqrt{300(300-150)(300-200)(300-250) \text{ cubits}^4} = \sqrt{(300)(150)(100)(50) \text{ cubits}^4}$$

$$= \sqrt{225,000,000 \text{ cubits}^4}$$

$$= 15,000 \text{ cubits}^2 \left( \frac{1 \text{ setat}}{10,000 \text{ cubits}^2} \right) = 1.5 \text{ setats}$$

25. a)  $d = 9$ cubits, $h = 10$ cubits

Construct a square whose sides are 9 units each, and inscribe a circle in the square. Thus, the diameter of the circle is 9 units. Now divide the sides of the square into three equal segments of 3 units each. Remove the triangular corners as shown and an irregular octagon is formed. The area of this octagon is then used as an approximation of the area of the circle: $A_{\text{circle}} \approx A_{\text{octagon}}$.

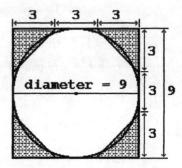

The area of the octagon is equal to the area of the large square minus the area of the four triangular corners. Rearranging these triangles, we can see that the area of two triangular corners is the same as the area of one square of side 3 units.

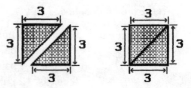

$$A_{\text{circle}} \approx A_{\text{octagon}} = A_{\text{square of side 9}} - 2A_{\text{square of side 3}}$$
$$= (9)^2 - 2(3)^2 = 81 - 18 = 63 \text{ square units}$$

25. Continued.

$$\sqrt{63} = 7.9372539 \approx 8 \text{ units. Therefore, we can say}$$

$$A_{\text{circle of diameter 9}} \approx 63 \approx 64 = A_{\text{square of side 8}}$$

The Eqyptians thought the area of the circle must be exactly the area of the square so they would use 64 square cubits as the area of the circle.

$$V = A_{\text{circle}} \bullet h = (64 \text{ cubits}^2)(10 \text{ cubits}) = 640 \text{ cubits}^3 \left( \frac{\frac{3}{2} \text{ khar}}{1 \text{ cubits}^3} \right) = 960 \text{ khar}$$

b) $r = 4.5$ cubits, $h = 10$ cubits

$$V = A_{\text{circle}} \bullet h = \pi r^2 h = \pi (4.5)^2 (10)$$

$$= 636.1725124 \text{ cubits}^3 \left( \frac{\frac{3}{2} \text{ khar}}{1 \text{ cubits}^3} \right) = 954.2587685 \text{ khar} \approx 954.2588 \text{ khar}$$

c) Error of Egyptian calculation relative to the calculator value.

$$V_{\text{Egyptian}} - V_{\text{Calculator}} = 960 - 954.2587685 = 5.74123147 \text{ khar}$$

$$\text{relative error} = \frac{5.74123147}{954.2587685} = 0.0060164 \approx 0.6\%$$

---

## Section 6.4                                                    The Greeks

1. The sides are proportional because the triangles are similar.

Compare the large triangle to the small triangle to find $y$:

$$\frac{y}{4} = \frac{90}{6}, \text{ then } y = \frac{90(4)}{6} = \frac{360}{60} = 60$$

Compare the small triangle to the large triangle to find $x$:

$$\frac{x}{75} = \frac{6}{90}, \text{ then } x = \frac{6(75)}{90} = \frac{450}{90} = 5$$

5. Use similar triangles. $x$ = height of tree.

$$\frac{x}{6} = \frac{21}{3.5}$$

$$x = \frac{21(6)}{3.5} = \frac{126}{3.5} = 36$$

The tree is 36 feet tall.

 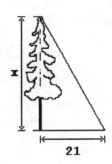

9. Rectangular Box: $L = 4$ ft, $w = 3$ ft, $h = 2$ ft

$c$ = diagonal on bottom of box
$d$ = diagonal of the box

$$c^2 = L^2 + w^2 = (4 \text{ ft})^2 + (3 \text{ ft})^2$$
$$= 16 \text{ ft}^2 + 9 \text{ ft}^2 = 25 \text{ ft}^2$$

$$c = \sqrt{25 \text{ ft}^2} = 5 \text{ ft}$$

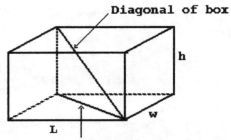

Diagonal of box

Diagonal of bottom

$$d^2 = c^2 + h^2 = (5 \text{ ft})^2 + (2 \text{ ft})^2 = 25 \text{ ft}^2 + 4 \text{ ft}^2 = 29 \text{ ft}^2$$

$$d = \sqrt{29 \text{ ft}^2} = \sqrt{29} \text{ ft} = 5.385 \text{ ft} \approx 5.3 \text{ ft}$$

The length of the longest object that will fit inside the box is 5.3 feet. Note: It is appropriate in this situation to not round up so that the object will fit inside the box.

13. Given: $AD = CD$
      $AB = CB$

Prove: $\angle DBA = \angle DBC$

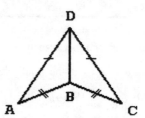

| Statements | Reasons |
|---|---|
| 1. $AD = CD$ | 1. Given |
| 2. $AB = CB$ | 2. Given |
| 3. $DB = DB$ | 3. Anything equals itself |
| 4. $\triangle ADB \cong \triangle CDB$ | 4. SSS |
| 5. $\angle DBA \cong \angle DBC$ | 5. Corresponding parts of congruent triangles are equal |

17.    Given: $AE = CE$
$\quad\quad\quad\quad AB = CB$

Prove: $\angle ADB = \angle CDB$

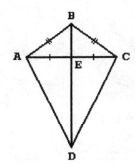

| Statements | Reasons |
|---|---|
| 1. $AE = CE$ | 1. Given |
| 2. $AB = CB$ | 2. Given |
| 3. $BE = BE$ | 3. Anything equals itself |
| 4. $\triangle ABE \cong \triangle CBE$ | 4. SSS |
| 5. $\angle ABD = \angle CBD$ | 5. Corresponding parts of congruent triangles are equal |
| 6. $BD = BD$ | 6. Anything equals itself |
| 7. $\triangle ABD \cong \triangle CBD$ | 7. SAS |
| 8. $\angle ADB = \angle CDB$ | 8. Corresponding parts of congruent triangles are equal |

21.    a)    $s = \left( \sqrt{2 - \sqrt{2 + \sqrt{2}}} \right) r = \left( \sqrt{2 - \sqrt{2 + 1.41421356}} \right) r$

$$= \left( \sqrt{2 - \sqrt{3.41421356}} \right) r = \left( \sqrt{2 - 1.84775907} \right) r = \sqrt{0.15224094} \; r = 0.39018064r$$

$C_{\text{circle}} \approx P_{\text{inscribed polygon with 16 sides}}$

$\quad 2\pi r \approx 16s$

$\quad 2\pi r \approx 16(0.39018064r\,)$

$\quad\quad \pi \approx \dfrac{16(0.39018064)r}{2r} = 8(0.39018064) = 3.121445152$

b)    $s = \left( \dfrac{2\sqrt{2 - \sqrt{2}}}{2 + \sqrt{2 + \sqrt{2}}} \right) r = \left( \dfrac{2\sqrt{2 - 1.41421356}}{2 + \sqrt{2 + 1.41421356}} \right) r$

$$= \left( \dfrac{2\sqrt{0.58578644}}{2 + \sqrt{3.41421356}} \right) r = \left( \dfrac{2(0.76536686)}{2 + 1.84775907} \right) r = \dfrac{1.53073373}{3.84775907} \; r = 0.39782473r$$

21. Continued.

$$C_{\text{circle}} \approx P_{\text{circumscribed polygon with 16 sides}}$$

$$2\pi r \approx 16s$$

$$2\pi r \approx 16(0.39782473r)$$

$$\pi \approx \frac{16(0.39782473)r}{2r} = 8(0.39782473) = 3.182597878$$

Combining (b) with (a), we have $3.121445152 < \pi < 3.182597878$

---

## Section 6.5                              Right Triangle Trigonometry

1.   opposite $= x$, adjacent $= y$, hypotenuse $= 6$

$\theta + 30° + 90° = 180°$                              Sum of angles is $180°$.
$\qquad \theta = 180° - 30° - 90° = 60°$

$\sin 30° = \dfrac{\text{opp}}{\text{hyp}} = \dfrac{x}{6}$, then $x = 6(\sin 30°) = 6\left(\dfrac{1}{2}\right) = 3$

$\cos 30° = \dfrac{\text{adj}}{\text{hyp}} = \dfrac{y}{6}$, then $y = 6(\cos 30°) = 6\left(\dfrac{\sqrt{3}}{2}\right) = 3\sqrt{3}$

5.   $x = \text{opp}, y = \text{hyp}, 3 = \text{adj}$

$\theta + 45° + 90° = 180°$                              Sum of angles is $180°$.
$\qquad \theta = 180° - 45° - 90° = 45°$

$\tan 45° = \dfrac{\text{opp}}{\text{adj}} = \dfrac{x}{3}$, then $x = 3(\tan 45°) = 3(1) = 3$

$\cos 45° = \dfrac{\text{adj}}{\text{hyp}} = \dfrac{3}{y}$, then $y = \dfrac{3}{\cos 45°} = \dfrac{3}{\left(\frac{1}{\sqrt{2}}\right)} = 3\sqrt{2}$

9.  $a = 12.0, A = 37°$

$B = 180° - 90° - 37° = 53°$

$\tan 37° = \dfrac{\text{opp}}{\text{adj}} = \dfrac{12.0}{b}$

$b(\tan 37°) = 12.0$

$b = \dfrac{12.0}{\tan 37°} = \dfrac{12.0}{0.7535...} = 15.924... \approx 15.9$

$\sin 37° = \dfrac{\text{opp}}{\text{hyp}} = \dfrac{12.0}{c}$

$c(\sin 37°) = 12.0$

$c = \dfrac{12.0}{\sin 37°} = \dfrac{12.0}{0.6018...} = 19.939... \approx 19.9$

13.  $c = 0.92, B = 49.9°$

$A = 180° - 90° - 49.9° = 40.1°$

$\cos 49.9° = \dfrac{\text{adj}}{\text{hyp}} = \dfrac{a}{0.92}$

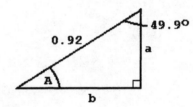

$a = 0.92(\cos 49.9°) = 0.92(0.6441) = 0.59259 \approx 0.59$

$\sin 49.0° = \dfrac{\text{opp}}{\text{hyp}} = \dfrac{b}{0.92}$

$b = 0.92(\sin 49.9°) = 0.92(0.7649) = 0.70372... \approx 0.70$

17.    $c = 54.40, A = 53.125°$

$B = 180° - 90° - 53.125° = 36.875°$

$\sin 53.125° = \dfrac{\text{opp}}{\text{hyp}} = \dfrac{a}{54.40}$

$a = 54.40(\sin 53.125°)$

$= 54.40(0.79995) = 43.51709 \approx 43.52$

$\cos 53.125° = \dfrac{\text{adj}}{\text{hyp}} = \dfrac{b}{54.40}$

$b = 54.40(\cos 53.125°)$

$= 54.40(0.60007) = 32.6439 \approx 32.64$

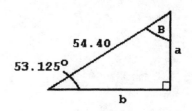

21.    $a = 15.0, c = 23.0$

$a^2 + b^2 = c^2$

$(15.0)^2 + b^2 = (23.0)^2$

$b^2 = (23.0)^2 - (15.0)^2 = 529 - 225 = 304$

$b = \sqrt{304} = 17.43559 \approx 17.4$

$A = \sin^{-1}\left(\dfrac{\text{opp}}{\text{hyp}}\right) = \sin^{-1}\left(\dfrac{15.0}{23.0}\right) = \sin^{-1}(0.65217...) = 40.7°$

$B = 180° - 90° - 40.7° = 49.3°$

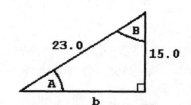

25.    $b = 0.123, c = 0.456$

$a^2 + b^2 = c^2$

$a^2 + (0.123)^2 = (0.456)^2$

$a^2 = (0.456)^2 - (0.123)^2$

$= 0.207936 - 0.015129 = 0.192807$

$a = \sqrt{0.1928007} = 0.43909979 \approx 0.439$

$B = \sin^{-1}\left(\dfrac{\text{opp}}{\text{hyp}}\right) = \sin^{-1}\left(\dfrac{0.123}{0.456}\right) = \sin^{-1}(0.269736) = 15.6°$

$A = 180° - 90° - 15.6° = 74.4°$

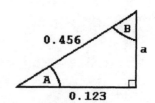

29.    tower = 20 ft, angle of depression = 7.6°

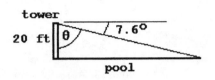

$\theta = 90° - 7.6° = 82.4°$

$\tan 82.4° = \dfrac{\text{pool}}{\text{tower}} = \dfrac{\text{pool}}{20}$

pool = 20(tan 82.4°) = 20(7.49465) = 149.893 ≈ 150 ft

33.    bell tower = 48.5 ft, θ = angle of elevation

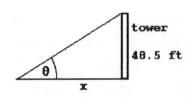

a)    $\theta = 15.4°$

$\tan 15.4° = \dfrac{\text{tower}}{x} = \dfrac{48.5}{x}$

$x(\tan 15.4°) = 48.5$

$x = \dfrac{48.5}{\tan 15.4°} = \dfrac{48.5}{0.27544} = 176.078 \approx 176.1 \text{ feet}$

b)    $\theta = 61.2°$

$x = \dfrac{48.5}{\tan 61.2°} = \dfrac{48.5}{1.81899} = 26.663 \approx 26.7 \text{ feet}$

37.    $\tan \theta = \dfrac{\text{opp}}{\text{adj}}$, then opp = adj(tan θ)

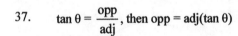

small triangle:
     opp = $h$, adj = $x$, $\theta = 61.0°$

     $h = x(\tan 61.0°)$

large triangle:
     opp = $h$, adj = $x + 50$, $\theta = 49.5°$

     $h = (x + 50)(\tan 49.5°)$

$h_{\text{small triangle}} = h_{\text{large triangle}}$

$x(\tan 61.0°) = (x + 50)(\tan 49.5°)$
$x(\tan 61.0°) = x(\tan 49.5°) + 50(\tan 49.5°)$
$x(\tan 61.0°) - x(\tan 49.5°) = 50(\tan 49.5°)$
$x(\tan 61.0° - \tan 49.5°) = 50(\tan 49.5°)$

$x = \dfrac{50(\tan 49.5°)}{\tan 61.0° - \tan 49.5°} = \dfrac{50(1.17085)}{1.80405 - 1.17085} = 92.455$

$h = x(\tan 61.0°) = 92.455(1.80405) = 166.7936 \approx 167 \text{ feet}$

41.   adj $= 900$ feet, opp $= h$, angle of elevation $= 58.24°$

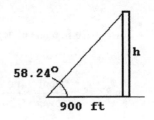

$$\tan 58.24° = \frac{\text{opp}}{\text{adj}} = \frac{h}{900}$$

$h = 900(\tan 58.24°) = 900(1.61535) = 1453.817 \approx 1454$ ft

---

| **Section 6.6** | **Conic Sections and Analytic Geometry** |
|---|---|

1.   $x^2 + y^2 = 1^2$

$(x - 0)^2 + (y - 0)^2 = 1^2$

Center is at $(0, 0)$, $r = 1$

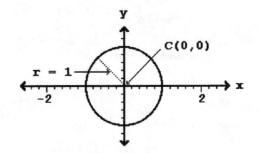

5.   $x^2 + y^2 - 10x + 4y + 13 = 0$

$(x^2 - 10x) + (y^2 + 4y) = -13$

To complete $x$:  $\left(\dfrac{10}{2}\right)^2 = 5^2 = 25$

To complete $y$:  $\left(\dfrac{4}{2}\right)^2 = 2^2 = 4$

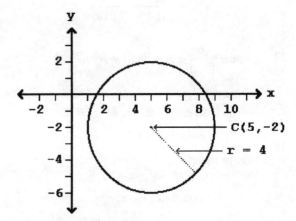

$(x^2 - 10x + 25) + (y^2 + 4y + 4) = -13 + 25 + 4$

$\qquad\quad (x - 5)^2 + (y + 2)^2 = 16$    Factor grouped terms

$\qquad\quad (x - 5)^2 + (y - (-2))^2 = 4^2$    Rewrite to match formula with minus signs and $r^2$.

Center is at $(5, -2)$, $r = 4$

9.    $y = x^2$, Vertex is at $(0,0)$

| $x$ | $y = x^2$ |
|-----|-----------|
| 0   | 0 |
| $\pm 1$ | 1 |
| $\pm 2$ | 4 |
| $\pm 3$ | 9 |

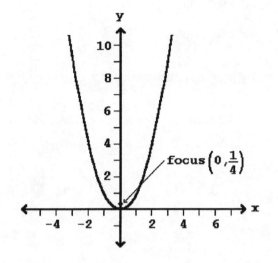

Find the focus:
  Use the point $Q(1, 1)$ which is
  on the parabola

$$4py = x^2$$
$$4p(1) = (1)^2$$
$$4p = 1$$

$$p = \frac{1}{4}, \text{ The focus is at } \left(0, \frac{1}{4}\right).$$

13.    Draw a parabola with its vertex at the origin.  $Q(4.5, 1.75)$ is a point on the parabola.

$$4py = x^2$$
$$4p(1.75) = (4.5)^2$$
$$7p = 20.25$$
$$p = 2.89286 \approx 2.9 \text{ ft}$$

Place the water container at the focus of the parabola which would be centered approximately 2.9 feet above the bottom of the reflector dish.

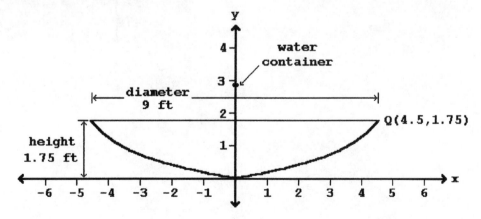

17.　$4x^2 + 9y^2 = 36$

$$\frac{4x^2}{36} + \frac{9y^2}{36} = \frac{36}{36}$$

$$\frac{x^2}{9} + \frac{y^2}{4} = 1$$

<u>*x*-intercept (*y* = 0)</u>

$$\frac{x^2}{9} + \frac{(0)^2}{4} = 1$$

$$\frac{x^2}{9} = 1$$

$$x^2 = 9$$

$$x = \pm 3$$

<u>*y*-intercept (*x* = 0)</u>

$$\frac{(0)^2}{9} + \frac{y^2}{4} = 1$$

$$\frac{y^2}{4} = 1$$

$$y^2 = 4$$

$$y = \pm 2$$

Finding the foci:

$$a^2 = 9, \, b^2 = 4$$

$$c^2 = a^2 - b^2 = 9 - 4 = 5$$

$$c = \pm\sqrt{5} \approx \pm 2.236$$

Foci are located on the *x*-axis at

$$\left(-\sqrt{5}, 0\right) \text{ and } \left(\sqrt{5}, 0\right).$$

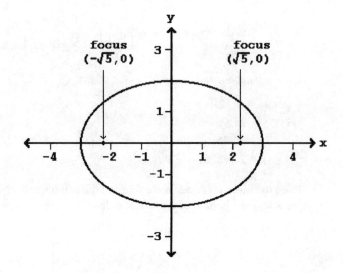

21.    $16x^2 + 9y^2 = 144$

$$\frac{16x^2}{144} + \frac{9y^2}{144} = \frac{144}{144}$$

$$\frac{x^2}{9} + \frac{y^2}{16} = 1$$

**_x_-intercept (_y_ = 0)**

$$\frac{x^2}{9} + \frac{(0)^2}{16} = 1$$

$$\frac{x^2}{9} = 1$$

$$x^2 = 9$$

$$x = \pm 3$$

**_y_-intercept (_x_ = 0)**

$$\frac{(0)^2}{9} + \frac{y^2}{16} = 1$$

$$\frac{y^2}{16} = 1$$

$$y^2 = 16$$

$$y = \pm 4$$

Finding the foci:

$$a^2 = 9, \, b^2 = 16$$

$$c^2 = b^2 - a^2 = 16 - 9 = 7$$

$$c = \pm\sqrt{7} \approx \pm 2.646$$

Foci are located on the _y_-axis at $\left(0, -\sqrt{7}\right)$ and $\left(0, \sqrt{7}\right)$.

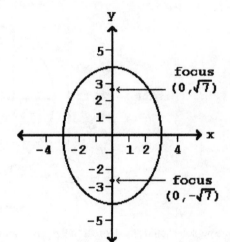

25.   Center an ellipse at the origin on a rectangular coordinate system so that the x-axis is 30 feet long and the y-axis is 24 feet long. This makes a = 15 and b = 12.

$$\frac{x^2}{a^2}+\frac{y^2}{b^2}=\frac{x^2}{15^2}+\frac{y^2}{12^2}=\frac{x^2}{225}+\frac{y^2}{144}=1$$

Finding the foci:
$$a^2 = 225, b^2 = 144$$

$$c^2 = a^2 - b^2 = 225 - 144 = 81$$

$$c = \pm\sqrt{81} = \pm 9$$

Foci are located on the x-axis at (−9, 0) and (9, 0).

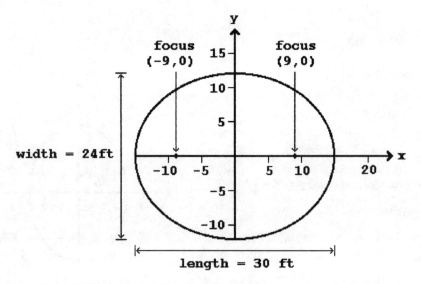

One person should stand 9 feet to the left from the center of the room along the longer axis which is the location of one focus. The other person should stand 9 feet to the right of the center at the other focus.

29.　　a)　　$x^2 - y^2 = 1$, $a^2 = 1$, $b^2 = 1$

**x-intercept (y = 0)**　　　　　　　　**y-intercept (x = 0)**

$x^2 - (0)^2 = 1$　　　　　　　　　　$(0)^2 - y^2 = 1$

$\qquad x^2 = 1$　　　　　　　　　　　$-y^2 = 1$

$\qquad x = \pm 1$　　　　　　　　　no solution; no y-intercepts

Locate $a = \pm 1$ on the x-axis and $b = \pm 1$ on the y-axis. Draw a rectangle with sides parallel to the axes going through these points. Now draw the diagonals of the rectangle. Then draw the branches of the hyperbola opening left and right.

Finding the foci:

$$a^2 = 1, b^2 = 1$$

$$c^2 = a^2 + b^2 = 1 + 1 = 2$$

$$c = \pm\sqrt{2} \approx \pm 1.41429$$

Foci are located on the x-axis at $\left(-\sqrt{2}, 0\right)$ and $\left(\sqrt{2}, 0\right)$.

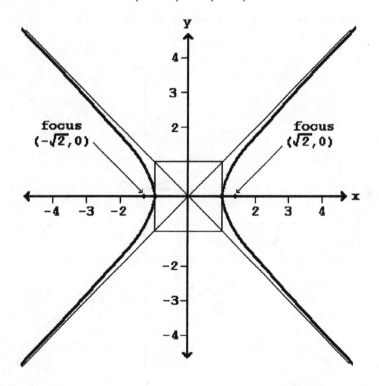

29. Continued.

b)  $y^2 - x^2 = 1$, $a^2 = 1$, $b^2 = 1$

<u>**x-intercept (y = 0)**</u>          <u>**y-intercept (x = 0)**</u>

$(0)^2 - x^2 = 1$                $y^2 - (0)^2 = 1$

$-x^2 = 1$                      $y^2 = 1$

no solution; no x-intercepts        $y = \pm 1$

Locate $a = \pm 1$ on the x-axis and $b = \pm 1$ on the y-axis. Draw a rectangle with sides parallel to the axes going through these points. Now draw the diagonals of the rectangle. Then draw the branches of the hyperbola opening up and down.

Finding the foci:

$a^2 = 1$, $b^2 = 1$

$c^2 = a^2 + b^2 = 1 + 1 = 2$

$c = \pm\sqrt{2} = \approx \pm 1.414$

Foci are located on the y-axis at $\left(0, -\sqrt{2}\right)$ and $\left(0, \sqrt{2}\right)$.

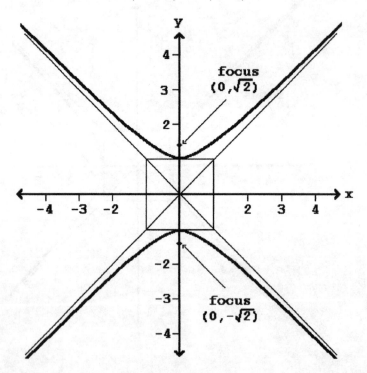

33. $25y^2 - 4x^2 = 100$

$$\frac{25y^2}{100} - \frac{4x^2}{100} = \frac{100}{100}$$

$$\frac{y^2}{4} - \frac{x^2}{25} = 1, \quad a^2 = 25, b^2 = 4$$

**x-intercept (y = 0)**

$$\frac{(0)^2}{4} - \frac{x^2}{25} = 1$$

$$-\frac{x^2}{25} = 1$$

$$-x^2 = 25$$

no solution; no x-intercepts

**y-intercept (x = 0)**

$$\frac{y^2}{4} - \frac{(0)^2}{25} = 1$$

$$\frac{y^2}{4} = 1$$

$$y^2 = 4$$

$$y = \pm 2$$

Locate $a = \pm 5$ on the x-axis and $b = \pm 2$ on the y-axis. Draw a rectangle with sides parallel to the axes going through these points. Now draw the diagonals of the rectangle. Then draw the branches of the hyperbola opening up and down.

Finding the foci:

$$a^2 = 25, b^2 = 4$$

$$c^2 = a^2 + b^2 = 25 + 4 = 29$$

$$c = \pm\sqrt{29} \approx \pm 5.385$$

Foci are located on the y-axis at $\left(0, -\sqrt{29}\right)$ and $\left(0, \sqrt{29}\right)$.

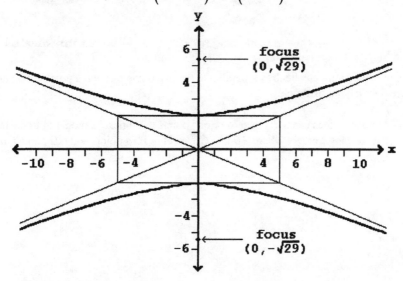

| **Section 6.7** | **Non-Euclidean Geometry** |
|---|---|

1.    Only one line parallel to the given line can be drawn through a point not on the given line.

5.    A pair of distinct lines can intersect in only zero or one points.

9.    A triangle can have zero to three right angles. (The sum of the three angles can be greater than 180°.) See the following drawings to visualize the different possible triangles.

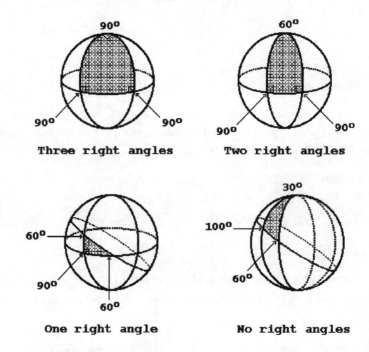

13.   Infinitely many lines can be drawn parallel to the given line through a point not on the given line. See Figure 6.99 in your textbook.

17.   A pair of distinct lines can meet in zero or one points in Poincaré's model of Lobachevskian geometry. See Figure 6.99 where *AB* and *CD* meet in zero points and *CD* and *EF* meet in one point.

| **Section 6.8** | **Fractal Geometry** |
| --- | --- |

1.    Follow the directions for drawing the Sierpinski gasket in the text.  It should look similar to Figure 6.107.

5.    $s = 5, n = 25$.

$s^d = n$              Calculate $d$
$5^d = 25$

$d = 2$              Since $5^2 = 25$

9.    To determine the scale and the number of new objects, draw one gasket and then draw a second step with double the length of the pattern in both directions as shown in the diagram.  Since the size of the object stays the same, this gives a valid scale of 2 with a resulting number of 3.

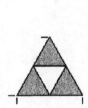 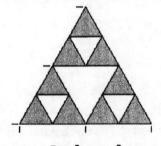

Scale = 1               Scale = 2
Number = 1              Number = 3

$s^d = n$        Calculate $d$

$2^d = 3$        Substituting

Find $d$ by trial and error:
$2^1 = 2$              which is less than 3
$2^2 = 4$              which is more than 3
$2^{1.5} = 2.828...$        which is less than 3, but close
$2^{1.6} = 3.031...$        which is more than 3, but closer than $2^{1.5}$

$d = 1.6$    Since $2^{1.6}$ gives the closest value rounded to the nearest tenth.

13.   To determine the scale and the number of new objects, draw one snowflake and then draw a second
      step with double the length of the pattern in both directions as shown in the diagram. Since that
      would not complete a complete pattern, try a scale of 3, which also does not complete the pattern.
      Trying a scale of 4 would complete a pattern in the same size as the original object. This gives a
      valid scale of 4 with a resulting number of 8.

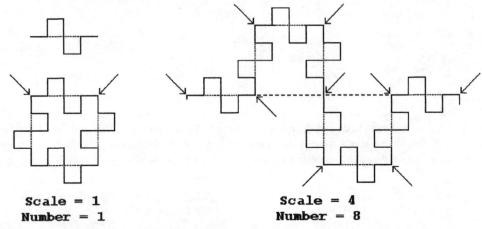

                  **Scale = 1**                              **Scale = 4**
                  **Number = 1**                             **Number = 8**

      $s^d = n$          Calculate $d$

      $4^d = 8$          Substituting

                  Find $d$ by trial and error:
                        $4^1 = 4$          which is less than 8
                        $4^2 = 16$         which is more than 8
                        $4^{1.5} = 8...$   which is 8

      $d = 1.5$     Since $4^{1.5} = 8$

17.   To determine the scale and the number of new objects, draw one rectangle and then draw a second
      step with double the length of the pattern in both directions as shown in the diagram. That does
      complete a pattern of the same size object. Further, trying a scale of 3, also completes the pattern in
      the same size as the original object. This gives a valid scale of 2 with a resulting number of 4 or a
      scale of 3 with a number of 9.

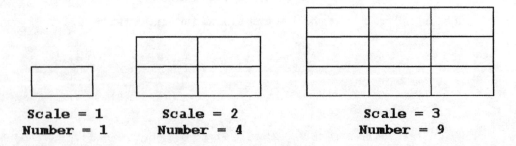

            **Scale = 1**           **Scale = 2**                   **Scale = 3**
            **Number = 1**          **Number = 4**                  **Number = 9**

17. Continued.

$s = 2, n = 4.$
$s^d = n$       Calculate $d$
$2^d = 4$
$d = 2$       Since $2^2 = 4$

Similarly, $s = 3, n = 9.$
$s^d = n$       Calculate $d$
$3^d = 25$
$d = 2$       Since $3^2 = 9$

## Section 6.9       The Perimeter and Area of a Fractal

1.   a)    Column 1: The number of the step in the process of drawing the gasket. Step 1 is the original triangle and each succeeding step is a part of the construction.

Column 2: The number of filled-in triangles when the step is completed. The number of triangles increase by a factor of 3 in each step.

Column 3: The length of one side of the triangle when the step is completed. The length is one-half of the previous length.

Column 4: The perimeter of one triangle when the step is completed. The perimeter is 3 times the length of one side from Column 3.

Column 5: The total perimeter of all the filled-in triangles in the gasket. The total perimeter is the product of multiplying the number of triangles (Column 2) by the perimeter of one triangle (Column 4).

b)

| Step | Number of triangles | Length of each side | Perimeter of one triangle | Total perimeter of all triangles |
|---|---|---|---|---|
| 1 | 1 | 1 ft | $3 \bullet 1$ ft $= 3$ ft | $1 \bullet 3$ ft $= 3$ ft |
| 2 | 3 | $\frac{1}{2} \bullet 1$ ft $= \frac{1}{2}$ ft | $3 \bullet \frac{1}{2}$ ft $= \frac{3}{2}$ ft | $3 \bullet \frac{3}{2}$ ft $= \frac{9}{2}$ ft |
| 3 | $3 \bullet 3 = 9$ | $\frac{1}{2} \bullet \frac{1}{2}$ ft $= \frac{1}{4}$ ft | $3 \bullet \frac{1}{4}$ ft $= \frac{3}{4}$ ft | $9 \bullet \frac{3}{4}$ ft $= \frac{27}{4}$ ft |
| 4 | $3 \bullet 9 = 27$ | $\frac{1}{2} \bullet \frac{1}{4}$ ft $= \frac{1}{8}$ ft | $3 \bullet \frac{1}{8}$ ft $= \frac{3}{8}$ ft | $27 \bullet \frac{3}{8}$ ft $= \frac{81}{8}$ ft |
| 5 | $3 \bullet 27 = 81$ | $\frac{1}{2} \bullet \frac{1}{8}$ ft $= \frac{1}{16}$ ft | $3 \bullet \frac{1}{16}$ ft $= \frac{3}{16}$ ft | $81 \bullet \frac{3}{16}$ ft $= \frac{243}{16}$ ft |
| 6 | $3 \bullet 81 = 243$ | $\frac{1}{2} \bullet \frac{1}{16}$ ft $= \frac{1}{32}$ ft | $3 \bullet \frac{1}{32}$ ft $= \frac{3}{32}$ ft | $243 \bullet \frac{3}{32}$ ft $= \frac{729}{32}$ ft |

c)    In each step the number of triangles increase by a factor of 3 because each previous triangle is replaced by 3 new triangles.

d)    The length of each side is decreased by a factor of $\frac{1}{2}$ because each side is divided into two equal pieces.

1. Continued.

e) The perimeter of one triangle changes by $\frac{1}{2}$ because taking $\frac{1}{2}$ of each piece and adding is the same as taking $\frac{1}{2}$ of the sum of the pieces. The perimeter of one triangle is decreasing because the sides are getting shorter.

f) The total perimeter of all the triangles changes by a factor of $\frac{3}{2}$ because there are 3 times as many triangles and the perimeter of each triangle is $\frac{1}{2}$ as long. The total perimeter of all triangles is increasing because $\frac{3}{2}$ is larger than 1.

g)
$$\text{Total Perimeter}_{\text{Step}} = \left(\begin{array}{c}\text{Perimeter}\\\text{in Step 1}\end{array}\right)\left(\begin{array}{c}\text{Factor for}\\\text{new triangles}\end{array}\right)^{\text{Step - 1}}\left(\begin{array}{c}\text{Factor for}\\\text{length of sides}\end{array}\right)^{\text{Step - 1}}$$

$$P_n = P_1 (3)^{n-1}\left(\frac{1}{2}\right)^{n-1}$$

$$P_n = P_1 \left(\frac{3}{2}\right)^{n-1}$$

h) Find the total perimeter for step 100:

$$P_n = P_1\left(\frac{3}{2}\right)^{n-1}$$

$$P_{100} = (3 \text{ ft})\left(\frac{3}{2}\right)^{100-1} = (3 \text{ ft})\left(\frac{3}{2}\right)^{99} = (3 \text{ ft})(2.710...\times 10^{17}) \approx 8.1 \times 10^{17} \text{ ft}$$

Since this is a very large number and would continue to increase in subsequent steps, the total perimeter of a Sierpinski gasket would be infinitely large.

5.  a)

| Step | Number of triangles | Length of each side of new triangle | Perimeter of one triangle | Total perimeter of all triangles |
|------|---------------------|-------------------------------------|---------------------------|----------------------------------|
| 1 | 1 | 1 ft | $3 \bullet 1 \text{ ft} = 3 \text{ ft}$ | $1 \bullet 3 \text{ ft} = 3 \text{ ft}$ |
| 2 | 6 | $\frac{1}{3} \bullet 1 \text{ ft} = \frac{1}{3} \text{ ft}$ | $3 \bullet \frac{1}{3} \text{ ft} = \frac{3}{3} \text{ ft}$ | $6 \bullet \frac{3}{3} \text{ ft} = \frac{18}{3} \text{ ft} = 6 \text{ ft}$ |
| 3 | $6 \bullet 6 = 36$ | $\frac{1}{3} \bullet \frac{1}{3} \text{ ft} = \frac{1}{9} \text{ ft}$ | $3 \bullet \frac{1}{9} \text{ ft} = \frac{3}{9} \text{ ft}$ | $36 \bullet \frac{3}{9} \text{ ft} = \frac{108}{9} \text{ ft} =$ 12 ft |
| 4 | $6 \bullet 36 = 216$ | $\frac{1}{3} \bullet \frac{1}{9} \text{ ft} = \frac{1}{27} \text{ ft}$ | $3 \bullet \frac{1}{27} \text{ ft} = \frac{3}{27} \text{ ft}$ | $216 \bullet \frac{3}{27} \text{ ft} = \frac{648}{27} \text{ ft} =$ 24 ft |

5. a) Continued.

$$\text{Total Perimeter}_{\text{Step}} = \begin{pmatrix}\text{Perimeter} \\ \text{in Step 1}\end{pmatrix}\begin{pmatrix}\text{Factor for} \\ \text{new triangles}\end{pmatrix}^{\text{Step - 1}}\begin{pmatrix}\text{Factor for} \\ \text{length of sides}\end{pmatrix}^{\text{Step - 1}}$$

$$P_n = P_1(6)^{n-1}\left(\frac{1}{3}\right)^{n-1} = P_1\left(\frac{6}{3}\right)^{n-1} = P_1(2)^{n-1}$$

b)    Find the total perimeter for step 100:

$$P_n = P_1(2)^{n-1}$$

$$P_{100} = (3\text{ ft})(2)^{100-1} = (3\text{ ft})(2)^{99} = (3\text{ ft})(6.338...\times10^{29}) \approx 1.9\times10^{30}\text{ ft}$$

Since this is a very large number and would continue to increase in subsequent steps, the total perimeter of a Mitsubishi gasket would be infinitely large.

c)

| Step | Number of triangles | Length of each side of new triangle | Height of new triangle | Area of one triangle | Total area of all triangles |
|------|------|------|------|------|------|
| 1 | 1 | 1 ft | $\frac{\sqrt{3}}{2}$ | $\frac{1}{2}(1\text{ ft})\left(\frac{\sqrt{3}}{2}\text{ ft}\right)$ $= \frac{\sqrt{3}}{4}\text{ ft}^2$ | $1\bullet\frac{\sqrt{3}}{4}\text{ ft}^2 = \frac{\sqrt{3}}{4}\text{ ft}^2$ |
| 2 | 6 | $\frac{1}{3}$ ft | $\frac{1}{3}\bullet\frac{\sqrt{3}}{2}$ $= \frac{\sqrt{3}}{6}$ | $\frac{1}{2}(\frac{1}{3}\text{ ft})\left(\frac{\sqrt{3}}{6}\text{ ft}\right)$ $= \frac{\sqrt{3}}{36}\text{ ft}^2$ | $6\bullet\frac{\sqrt{3}}{36}\text{ ft}^2 = \frac{\sqrt{3}}{6}\text{ ft}^2$ |
| 3 | 36 | $\frac{1}{9}$ ft | $\frac{1}{3}\bullet\frac{\sqrt{3}}{6}$ $= \frac{\sqrt{3}}{18}$ | $\frac{1}{2}(\frac{1}{9}\text{ ft})\left(\frac{\sqrt{3}}{18}\text{ ft}\right)$ $= \frac{\sqrt{3}}{324}\text{ ft}^2$ | $36\bullet\frac{\sqrt{3}}{324}\text{ ft}^2$ $= \frac{\sqrt{3}}{9}\text{ ft}^2$ |
| 4 | 216 | $\frac{1}{27}$ ft | $\frac{1}{3}\bullet\frac{\sqrt{3}}{18}$ $= \frac{\sqrt{3}}{54}$ | $\frac{1}{2}(\frac{1}{27}\text{ ft})\left(\frac{\sqrt{3}}{54}\text{ ft}\right)$ $= \frac{\sqrt{3}}{2916}\text{ ft}^2$ | $216\bullet\frac{\sqrt{3}}{2916}\text{ ft}^2$ $= \frac{2\sqrt{3}}{27}\text{ ft}^2$ |

$$\text{Total Area}_{\text{Step}} = \begin{pmatrix}\text{Area in} \\ \text{Step 1}\end{pmatrix}\begin{pmatrix}\text{Factor for} \\ \text{triangles}\end{pmatrix}^{\text{Step - 1}}\begin{pmatrix}\text{Factor for} \\ \text{sides}\end{pmatrix}^{\text{Step - 1}}\begin{pmatrix}\text{Factor for} \\ \text{height}\end{pmatrix}^{\text{Step - 1}}$$

$$A_n = A_1(6)^{n-1}\left(\frac{1}{3}\right)^{n-1}\left(\frac{1}{3}\right)^{n-1} = A_1\left(\frac{6}{9}\right)^{n-1} = A_1\left(\frac{2}{3}\right)^{n-1}$$

5. Continued.

  d)     Find the total area for step 100:

$$A_n = A_1\left(\frac{2}{3}\right)^{n-1}$$

$$A_{100} = \left(\frac{\sqrt{3}}{4}\ \text{ft}^2\right)\left(\frac{2}{3}\right)^{100-1} = \left(\frac{\sqrt{3}}{4}\ \text{ft}^2\right)\left(\frac{2}{3}\right)^{99} = \left(\frac{\sqrt{3}}{4}\ \text{ft}^2\right)(3.689...\times10^{-18}) \approx 1.6\times10^{-18}\ \text{ft}^2$$

  Since this is a very small number and would continue to decrease in subsequent steps, the total area of a Mitsubishi gasket would be zero.

---

| **Chapter 6** | **Review** |
|---|---|

1.    Rectangle: $L = x$ ft, $w = (x - 2)$ ft

  $A = L \bullet w = x(x - 2) = (x^2 - 2x)\ \text{ft}^2$

  $P = x + (x - 2) + x + (x - 2) = (4x - 4)$ ft

5.    a)    Extending the line that is 6 yards long to the base creates a
         small square ($s = 2$ yd) and a trapezoid ($b_1 = 6 + 2 = 8$ yd,
         $b_2 = 5$ yd and $h = 12 - 2 = 10$yd).

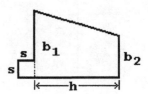

  $A = A_{\text{square}} + A_{\text{trapezoid}}$

$$= s^2 + \frac{b_1 + b_2}{2}h = (2\ \text{yd})^2 + \frac{8\ \text{yd} + 5\ \text{yd}}{2}(10\ \text{yd})$$

$$= 4\ \text{yd}^2 + (13\ \text{yd})(5\ \text{yd}) = (4 + 65)\ \text{yd}^2 = 69\ \text{yd}^2$$

  b)    Draw a line across from the top of the 5 yard side to the 6
         yard side creating a triangle so that the length of the
         hypotenuse (the unmarked side) can be found. The base
         of the triangle $= b = 10$ yd, the height $= h = (6 + 2 - 5)$ yd
         $= 3$ yd.

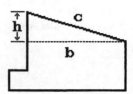

  $c^2 = b^2 + h^2 = 10^2 + 3^2 = 100 + 9 = 109,$

  $c = \sqrt{109}$

5. b)  Continued.

To find the perimeter, start at one corner and add the lengths.

$$P = (5 + 12 + 2 + 2 + 6 + \sqrt{109}) \text{ yd}$$

$$= (27 + \sqrt{109}) \text{ yd} = (27 + 10.4403) \text{ yd} = 37.4403 \text{ yd} \approx 37.4 \text{ yd}$$

9.  Rectangular box:  $L = 14$ in., $w = 8$ in., $h = 10$ in.
Small cube:  $s = 4$ in.

$$V = V_{\text{rectangular box}} - V_{\text{small cube}}$$
$$= L \bullet w \bullet h - s^3$$
$$= (14 \text{ in.})(8 \text{ in.})(10 \text{ in.}) - (4 \text{ in.})^3 = (1120 - 64) \text{ in.}^3 = 1056 \text{ in.}^3$$

To find the surface area, note that if the missing cube's sides are "popped out" it completes the

rectangular box.

$$A = 2(A_{\text{bottom}}) + 2(A_{\text{side}}) + 2(A_{\text{front}})$$
$$= 2(L \bullet w) + 2(w \bullet h) + 2(L \bullet h)$$
$$= 2(14 \text{ in.})(8 \text{ in.}) + 2(8 \text{ in.})(10 \text{ in.}) + 2(14 \text{ in.})(10 \text{ in.}) = 224 \text{ in.}^2 + 160 \text{ in.}^2 + 280 \text{ in.}^2 = 664 \text{ in.}^2$$

13.  $L = (18 - 2(2))$ in. $= (18 - 4)$ in. $= 14$ in.
$w = (12 - 2(2))$ in. $= (12 - 4)$ in. $= 8$ in.
$h = 2$ in.

a)  $V = L \bullet w \bullet h$
$$= (14 \text{ in.})(8 \text{ in.})(2 \text{ in.}) = 224 \text{ in.}^3$$

b)  $A = A_{\text{bottom}} + 2(A_{\text{front}}) + 2(A_{\text{side}})$
$$= L \bullet w + 2(L \bullet h) + 2(w \bullet h)$$
$$= (14 \text{ in.})(8 \text{ in.}) + 2(14 \text{ in.})(2 \text{ in.}) + 2(8 \text{ in.})(2 \text{ in.}) = 112 \text{ in.}^2 + 56 \text{ in.}^2 + 32 \text{ in.}^2 = 200 \text{ in.}^2$$

17.  Sphere:  $r = 1$ cubit

a)  Egyptian approximation of pi  $\left( \pi = \dfrac{256}{81} \right)$

$$V = \frac{4}{3} \pi r^3 = \frac{4}{3} \bullet \frac{256}{81} (1 \text{ cubit})^3 = \frac{1024}{243} (1 \text{ cubit})^3 = 4.21399177 \text{ cubits}^3 \approx 4.2140 \text{ cubits}^3$$

b)  Calculator value of pi.

$$V = \frac{4}{3} \pi r^3 = \frac{4}{3} \pi (1 \text{ cubit})^3 = \frac{4}{3} \pi \text{ cubits}^3 = 4.188790205 \text{ cubits}^3 \approx 4.1888 \text{ cubits}^3$$

17. Continued.

   c)   Error of Egyptian calculation relative to the calculator value.

   $$V_{\text{Egyptian}} - V_{\text{Calculator}} = 4.21399177 - 4.188790205 = 0.025201564 \text{ cubits}^3$$

   $$\text{relative error} = \frac{0.025201564 \text{ cubits}^3}{4.188790205 \text{ cubits}^3} = 0.006016 \approx 0.6\%$$

21.   Rectangular Box:
   $L = 5$ ft,
   $w = 3.5$ ft,
   $h = 2.5$ ft

   $c$ = diagonal on bottom of box

   $d$ = diagonal of the box

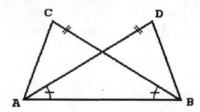

   Diagonal of box

   Diagonal of bottom

   $$c^2 = L^2 + w^2$$

   $$c^2 = (5 \text{ ft})^2 + (3.5 \text{ ft})^2 = 25 \text{ ft}^2 + 12.25 \text{ ft}^2 = 37.25 \text{ ft}^2$$

   $$d^2 = c^2 + h^2 = 37.25 \text{ ft}^2 + (2.5 \text{ ft})^2 = 37.25 \text{ ft}^2 + 6.25 \text{ ft}^2 = 43.5 \text{ ft}^2$$

   $$d = \sqrt{43.5 \text{ ft}^2} = 6.59545 \text{ ft} \approx 6.5 \text{ ft}$$

   The length of the longest object that will fit inside the box is 6.5 feet.  Note:  It is appropriate in this situation to not round up so that the object will fit inside the box.

25.   Given:  $\angle CBA = \angle DAB$
   $BC = AD$

   Prove:  $\angle CAD = \angle DBC$

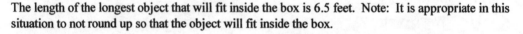

| Statements | Reasons |
|---|---|
| 1. $\angle CBA = \angle DAB$ | 1. Given |
| 2. $BC = AD$ | 2. Given |
| 3. $AB = AB$ | 3. Anything equals itself |
| 4. $\triangle ACB \cong \triangle BDA$ | 4. SAS |
| 5. $\angle CAB = \angle DBA$ | 5. Corresponding parts of congruent triangles are equal |
| 6. $\angle CAD = \angle DBC$ | 6. Equals subtracted from equals are equal |

29.    a)    $9x^2 - 4y^2 = 36$

$$\frac{9x^2}{36} - \frac{4y^2}{36} = \frac{36}{36}$$

$$\frac{x^2}{4} - \frac{y^2}{9} = 1, \quad a^2 = 4, b^2 = 9$$

**x-intercept (y = 0)**

$$\frac{x^2}{4} - \frac{(0)^2}{9} = 1$$

$$\frac{x^2}{4} = 1$$

$$x^2 = 4$$

$$x = \pm 2$$

**y-intercept (x = 0)**

$$\frac{(0)^2}{4} - \frac{y^2}{9} = 1$$

$$-\frac{y^2}{9} = 1$$

$$-y^2 = 9$$

no solution; no y-intercepts

Locate $a = \pm 2$ on the x-axis and $b = \pm 3$ on the y-axis. Draw a rectangle with sides parallel to the axes going through these points. Now draw the diagonals of the rectangle. Then draw the branches of the hyperbola opening left and right.

Finding the foci:
$$a^2 = 4, b^2 = 9$$

$$c^2 = a^2 + b^2 = 4 + 9 = 13$$

$$c = \pm\sqrt{13} \approx \pm 3.606$$

Foci are located on the x-axis at $\left(-\sqrt{13}, 0\right)$ and $\left(\sqrt{13}, 0\right)$.

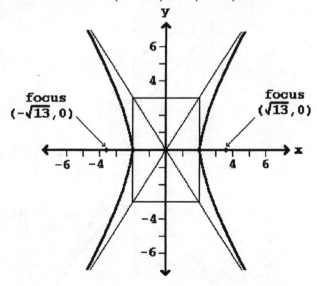

focus $(-\sqrt{13}, 0)$

focus $(\sqrt{13}, 0)$

29. Continued.

b)    $4y^2 - 9x^2 = 36$

$$\frac{4y^2}{36} - \frac{9x^2}{36} = \frac{36}{36}$$

$$\frac{y^2}{9} - \frac{x^2}{4} = 1, \quad a^2 = 4, b^2 = 9$$

| **x-intercept ($y = 0$)** | **y-intercept ($x = 0$)** |
|---|---|
| $\dfrac{(0)^2}{9} - \dfrac{x^2}{4} = 1$ | $\dfrac{y^2}{9} - \dfrac{(0)^2}{4} = 1$ |
| $-\dfrac{x^2}{4} = 1$ | $\dfrac{y^2}{9} = 1$ |
| $-x^2 = 4$ | $y^2 = 9$ |
| no solution; no $x$-intercepts | $y = \pm 3$ |

Locate $a = \pm 2$ on the $x$-axis and $b = \pm 3$ on the $y$-axis.  Draw a rectangle with sides parallel to the axes going through these points.  Now draw the diagonals of the rectangle.  Then draw the branches of the hyperbola opening up and down.

Finding the foci:

$$a^2 = 4, b^2 = 9$$

$$c^2 = a^2 + b^2 = 4 + 9 = 13$$

$$c = \pm\sqrt{13} \approx \pm 3.606$$

Foci are located on the $y$-axis at $\left(0, -\sqrt{13}\right)$ and $\left(0, \sqrt{13}\right)$.

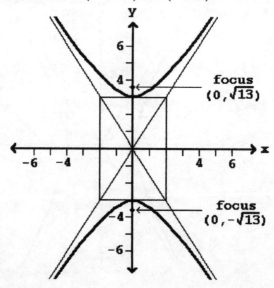

33.     $B = 180° - 90° - 25.4° = 64.6°$

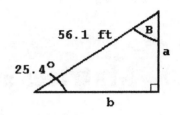

$$\sin 25.4° = \frac{\text{opp}}{\text{hyp}} = \frac{a}{56.1}$$

$a = 56.1(\sin 25.4°) = 56.1(0.42894) = 24.063 \approx 24.1 \text{ ft}$

$$\cos 25.4° = \frac{\text{adj}}{\text{hyp}} = \frac{b}{56.1}$$

$b = 56.1(\cos 25.4°) = 56.1(0.90334) = 50.677 \approx 50.7 \text{ ft}$

37.     a)     The Parallel Postulate is:  Given a line and a point not on that line, there is one and only one line through the point parallel to the original line.

        b)     Alternative 1:  Given a line and a point not on the line, there are no lines through the point parallel to the original line (parallels do not exist).

               Alternative 2:  Given a line and a point not on the line, there are at least two lines through the point parallel to the original line.

        c)     Alternative 1:  Riemannian geometry uses the sphere as a model where lines are defined as great circles that encompass the sphere.

               Alternative 2:  Lobachevskian geometry uses the interior of a circle as a model and defines lines as either the diameter of the disk or a circular arc connecting two points on the boundary of the disk.

        d)     **Alternative 1**

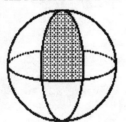

**Riemannian Geometry**

       **Alternative 2**

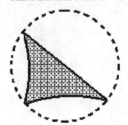

**Lobachevskian Geometry**

# 7  Matrices and Markov Chains

| Section 7.0 | Review of Matrices |
|---|---|

1.  a)  3 rows and 2 columns, so the dimensions of matrix $A$ are $3 \times 2$.
    b)  Matrix $A$ is neither a row matrix, a column matrix, nor a square matrix.

5.  a)  1 row and 2 columns, so the dimensions of matrix $E$ are $1 \times 2$.
    b)  Matrix $E$ is a row matrix.

9.  a)  3 rows and 1 column, so the dimensions of matrix $J$ are $3 \times 1$.
    b)  Matrix $J$ is a column matrix.

13.  $c_{21} = 41$   row 2, column 1 of matrix $C$ (Exercise 3)

17.  $g_{12} = -11$   row 1, column 2 of matrix $G$ (Exercise 7)

21.  a)  $AC = \begin{bmatrix} 5 & 0 \\ 22 & -3 \\ 18 & 9 \end{bmatrix} \begin{bmatrix} 23 \\ 41 \end{bmatrix} = \begin{bmatrix} 5 \cdot 23 + 0 \cdot 41 \\ 22 \cdot 23 + -3 \cdot 41 \\ 18 \cdot 23 + 9 \cdot 41 \end{bmatrix} = \begin{bmatrix} 115 \\ 383 \\ 783 \end{bmatrix}$

   b)  $CA = \begin{bmatrix} 23 \\ 41 \end{bmatrix} \begin{bmatrix} 5 & 0 \\ 22 & -3 \\ 18 & 9 \end{bmatrix} =$ does not exist

   $\quad\quad 2 \times 1 \quad 3 \times 2$

   *different, so we can't multiply*

25.  a)  $CG = \begin{bmatrix} 23 \\ 41 \end{bmatrix} \begin{bmatrix} 12 & -11 & 5 \\ -9 & 4 & 0 \\ 1 & 9 & 5 \end{bmatrix} =$ does not exist

   $\quad\quad 2 \times 1 \quad\quad 3 \times 3$

   *different, so we can't multiply*

25.  Continued.

b)    $GC = \begin{bmatrix} 12 & -11 & 5 \\ -9 & 4 & 0 \\ 1 & 9 & 5 \end{bmatrix}\begin{bmatrix} 23 \\ 41 \end{bmatrix}$ = does not exist

$$3 \times 3 \qquad 2 \times 1$$
$$\uparrow \qquad \uparrow$$

*different, so we can't multiply*

29.  a)    $AF = \begin{bmatrix} 5 & 0 \\ 22 & -3 \\ 18 & 9 \end{bmatrix}\begin{bmatrix} -2 & 10 \\ 4 & -3 \end{bmatrix} = \begin{bmatrix} 5\cdot-2+0\cdot4 & 5\cdot10+0\cdot-3 \\ 22\cdot-2+-3\cdot4 & 22\cdot10+-3\cdot-3 \\ 18\cdot-2+9\cdot4 & 18\cdot10+9\cdot-3 \end{bmatrix} = \begin{bmatrix} -10 & 50 \\ -56 & 229 \\ 0 & 153 \end{bmatrix}$

b)    $FA = \begin{bmatrix} -2 & 10 \\ 4 & -3 \end{bmatrix}\begin{bmatrix} 5 & 0 \\ 22 & -3 \\ 18 & 9 \end{bmatrix}$ = does not exist

$$2 \times 2 \qquad 3 \times 2$$
$$\uparrow \qquad \uparrow$$

*different, so we can't multiply*

33.    Table could be rewritten as the matrix $T$ and the purchase as matrix $P$.  Find $PT$, not $TP$, so that the dimensions match.

$$T = \begin{bmatrix} 1.25 & 1.30 \\ 0.95 & 1.10 \end{bmatrix}, \ P = \begin{bmatrix} 2 & 1 \end{bmatrix}$$

$$PT = \begin{bmatrix} 2 & 1 \end{bmatrix}\begin{bmatrix} 1.25 & 1.30 \\ 0.95 & 1.10 \end{bmatrix} = \begin{bmatrix} 2\cdot1.25+1\cdot0.95 & 2\cdot1.30+1\cdot1.10 \end{bmatrix} = \begin{bmatrix} 3.45 & 3.70 \end{bmatrix}$$

The price at Blondie's would be \$3.45 and the price at SliceMan's would be \$3.70.

37.    $BC = \begin{bmatrix} -3 & 0 & -1 \\ 1 & 4 & -2 \end{bmatrix}\begin{bmatrix} 0 \\ 5 \\ -1 \end{bmatrix} = \begin{bmatrix} -3\cdot0+0\cdot5+-1\cdot-1 \\ 1\cdot0+4\cdot5+-2\cdot-1 \end{bmatrix} = \begin{bmatrix} 1 \\ 22 \end{bmatrix}$

$A(BC) = \begin{bmatrix} -4 & 5 \\ 2 & 3 \end{bmatrix}\begin{bmatrix} 1 \\ 22 \end{bmatrix} = \begin{bmatrix} -4\cdot1+5\cdot22 \\ 2\cdot1+3\cdot22 \end{bmatrix} = \begin{bmatrix} 106 \\ 68 \end{bmatrix}$

37. Continued.

$$AB = \begin{bmatrix} -4 & 5 \\ 2 & 3 \end{bmatrix}\begin{bmatrix} -3 & 0 & -1 \\ 1 & 4 & -2 \end{bmatrix} = \begin{bmatrix} -4\cdot-3+5\cdot1 & -4\cdot0+5\cdot4 & -4\cdot-1+5\cdot-2 \\ 2\cdot-3+3\cdot1 & 2\cdot0+3\cdot4 & 2\cdot-1+3\cdot-2 \end{bmatrix} = \begin{bmatrix} 17 & 20 & -6 \\ -3 & 12 & -8 \end{bmatrix}$$

$$(AB)C = \begin{bmatrix} 17 & 20 & -6 \\ -3 & 12 & -8 \end{bmatrix}\begin{bmatrix} 0 \\ 5 \\ -1 \end{bmatrix} = \begin{bmatrix} 17\cdot0+20\cdot5+-6\cdot-1 \\ -3\cdot0+12\cdot5+-8\cdot-1 \end{bmatrix} = \begin{bmatrix} 106 \\ 68 \end{bmatrix}$$

41. $\begin{bmatrix} 1 & 0 & 0 \\ 0 & 1 & 0 \\ 0 & 0 & 1 \end{bmatrix}\begin{bmatrix} 4 & 1 & -1 \\ 5 & 12 & 3 \end{bmatrix}$ = does not exist

$\quad\quad 3 \times 3 \quad\quad 2 \times 3$
$\quad\quad\quad \uparrow \quad\quad\quad \uparrow$

*different, so we can't multiply*

45. Omitted.

49. Omitted.

53. Omitted.

57. a) $\quad EF = \begin{bmatrix} 5 & -31 \\ 83 & -33 \\ 60 & 0 \end{bmatrix}\begin{bmatrix} 93 & -11 & 39 \\ 53 & 66 & 83 \end{bmatrix} = \begin{bmatrix} -1178 & -2101 & -2378 \\ 5970 & -3091 & 498 \\ 5580 & -660 & 2340 \end{bmatrix}$

b) $\quad FE = \begin{bmatrix} 93 & -11 & 39 \\ 53 & 66 & 83 \end{bmatrix}\begin{bmatrix} 5 & -31 \\ 83 & -33 \\ 60 & 0 \end{bmatrix} = \begin{bmatrix} 1,892 & -2,520 \\ 10,723 & -3,821 \end{bmatrix}$

61. a) $\quad C^2 = \begin{bmatrix} 8 & -12 & 13 \\ 52 & 17 & -31 \\ 72 & 28 & -15 \end{bmatrix}^2 = \begin{bmatrix} 376 & 64 & 281 \\ -932 & -1203 & 614 \\ 952 & -808 & 293 \end{bmatrix}$

b) $\quad C^5 = \begin{bmatrix} 8 & -12 & 13 \\ 52 & 17 & -31 \\ 72 & 28 & -15 \end{bmatrix}^5 = \begin{bmatrix} 4,599,688 & -2,586,492 & 9,810,101 \\ -1,887,820 & -24,086,567 & 2,293,357 \\ 42,244,296 & -7,140,820 & -4,471,911 \end{bmatrix}$

## Section 7.1                                          **Markov Chains**

1.  a)    Given:  $p$(current purchase is KickKola) = 0.14
          Complement:  $p$(current purchase is not KickKola) = 1 − 0.14 = 0.86

    b)              K      K'
          $P = \begin{bmatrix} 0.14 & 0.86 \end{bmatrix}$

5.  a)    Given:  $p$(Silver's Gym) = 0.48
                  $p$(Fitness Lab) = 0.37

          Complement:  $p$(ThinNFit) = 1 − (0.48 + 0.37) = 0.15

    b)              S       F       T
          $P = \begin{bmatrix} 0.48 & 0.37 & 0.15 \end{bmatrix}$

9.  a)    Given:  $p$(will buy home | currently rents) = 0.12
                  $p$(will sell and then rent | currently owns) = 0.03

          Complements:  $p$(will not buy home | currently rents) = 1 − 0.12 = 0.88
                        $p$(will not sell home | currently owns) = 1 − 0.03 = 0.97

    b)              H       H'
          $T = \begin{bmatrix} 0.97 & 0.03 \\ 0.12 & 0.88 \end{bmatrix} \begin{matrix} H \\ H' \end{matrix}$

13.  Refer to Exercises 1 and 7.

    a)
    | current purchase | 1st following purchase | probabilities |
    |---|---|---|
    | 0.14 **KickKola** | 0.63 **KickKola** | (0.14)(0.63) = 0.0882 |
    |  | 0.37 **not KickKola** | (0.14)(0.37) = 0.0518 |
    | 0.86 **not KickKola** | 0.12 **KickKola** | (0.86)(0.12) = 0.1032 |
    |  | 0.88 **not KickKola** | (0.86)(0.88) = 0.7568 |

    $p$(1st following purchase is KickKola) = (0.14)(0.63) + (0.86)(0.12)
                                     = 0.0882 + 0.1032 = 0.1914

    KickKola's market share at the next following purchase will be 19% if current trends continue.

13. Continued.

b)

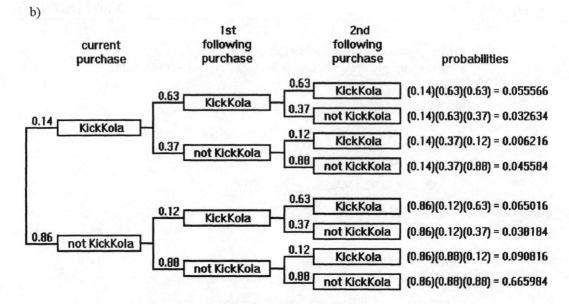

$p$(2nd following purchase is KickKola) = 0.055566 + 0.006216 + 0.065016 + 0.090816
$$= 0.217614$$

KickKola's market share at the second following purchase will be 22% if current trends continue.

c)
$$\begin{array}{cc} \text{K} & \text{K'} \end{array}$$
$$P = \begin{bmatrix} 0.14 & 0.86 \end{bmatrix}$$

$$\begin{array}{cc} \text{K} & \text{K'} \end{array}$$
$$T = \begin{bmatrix} 0.63 & 0.37 \\ 0.12 & 0.88 \end{bmatrix} \begin{array}{c} \text{K} \\ \text{K'} \end{array}$$

$$PT^1 = \begin{bmatrix} 0.14 & 0.86 \end{bmatrix}\begin{bmatrix} 0.63 & 0.37 \\ 0.12 & 0.88 \end{bmatrix}$$

$$= \begin{bmatrix} 0.14 \cdot 0.63 + 0.86 \cdot 0.12 & 0.14 \cdot 0.37 + 0.86 \cdot 0.88 \end{bmatrix} = \begin{bmatrix} 0.1914 & 0.8086 \end{bmatrix}$$

KickKola's market share at the next following purchase will be 19% if current trends continue.

13. Continued.

d) $\quad PT^2 = (PT)T = [0.1914 \quad 0.8086]\begin{bmatrix} 0.63 & 0.37 \\ 0.12 & 0.88 \end{bmatrix}$

$\quad = [0.1914 \cdot 0.63 + 0.8086 \cdot 0.12 \quad 0.1914 \cdot 0.37 + 0.8086 \cdot 0.88]$

$\quad = [0.217614 \quad 0.782386]$

KickKola's market share at the second following purchase will be 22% if current trends continue.

e)   Omitted.

17.       S     S'

$P = [0.41 \quad 0.59]$

$\qquad$ S $\qquad$ S'

$T = \begin{bmatrix} 0.12 & 0.88 \\ 0.31 & 0.69 \end{bmatrix} \begin{matrix} \text{S} \\ \text{S'} \end{matrix}$

Find the second following purchase. $\left( \dfrac{4 \text{ years}}{2 \text{ years}} = 2 \right)$

$PT^1 = [0.41 \quad 0.59]\begin{bmatrix} 0.12 & 0.88 \\ 0.31 & 0.69 \end{bmatrix} = [0.2321 \quad 0.7679]$

$PT^2 = (PT)T = [0.2321 \quad 0.7679]\begin{bmatrix} 0.12 & 0.88 \\ 0.31 & 0.69 \end{bmatrix} = [0.265901 \quad 0.734099]$

Sierra Cruiser's market share in four years, at the second following purchase, will be 27%.

---

| **Section 7.2** | **Systems of Linear Equations** |
| --- | --- |

1.   Substitute (4, 1) into the system of equations:

$\qquad 3x - 5y = 3(4) - 5(1) = 12 - 5 = 7$

$\qquad 2x + 2y = 2(4) + 2(1) = \;\; 8 + 2 = 10$

Both equations are satisfied, so (4, 1) is a solution.

5.   Substitute (4, −1, 2) into the system of equations:

$\qquad 2x + 3y - z = 2(4) + 3(-1) - (2) = 8 - 3 - 2 = 3$

$\qquad x + y + z = (4) + (-1) + (2) = 4 - 1 + 2 = 5$

$\qquad 10x - 2y = 10(4) - 2(-1) = 40 + 2 = 42 \qquad \text{not } 3$

The third equation is not satisfied, so (4, −1, 2) is not a solution for the system.

9.   a)    $4x + 3y = 12$                    $8x + 6y = 24$

$3y = -4x + 12$                  $6y = -8x + 24$

$y = -\dfrac{4}{3}x + 4$              $y = -\dfrac{8}{6}x + \dfrac{24}{6} = -\dfrac{4}{3}x + 4$

The slope of each equation is $-\dfrac{4}{3}$ and the $y$-intercept is 4.

b)    Therefore, the equations are the same and there are an infinite number of solutions.

13.   This system could have a unique solution since it has as many unique equations as it has unknowns.

17.   This system will not have a single solution, it will have either no solution or an infinite number of solutions since it has fewer equations (2) than unknowns (3).

21.   $3x - 7y = 27$          multiply by 4          $12x - 28y = 108$
$4x - 5y = 23$          multiply by –3          $\underline{-12x + 15y = -69}$
                                    add together          $0x - 13y = 39$

solve for $y$          $y = \dfrac{39}{-13} = -3$

Find $x$ by substitution:     $3x - 7(-3) = 27$
$3x + 21 = 27$
$3x = 6$
$x = 2$

Solution is (2, –3)

*Check solution* (2, –3):     $3x - 7y = 3(2) - 7(-3) = 27$
$4x - 5y = 4(2) - 5(-3) = 23$

25.   Omitted.

29.   Omitted.

33.   Solve each equation for $y$:

$5x - 9y = -12$                    $3x + 7y = -1$

$-9y = -5x - 12$                  $7y = -3x - 1$

$\dfrac{-9y}{-9} = \dfrac{-5x}{-9} - \dfrac{12}{-9}$          $\dfrac{7y}{7} = \dfrac{-3x}{7} - \dfrac{1}{7}$

$y = \dfrac{5}{9}x + \dfrac{4}{3}$              $y = -\dfrac{3}{7}x - \dfrac{1}{7}$

**33.** Continued.

Enter results on graphing calculator as $Y_1$ and $Y_2$ and then graph and find intersection point.

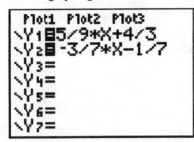 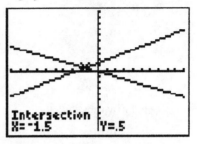

*Check* by substituting the result $(-1.5, 0.5)$ into each of the original equations.

$$5x - 9y = 5(-1.5) - 9(0.5) = -7.5 - 4.5 = -12$$
$$3x + 7y = 3(-1.5) + 7(0.5) = -4.5 + 3.5 = -1$$

| **Section 7.3** | **Long-Range Predictions with Markov Chains** |
|---|---|

**1.**    **Step 1:** Create the transition matrix $T$.

$$T = \begin{bmatrix} 0.1 & 0.9 \\ 0.2 & 0.8 \end{bmatrix}$$

**Step 2:** Create the equilibrium matrix $L$.

$$L = \begin{bmatrix} x & y \end{bmatrix}$$

**Step 3:** Find and simplify the system of equations described by $LT = L$.

$$\begin{bmatrix} x & y \end{bmatrix}\begin{bmatrix} 0.1 & 0.9 \\ 0.2 & 0.8 \end{bmatrix} = \begin{bmatrix} x & y \end{bmatrix}$$

$$\begin{bmatrix} 0.1x + 0.2y & 0.9x + 0.8y \end{bmatrix} = \begin{bmatrix} x & y \end{bmatrix}$$

Yields:    $0.1x + 0.2y = x$
$0.9x + 0.8y = y$

Combining like terms gives:    $-0.9x + 0.2y = 0$
$0.9x - 0.2y = 0$

**Step 4:** Discard any redundant equations and include $x + y = 1$.
The Step 3 equations are the same so use only once.
The system to be solved is:

$$x + y = 1$$
$$0.9x - 0.2y = 0$$

1. Continued.

Step 5: Solve the resulting system by elimination.

$$x + y = 1 \qquad \text{times 0.2} \qquad 0.2x + 0.2y = 0.2$$
$$0.9x - 0.2y = 0 \qquad\qquad\qquad \underline{0.9x - 0.2y = \ \ 0}$$
$$\text{add together} \qquad 1.1x \qquad\quad = 0.2$$

$$\text{solve for } x \qquad\qquad x = \frac{0.2}{1.1} = 0.1818$$

Find $y$ by substitution:

$$0.1818 + y = 1$$
$$y = 1 - 0.1818 = 0.8182$$

Resulting matrix is    $L = \begin{bmatrix} 0.1818 & 0.8182 \end{bmatrix} \approx \begin{bmatrix} 0.18 & 0.82 \end{bmatrix}$

Step 6: *Check the work by verifying that LT = L.*

$$LT = \begin{bmatrix} 0.1818 & 0.8182 \end{bmatrix} \begin{bmatrix} 0.1 & 0.9 \\ 0.2 & 0.8 \end{bmatrix} = \begin{bmatrix} 0.18182 & 0.81818 \end{bmatrix} \approx \begin{bmatrix} 0.1818 & 0.8182 \end{bmatrix}$$

5.    Refer to Exercise 13 in Section 7.1.

Step 1: Create the transition matrix $T$.

$$T = \begin{bmatrix} 0.63 & 0.37 \\ 0.12 & 0.88 \end{bmatrix}$$

Step 2: Create the equilibrium matrix $L$.

$$L = \begin{bmatrix} x & y \end{bmatrix}$$

Step 3: Find and simplify the system of equations described by $LT = L$.

$$\begin{bmatrix} x & y \end{bmatrix} \begin{bmatrix} 0.63 & 0.37 \\ 0.12 & 0.88 \end{bmatrix} = \begin{bmatrix} x & y \end{bmatrix}$$
$$\begin{bmatrix} 0.63x + 0.12y & 0.37x + 0.88y \end{bmatrix} = \begin{bmatrix} x & y \end{bmatrix}$$

Yields:    $0.63x + 0.12y = x$
$\qquad\qquad 0.37x + 0.88y = y$

Combining like terms:    $-0.37x + 0.12y = 0$
$\qquad\qquad\qquad\qquad\qquad 0.37x - 0.12y = 0$

Step 4: Discard any redundant equations and include $x + y = 1$.
The Step 3 equations are the same so use only once.
The system to be solved is:

$$x + y = 1$$
$$0.37x - 0.12y = 0$$

5. Continued.

**Step 5:** Solve the resulting system by elimination.

$$x + y = 1 \qquad \text{times } 0.12 \qquad 0.12x + 0.12y = 0.12$$
$$0.37x - 0.12y = 0 \qquad\qquad\qquad \underline{0.37x - 0.12y = \quad 0}$$
$$\text{add together} \qquad 0.49x \qquad = 0.12$$

$$\text{solve for } x \qquad x = \frac{0.12}{0.49} = 0.2449$$

Find $y$ by substitution:

$$0.2449 + y = 1$$
$$y = 1 - 0.2449 = 0.7551$$

Resulting matrix is $L = \begin{bmatrix} 0.2449 & 0.7551 \end{bmatrix}$

**Step 6:** *Check the work by verifying that LT = L.*

$$LT = \begin{bmatrix} 0.2449 & 0.7551 \end{bmatrix} \begin{bmatrix} 0.63 & 0.37 \\ 0.12 & 0.88 \end{bmatrix} = \begin{bmatrix} 0.244899 & 0.755101 \end{bmatrix} \approx \begin{bmatrix} 0.2449 & 0.7551 \end{bmatrix} = L$$

KickKola's long-range market share should stabilize at 24%.

9.    Refer to Exercise 17 in Section 7.1.

**Step 1:** Create the transition matrix $T$.

$$T = \begin{bmatrix} 0.12 & 0.88 \\ 0.31 & 0.69 \end{bmatrix}$$

**Step 2:** Create the equilibrium matrix $L$.

$$L = \begin{bmatrix} x & y \end{bmatrix}$$

**Step 3:** Find and simplify the system of equations described by $LT = L$.

$$\begin{bmatrix} x & y \end{bmatrix} \begin{bmatrix} 0.12 & 0.88 \\ 0.31 & 0.69 \end{bmatrix} = \begin{bmatrix} x & y \end{bmatrix}$$

$$\begin{bmatrix} 0.12x + 0.31y & 0.88x + 0.69y \end{bmatrix} = \begin{bmatrix} x & y \end{bmatrix}$$

Yields:    $0.12x + 0.31y = x$
$$0.88x + 0.69y = y$$

Combining like terms:    $-0.88x + 0.31y = 0$
$$0.88x - 0.31y = 0$$

**9. Continued.**

**Step 4:** Discard any redundant equations and include $x + y = 1$.
The Step 3 equations are the same so use only once.
The system to be solved is:

$$x + \quad y = 1$$
$$0.88x - 0.31y = 0$$

**Step 5:** Solve the resulting system by elimination.

| | | |
|---|---|---|
| $x + \quad y = 1$ | times 0.31 | $0.31x + 0.31y = 0.31$ |
| $0.88x - 0.31y = 0$ | | $\underline{0.88x - 0.31y = \quad 0}$ |
| | add together | $1.19x \qquad = 0.31$ |
| | solve for $x$ | $x = \dfrac{0.31}{1.19} = 0.2605$ |

Find $y$ by substitution:

$$0.2605 + y = 1$$
$$y = 1 - 0.2605 = 0.7395$$

Resulting matrix is $L = \begin{bmatrix} 0.2605 & 0.7395 \end{bmatrix}$

**Step 6:** *Check the work by verifying that LT = L.*

$$LT = \begin{bmatrix} 0.2605 & 0.7395 \end{bmatrix} \begin{bmatrix} 0.12 & 0.88 \\ 0.31 & 0.69 \end{bmatrix} = \begin{bmatrix} 0.260505 & 0.739495 \end{bmatrix} \approx \begin{bmatrix} 0.2605 & 0.7395 \end{bmatrix} = L$$

Sierra Cruiser's ultimate market share should stabilize at 26%.

---

| | |
|---|---|
| **Section 7.4** | **Solving Larger Systems of Equations** |

1.     (1)     $5x + 1y - 1z = 17$
        (2)     $2x + 5y + 2z = 0$
        (3)     $3x + 1y + 1z = 11$

Eliminate $z$ from equation (1) and equation (2):

      (1)     $10x + 2y - 2z = 34$      2 times equation (1)
      (2)     $\underline{2x + 5y + 2z = \quad 0}$      no change
      (4)     $12x + 7y + 0z = 34$      sum = equation (4)

Eliminate $z$ from equation (1) and equation (3):

      (1)     $5x + 1y - 1z = 17$      no change
      (3)     $\underline{3x + 1y + 1z = 11}$      no change
      (5)     $8x + 2y + 0z = 28$      sum = equation (5)

1. **Continued.**

Eliminate $y$ from equation (4) and equation (5):

| | | |
|---|---|---|
| (4) | $24x + 14y = \phantom{-}68$ | 2 times equation (4) |
| (5) | $\underline{-56x - 14y = -196}$ | $-7$ times equation (5) |
| | $-32x + \phantom{0}0y = -128$ | sum |

Solve for $x$    $x = \dfrac{-128}{-32} = 4$

Find $y$ by substituting $x = 4$ into equation (5):

$$8x + 2y = 28$$
$$8(4) + 2y = 28$$
$$32 + 2y = 28$$
$$2y = -4$$
$$y = -2$$

Substitute $x = 4$ and $y = -2$ into equation (3):

$$3x + y + z = 11$$
$$3(4) + (-2) + z = 11$$
$$12 - 2 + z = 11$$
$$10 + z = 11$$
$$z = 1$$

Solution is $(4, -2, 1)$

*Check solution* $(4, -2, 1)$ in equations (1) and (2) since (3) was used to solve for $x$ and $y$:

| | |
|---|---|
| (1) | $5x + y - z = 5(4) + (-2) - (1) = 20 - 2 - 1 = 17$ |
| (2) | $2x + 5y + 2z = 2(4) + 5(-2) + 2(1) = 8 - 10 + 2 = 0$ |

5.  
| | |
|---|---|
| (1) | $1x - 1y + 4z = -13$ |
| (2) | $2x + 0y - 1z = 12$ |
| (3) | $3x + 1y + 0z = 25$ |

Eliminate $z$ from equation (1) and equation (2):

| | | |
|---|---|---|
| (1) | $1x - 1y + 4z = -13$ | no change |
| (2) | $\underline{8x + 0y - 4z = \phantom{0}48}$ | 4 times equation (2) |
| (4) | $9x - 1y + 0z = \phantom{0}35$ | sum = equation (4) |

Eliminate $y$ from equation (3) and equation (4):

| | | |
|---|---|---|
| (3) | $3x + 1y = 25$ | |
| (4) | $\underline{9x - 1y = 35}$ | |
| | $12x + 0y = 60$ | sum |

5. Continued.

Solve for $x$     $x = \dfrac{60}{12} = 5$

Find $y$ by substituting $x = 5$ into equation (3):

$$3x + y = 25$$
$$3(5) + y = 25$$
$$15 + y = 25$$
$$y = 10$$

Substitute $x = 5$ into equation (2):
$$2x - z = 12$$
$$2(5) - z = 12$$
$$10 - z = 12$$
$$-z = 2$$
$$z = -2$$

Solution is $(5, 10, -2)$

*Check solution* $(5, 10, -2)$ in equation (1) only since (2) and (3) were used to solve for $x$ and $z$:

(1)     $x - y + 4z = (5) - (10) + 4(-2) = 5 - 10 - 8 = -13$

9.     Write with all variables to the left and all coefficients showing.

$$2x - 1y = 5$$
$$3x - 1y = 41$$

Rewrite in matrix form

$$\begin{bmatrix} 2 & -1 & 5 \\ 3 & -1 & 41 \end{bmatrix}$$

13.     Write with all variables to the left and all coefficients showing.

$$5x + 0y + 1z = 2$$
$$3x - 1y + 0z = 15$$
$$2x + 2y + 2z = 53$$

Rewrite in matrix form

$$\begin{bmatrix} 5 & 0 & 1 & 2 \\ 3 & -1 & 0 & 15 \\ 2 & 2 & 2 & 53 \end{bmatrix}$$

17. $\begin{bmatrix} 1 & 3 & 12 \\ 0 & 5 & 7 \end{bmatrix}$

$$R2 \div 5 \begin{bmatrix} 1 & 3 & 12 \\ 0 & 1 & 7\!/\!5 \end{bmatrix}$$

$$-3R2 + R1 : R1 \begin{bmatrix} 1 & 0 & 39\!/\!5 \\ 0 & 1 & 7\!/\!5 \end{bmatrix}$$

21. To determine the solution, read down each column, take a right turn at 1, and read the number at the end of the row.

$$\begin{bmatrix} 1 & 0 & 9 \\ 0 & 1 & 2 \end{bmatrix} \begin{matrix} x = 9 \\ y = 2 \end{matrix} \qquad \text{The solution is } (9, 2).$$

25. **Step 1:** Write with all variables to the left and all coefficients showing.

$$1x + 1y = 3$$
$$2x - 3y = -4$$

**Step 2:** Rewrite in matrix form.

$$\begin{bmatrix} 1 & 1 & 3 \\ 2 & -3 & -4 \end{bmatrix}$$

**Step 3:** Row operations.

First column:

$$-2R1 + R2 : R2 \begin{bmatrix} 1 & 1 & 3 \\ 0 & -5 & -10 \end{bmatrix}$$

Second column:

$$R2 \div -5 : R2 \begin{bmatrix} 1 & 1 & 3 \\ 0 & 1 & 2 \end{bmatrix}$$

$$-1R2 + R1 : R1 \begin{bmatrix} 1 & 0 & 1 \\ 0 & 1 & 2 \end{bmatrix}$$

**Step 4:** Read the solution. $x = 1, y = 2$ or $(1, 2)$

**Step 5:** *Check the solution* $(1, 2)$:

$$x + y = (1) + (2) = 3$$
$$2x - 3y = 2(1) - 3(2) = 2 - 6 = -4$$

29.    **Step 1:** Write with all variables to the left and all coefficients showing.

$$5x + 2y = 19$$
$$3x - 4y = -25$$

**Step 2:** Rewrite in matrix form.

$$\begin{bmatrix} 5 & 2 & 19 \\ 3 & -4 & -25 \end{bmatrix}$$

**Step 3:** Row operations.

First column:

$$R1 \div 5 : R1 \begin{bmatrix} 1 & 0.4 & 3.8 \\ 3 & -4 & -25 \end{bmatrix}$$

$$-3R1 + R2 : R2 \begin{bmatrix} 1 & 0.4 & 3.8 \\ 0 & -5.2 & -36.4 \end{bmatrix}$$

Second column:

$$R2 \div -5.2 : R2 \begin{bmatrix} 1 & 0.4 & 3.8 \\ 0 & 1 & 7 \end{bmatrix}$$

$$-0.4R2 + R1 : R1 \begin{bmatrix} 1 & 0 & 1 \\ 0 & 1 & 7 \end{bmatrix}$$

**Step 4:** Read the solution. $x = 1, y = 7$ or $(1, 7)$

**Step 5:** *Check the solution* $(1, 7)$:

$$5x + 2y = 5(1) + 2(7) = 5 + 14 = 19$$
$$3x - 4y = 3(1) - 4(7) = 3 - 28 = -25$$

33.    **Step 1:** Write with all variables to the left and all coefficients showing.

$$1x + 1y + 1z = 14$$
$$3x - 2y + 1z = 3$$
$$5x + 1y + 2z = 29$$

**Step 2:** Rewrite in matrix form.

$$\begin{bmatrix} 1 & 1 & 1 & 14 \\ 3 & -2 & 1 & 3 \\ 5 & 1 & 2 & 29 \end{bmatrix}$$

**33.** Continued.

Step 3: Row operations.

First column:

$$-3R1 + R2 : R2 \quad \begin{bmatrix} 1 & 1 & 1 & 14 \\ 0 & -5 & -2 & -39 \\ 0 & -4 & -3 & -41 \end{bmatrix}$$
$$-5R1 + R3 : R3$$

Second column:

$$R2 \div -5 : R2 \quad \begin{bmatrix} 1 & 1 & 1 & 14 \\ 0 & 1 & 0.4 & 7.8 \\ 0 & -4 & -3 & -41 \end{bmatrix}$$

$$-R2 + R1 : R1 \quad \begin{bmatrix} 1 & 0 & 0.6 & 6.2 \\ 0 & 1 & 0.4 & 7.8 \\ 0 & 0 & -1.4 & -9.8 \end{bmatrix}$$
$$4R2 + R3 : R3$$

Third column:

$$R3 \div -1.4 : R3 \quad \begin{bmatrix} 1 & 0 & 0.6 & 6.2 \\ 0 & 1 & 0.4 & 7.8 \\ 0 & 0 & 1 & 7 \end{bmatrix}$$

$$-0.6R3 + R1 : R1 \quad \begin{bmatrix} 1 & 0 & 0 & 2 \\ 0 & 1 & 0 & 5 \\ 0 & 0 & 1 & 7 \end{bmatrix}$$
$$-0.4R3 + R2 : R2$$

Step 4: Read the solution. $x = 2, y = 5, z = 7$ or $(2, 5, 7)$

Step 5: *Check the solution* $(2, 5, 7)$:

$$x + y + z = (2) + (5) + (7) = 14$$
$$3x - 2y + z = 3(2) - 2(5) + (7) = 6 - 10 + 7 = 3$$
$$5x + y + 2z = 5(2) + (5) + 2(7) = 10 + 5 + 14 = 29$$

**37.** The first equation was ignored: $3x - 5y + z = 12$

Step 1: Write with all variables to the left and all coefficients showing. (The first equation is ignored)

$$2x + 1y + 1z = 3$$
$$5x - 4y + 1z = 0$$
$$1x + 1y + 1z = 4$$

37. Continued.

**Step 2:** Rewrite in matrix form.

$$\begin{bmatrix} 2 & 1 & 1 & 3 \\ 5 & -4 & 1 & 0 \\ 1 & 1 & 1 & 4 \end{bmatrix}$$

**Step 3:** Row operations.

First column:

$$\begin{matrix} \text{Move R3 : R1} \\ \text{Move R1 : R2} \\ \text{Move R2 : R3} \end{matrix} \begin{bmatrix} 1 & 1 & 1 & 4 \\ 2 & 1 & 1 & 3 \\ 5 & -4 & 1 & 0 \end{bmatrix}$$

$$\begin{matrix} \\ -2R1 + R2 : R2 \\ -5R1 + R3 : R3 \end{matrix} \begin{bmatrix} 1 & 1 & 1 & 4 \\ 0 & -1 & -1 & -5 \\ 0 & -9 & -4 & -20 \end{bmatrix}$$

Second column:

$$\begin{matrix} \\ R2 \div -1 : R2 \\ \\ \end{matrix} \begin{bmatrix} 1 & 1 & 1 & 4 \\ 0 & 1 & 1 & 5 \\ 0 & -9 & -4 & -20 \end{bmatrix}$$

$$\begin{matrix} -R2 + R1 : R1 \\ \\ 9R2 + R3 : R3 \end{matrix} \begin{bmatrix} 1 & 0 & 0 & -1 \\ 0 & 1 & 1 & 5 \\ 0 & 0 & 5 & 25 \end{bmatrix}$$

Third column:

$$\begin{matrix} \\ \\ R3 \div 5 : R3 \end{matrix} \begin{bmatrix} 1 & 0 & 0 & -1 \\ 0 & 1 & 1 & 5 \\ 0 & 0 & 1 & 5 \end{bmatrix}$$

$$\begin{matrix} \\ -R3 + R2 : R2 \\ \\ \end{matrix} \begin{bmatrix} 1 & 0 & 0 & -1 \\ 0 & 1 & 0 & 0 \\ 0 & 0 & 1 & 5 \end{bmatrix}$$

**Step 4:** Read the solution. $x = -1, y = 0, z = 5$ or $(-1, 0, 5)$

**Step 5:** *Check the solution* $(-1, 0, 5)$:

$$2x + y + z = 2(-1) + (0) + (5) = -2 + 5 = 3$$
$$5x - 4y + z = 5(-1) - 4(0) + (5) = -5 + 5 = 0$$
$$x + y + z = (-1) + (0) + (5) = -1 + 5 = 4$$

37. Continued.

   **Step 6:** *Check the solution in the first equation.*

$$3x - 5y + z = 3(-1) - 5(0) + (5) = -3 + 5 = 2 \text{ not } 12$$

   Hence this system has no solutions.

41.    **Step 1:** Write with all variables to the left and all coefficients showing.

$$1x + 1y + 1z = 3$$
$$2x - 1y + 1z = 2$$

   **Step 2:** Rewrite in matrix form.

$$\begin{bmatrix} 1 & 1 & 1 & 3 \\ 2 & -1 & 1 & 2 \end{bmatrix}$$

   **Step 3:** Row operations.

   First column:

$$-2R1 + R2 : R2 \quad \begin{bmatrix} 1 & 1 & 1 & 3 \\ 0 & -3 & -1 & -4 \end{bmatrix}$$

   Second column:

$$R2 \div -3 : R2 \quad \begin{bmatrix} 1 & 1 & 1 & 3 \\ 0 & 1 & \frac{1}{3} & \frac{4}{3} \end{bmatrix}$$

$$-R2 + R1 : R1 \quad \begin{bmatrix} 1 & 0 & \frac{2}{3} & \frac{5}{3} \\ 0 & 1 & \frac{1}{3} & \frac{4}{3} \end{bmatrix}$$

   **Step 4:** Write the new system of equations.

$$1x + 0y + \frac{2}{3}z = \frac{5}{3} \quad \text{or} \quad 3x + 2z = 5$$

$$0x + 1y + \frac{1}{3}z = \frac{4}{3} \quad \text{or} \quad 3y + z = 4$$

41. Continued.

**Step 5:** Let $z$ equal other numbers and find the related $x$ and $y$.

| $z = 0$ | $3x + 2z = 5$ <br> $3x + 2(0) = 5$ <br> $3x = 5$ <br> $x = \dfrac{5}{3}$ | $3y + z = 4$ <br> $3y + (0) = 4$ <br> $3y = 4$ <br> $y = \dfrac{4}{3}$ |
|---|---|---|
| $z = 1$ | $3x + 2z = 5$ <br> $3x + 2(1) = 5$ <br> $3x = 5 - 2$ <br> $3x = 3$ <br> $x = 1$ | $3y + z = 4$ <br> $3y + (1) = 4$ <br> $3y = 4 - 1$ <br> $3y = 3$ <br> $y = 1$ |
| $z = -1$ | $3x + 2z = 5$ <br> $3x + 2(-1) = 5$ <br> $3x = 5 + 2$ <br> $3x = 7$ <br> $x = \dfrac{7}{3}$ | $3y + z = 4$ <br> $3y + (-1) = 4$ <br> $3y = 4 + 1$ <br> $3y = 5$ <br> $y = \dfrac{5}{3}$ |

Some selected solutions would be: $\left(\dfrac{5}{3}, \dfrac{4}{3}, 0\right), (1, 1, 1), \left(\dfrac{7}{3}, \dfrac{5}{3}, -1\right)$.

45.   Omitted.

49.   Write the system of equations in standard form:

$$5.1x - 3.2y + 9.8z = 15$$
$$7.3x - 4.6y - 1.2z = 5$$
$$8.0x + 1.0y + 2.0z = 14$$

Rewrite the system in matrix form:

$$\begin{bmatrix} 5.1 & -3.2 & 9.8 & 15 \\ 7.3 & -4.6 & -1.2 & 5 \\ 8.0 & 1.0 & 2.0 & 14 \end{bmatrix}$$

Matrix is $3 \times 4$

Row Operations:

Column 1:   Divide Row1 by 5.1   or   Multiply by $\frac{1}{5.1}$

Add $-7.3 \bullet$ Row1 + Row2 : Row2

Add $-8 \bullet$ Row1 + Row3 : Row3

**49. Continued.**

Column 2:    Divide Row2 by $\frac{-1}{51}$ or Multiply by $-51$

Add $\frac{32}{51} \bullet$ Row2 + Row1 : Row1

Add $-\frac{307}{51} \bullet$ Row2 + Row3 : Row3

Column 3:    Divide Row3 by $-4688.2$ or Multiply by $-\frac{5}{23,441}$

Add $-489.2 \bullet$ Row3 + Row1 : Row1
Add $-776.6 \bullet$ Row3 + Row2 : Row2

Solution: $x = 1.37767...$, $y = 0.81745...5$, $z = 1.08058...$
     or approximately $(1.3777, 0.8174, 1.0806)$

*Check the solution by substituting into the original equations*:

$5.1x - 3.2y + 9.8z = 5.1(1.3777) - 3.2(0.8174) + 9.8(1.0806) = 15.0005 \approx 15$

$7.3x - 4.6y - 1.2z = 7.3(1.3777) - 4.6(0.8174) - 1.2(1.0806) = 5.0005 \approx 5$

$8.0x + 1.0y + 2.0z = 8.0(1.3777) + 1.0(0.8174) + 2.0(1.0806) = 14.0002 \approx 14$

**53.**    The system of equations is in standard form

$1.3x + 5.3y - 8.9z + 5.2w = 2.8$
$4.7x - 5.5y - 3.8z - 7.3w = 5.0$
$5.3x - 1.0y - 3.3z + 8.9w = 8.3$
$7.4x - 3.2y + 9.9z + 5.7w = 82.9$

Rewrite the system in matrix form:

$$\begin{bmatrix} 1.3 & 5.3 & -8.9 & 5.2 & 2.8 \\ 4.7 & -5.5 & -3.8 & -7.3 & 5.0 \\ 5.3 & -1.0 & -3.3 & 8.9 & 8.3 \\ 7.4 & -3.2 & 9.9 & 5.7 & 82.9 \end{bmatrix}$$

Matrix dimensions are $4 \times 5$

Row Operations:

Column 1:    Divide Row1 by 1.3  or  Multiply by $\frac{1}{1.3}$

Add $-4.7 \bullet$ Row1 + Row2:Row2
Add $-5.3 \bullet$ Row1 + Row3:Row3
Add $-7.4 \bullet$ Row1 + Row4:Row4

**53. Continued.**

Column 2:  Divide Row2 by $-\frac{1603}{65}$ or Multiply by $-\frac{65}{1603}$

Add $-\frac{53}{13}$ •Row2 + Row1:Row1

Add $+\frac{2939}{130}$ •Row2 + Row3:Row3

Add $+\frac{2169}{65}$ •Row2 + Row4:Row4

Column 3:  Divide Row3 by $-6.97096\ldots$ or Multiply by $-\frac{4580}{31,927}$

Add $+\frac{987}{458}$ •Row3 + Row1:Row1

Add $+\frac{527}{458}$ •Row3 + Row2:Row2

Add $-\frac{25,379}{1145}$ •Row3 + Row4:Row4

Column 4:  Divide Row4 by $-25.5519\ldots$ or Multiply by $\frac{1}{25.5519\ldots}$

Add $-3.27948\ldots$•Row4 + Row1:Row1

Add $-2.97742\ldots$•Row4 + Row2:Row2

Add $-1.66783\ldots$•Row4 + Row3:Row3

Solution: $x = 10.63446\ldots, y = 8.49327\ldots, z = 4.72179\ldots, w = -2.69516\ldots$
or approximately $(10.6345, 8.4933, 4.7218, -2.6952)$

*Check the solution by substituting into the original equations.*

$$1.3x + 5.3y - 8.9z + 5.2w = 1.3(10.6345) + 5.3(8.4933) - 8.9(4.7218) + 5.2(-2.6952)$$
$$= 2.8003 \approx 2.8$$

$$4.7x - 5.5y - 3.8z - 7.3w = 4.7(10.6345) - 5.5(8.4933) - 3.8(4.7218) - 7.3(-2.6952)$$
$$= 5.0011 \approx 5.0$$

$$5.3x - 1.0y - 3.3z + 8.9w = 5.3(10.6345) - 1.0(8.4933) - 3.3(4.7218) + 8.9(-2.6952)$$
$$= 8.3003 \approx 8.3$$

$$7.4x - 3.2y + 9.9z + 5.7w = 7.4(10.6345) - 3.2(8.4933) + 9.9(4.7218) + 5.7(-2.6952)$$
$$= 82.8999 \approx 82.9$$

---

| **Section 7.5** | **More on Markov Chains** |
|---|---|

1.  **Step 1:** The transition matrix $T$ was given.

$$T = \begin{bmatrix} 0.3 & 0.2 & 0.5 \\ 0.1 & 0.8 & 0.1 \\ 0.4 & 0.3 & 0.3 \end{bmatrix}$$

1. Continued.

Step 2: Create the equilibrium matrix $L$.

$$L = \begin{bmatrix} x & y & z \end{bmatrix}$$

Step 3: Find and simplify the system of equations described by $LT = L$.

$$\begin{bmatrix} x & y & z \end{bmatrix} \begin{bmatrix} 0.3 & 0.2 & 0.5 \\ 0.1 & 0.8 & 0.1 \\ 0.4 & 0.3 & 0.3 \end{bmatrix} = \begin{bmatrix} x & y & z \end{bmatrix}$$

$$\begin{bmatrix} 0.3x + 0.1y + 0.4z & 0.2x + 0.8y + 0.3z & 0.5x + 0.1y + 0.3z \end{bmatrix} = \begin{bmatrix} x & y & z \end{bmatrix}$$

Yields:  $0.3x + 0.1y + 0.4z = x$
$0.2x + 0.8y + 0.3z = y$
$0.5x + 0.1y + 0.3z = z$

Combining like terms gives:  $-0.7x + 0.1y + 0.4z = 0$
$0.2x - 0.2y + 0.3z = 0$
$0.5x + 0.1y - 0.7z = 0$

Step 4: Discard any redundant equations and include $x + y + z = 1$.

If the first two equations in Step 3 are added together they will equal the negative of the last equation. Therefore the last equation is discarded.

The system to be solved is:

$$\begin{aligned} x + \quad y + \quad z &= 1 \\ -0.7x + 0.1y + 0.4z &= 0 \\ 0.2x - 0.2y + 0.3z &= 0 \end{aligned}$$

Step 5: Solve the resulting system.

Rewrite in matrix form.

$$\begin{bmatrix} 1 & 1 & 1 & 1 \\ -0.7 & 0.1 & 0.4 & 0 \\ 0.2 & -0.2 & 0.3 & 0 \end{bmatrix}$$

Row operations.

First column:

$$\begin{array}{l} \\ 0.7R1 + R2 : R2 \\ -0.2R1 + R3 : R3 \end{array} \begin{bmatrix} 1 & 1 & 1 & 1 \\ 0 & 0.8 & 1.1 & 0.7 \\ 0 & -0.4 & 0.1 & -0.2 \end{bmatrix}$$

1. Continued.

Second column:

$$R2 \div 0.8 : R2 \begin{bmatrix} 1 & 1 & 1 & 1 \\ 0 & 1 & 1.375 & 0.875 \\ 0 & -0.4 & 0.1 & -0.2 \end{bmatrix}$$

$$\begin{array}{r} -R2 + R1 : R1 \\ \\ 0.4R2 + R3 : R3 \end{array} \begin{bmatrix} 1 & 0 & -0.375 & 0.125 \\ 0 & 1 & 1.375 & 0.875 \\ 0 & 0 & 0.65 & 0.15 \end{bmatrix}$$

Third column:

$$R3 \div 0.65 : R3 \begin{bmatrix} 1 & 0 & -0.375 & 0.125 \\ 0 & 1 & 1.375 & 0.875 \\ 0 & 0 & 1 & 0.2308 \end{bmatrix}$$

$$\begin{array}{r} 0.375R3 + R1 : R1 \\ -1.375R3 + R2 : R2 \\ \end{array} \begin{bmatrix} 1 & 0 & 0 & 0.2116 \\ 0 & 1 & 0 & 0.5577 \\ 0 & 0 & 1 & 0.2308 \end{bmatrix}$$

Resulting matrix is $\quad L \approx \begin{bmatrix} 0.2116 & 0.5577 & 0.2308 \end{bmatrix}$

**Step 6:** *Check the work by verifying that LT = L.*

$$LT = \begin{bmatrix} 0.2116 & 0.5577 & 0.2308 \end{bmatrix} \begin{bmatrix} 0.3 & 0.2 & 0.5 \\ 0.1 & 0.8 & 0.1 \\ 0.4 & 0.3 & 0.3 \end{bmatrix} = \begin{bmatrix} 0.2116 & 0.5577 & 0.2308 \end{bmatrix} = L$$

5.     **Step 1:** Create the transition matrix $T$.

$$T = \begin{bmatrix} 0.87 & 0.08 & 0.05 \\ 0.12 & 0.86 & 0.02 \\ 0.13 & 0.10 & 0.77 \end{bmatrix}$$

**Step 2:** Create the equilibrium matrix $L$.

$$L = \begin{bmatrix} x & y & z \end{bmatrix}$$

5. Continued.

**Step 3:** Find and simplify the system of equations described by $LT = L$.

$$\begin{bmatrix} x & y & z \end{bmatrix} \begin{bmatrix} 0.87 & 0.08 & 0.05 \\ 0.12 & 0.86 & 0.02 \\ 0.13 & 0.10 & 0.77 \end{bmatrix} = \begin{bmatrix} x & y & z \end{bmatrix}$$

$$\begin{bmatrix} 0.87x + 0.12y + 0.13z & 0.08x + 0.86y + 0.10z & 0.05x + 0.02y + 0.77z \end{bmatrix} = \begin{bmatrix} x & y & z \end{bmatrix}$$

Yields:  $0.87x + 0.12y + 0.13z = x$
$0.08x + 0.86y + 0.10z = y$
$0.05x + 0.02y + 0.77z = z$

Combining like terms gives:   $-0.13x + 0.12y + 0.13z = 0$
$0.08x - 0.14y + 0.10z = 0$
$0.05x + 0.02y - 0.23z = 0$

**Step 4:** Discard any redundant equations and include $x + y + z = 1$.

If the first two equations in Step 3 are added together they will equal the negative of the last equation. Therefore the last equation is discarded.

The system to be solved is:

$$x + y + z = 1$$
$$-0.13x + 0.12y + 0.13z = 0$$
$$0.08x - 0.14y + 0.10z = 0$$

**Step 5:** Solve the resulting system.

Rewrite in matrix form.

$$\begin{bmatrix} 1 & 1 & 1 & 1 \\ -0.13 & 0.12 & 0.13 & 0 \\ 0.08 & -0.14 & 0.10 & 0 \end{bmatrix}$$

Row operations.

First column:

$$\begin{array}{r} \\ 0.13R1 + R2 : R2 \\ -0.08R1 + R3 : R3 \end{array} \begin{bmatrix} 1 & 1 & 1 & 1 \\ 0 & 0.25 & 0.26 & 0.13 \\ 0 & -0.22 & 0.02 & -0.08 \end{bmatrix}$$

**5. Continued.**

Second column:

$$R2 \div 0.25 : R2 \begin{bmatrix} 1 & 1 & 1 & 1 \\ 0 & 1 & 1.04 & 0.52 \\ 0 & -0.22 & 0.02 & -0.08 \end{bmatrix}$$

$$\begin{array}{c} -R2 + R1 : R1 \\ \\ 0.22R2 + R3 : R3 \end{array} \begin{bmatrix} 1 & 0 & -0.04 & 0.48 \\ 0 & 1 & 1.04 & 0.52 \\ 0 & 0 & 0.2488 & 0.0344 \end{bmatrix}$$

Third column:

$$R3 \div 0.2488 : R3 \begin{bmatrix} 1 & 0 & -0.04 & 0.48 \\ 0 & 1 & 1.04 & 0.52 \\ 0 & 0 & 1 & 0.138264 \end{bmatrix}$$

$$\begin{array}{c} 0.04R3 + R1 : R1 \\ -1.04R3 + R2 : R2 \end{array} \begin{bmatrix} 1 & 0 & 0 & 0.485531 \\ 0 & 1 & 0 & 0.376206 \\ 0 & 0 & 1 & 0.138264 \end{bmatrix}$$

Resulting matrix is $\quad L \approx \begin{bmatrix} 0.4855 & 0.3762 & 0.1383 \end{bmatrix}$

**Step 6:** *Check the work by verifying that LT = L.*

$$LT = \begin{bmatrix} 0.4855 & 0.3762 & 0.1383 \end{bmatrix} \begin{bmatrix} 0.87 & 0.08 & 0.05 \\ 0.12 & 0.86 & 0.02 \\ 0.13 & 0.10 & 0.77 \end{bmatrix} = \begin{bmatrix} 0.4855 & 0.3762 & 0.1383 \end{bmatrix} = L$$

Safe Shop's long-range market share should stabilize at 49% and PayNEat's at 38%.

---

| **Chapter 7** | **Review** |

1.  The given matrix is neither a row matrix, a column matrix, nor a square matrix. The dimensions are $3 \times 2$.

5.  $\begin{bmatrix} 1 & 6 \\ 8 & 4 \end{bmatrix} \cdot \begin{bmatrix} -3 & 6 \\ 0 & -5 \end{bmatrix} = \begin{bmatrix} 1 \cdot -3 + 6 \cdot 0 & 1 \cdot 6 + 6 \cdot -5 \\ 8 \cdot -3 + 4 \cdot 0 & 8 \cdot 6 + 4 \cdot -5 \end{bmatrix} = \begin{bmatrix} -3 & -24 \\ -24 & 28 \end{bmatrix}$

9.  The product does not exist because the dimensions do not match. ($3 \times 4$ and $2 \times 3$).

13. $5x - 7y = -29$    multiply by 2      $10x - 14y = -58$

      $2x + 4y = 2$      multiply by $-5$     $\underline{-10x - 20y = -10}$

                         add together         $0x - 34y = -68$

                         solve for $y$             $y = \dfrac{-68}{-34} = 2$

Find $x$ by substitution:     $5x - 7(2) = -29$

                               $5x - 14 = -29$

                                   $5x = -15$

                                     $x = -3$

Solution is $(-3, 2)$

*Check solution* $(-3, 2)$:     $5x - 7y = 5(-3) - 7(2) = -15 - 14 = -29$

                                 $2x + 4y = 2(-3) + 4(2) = -6 + 8 = 2$

17.          T      T'

    $P = \begin{bmatrix} 0.12 & 0.88 \end{bmatrix}$

          T     T'

    $T = \begin{bmatrix} 0.62 & 0.38 \\ 0.16 & 0.84 \end{bmatrix} \begin{matrix} \text{T} \\ \text{T'} \end{matrix}$

a)     Find the first following purchase

        $PT^1 \quad = \begin{bmatrix} 0.12 & 0.88 \end{bmatrix} \begin{bmatrix} 0.62 & 0.38 \\ 0.16 & 0.84 \end{bmatrix} = \begin{bmatrix} 0.2152 & 0.7848 \end{bmatrix}$

        After three years, Toyonda Motors' market share is predicted to be 21.5%.

b)     Find the second following purchase. $\left( \dfrac{6 \text{ years}}{3 \text{ years}} \right) = 2$

        $PT^2 \quad = (PT)T = \begin{bmatrix} 0.2152 & 0.7848 \end{bmatrix} \begin{bmatrix} 0.62 & 0.38 \\ 0.16 & 0.84 \end{bmatrix} = \begin{bmatrix} 0.258992 & 0.741008 \end{bmatrix}$

        Toyonda Motors' market share in six years (the second following purchase) is predicted to be 25.9%

c)     **Step 1:** Create the transition matrix $T$.

        $T = \begin{bmatrix} 0.62 & 0.38 \\ 0.16 & 0.84 \end{bmatrix}$

        **Step 2:** Create the equilibrium matrix $L$.

        $L = \begin{bmatrix} x & y \end{bmatrix}$

**17. Continued.**

**Step 3:** Find and simplify the system of equations described by $LT = L$.

$$[x \quad y] \begin{bmatrix} 0.62 & 0.38 \\ 0.16 & 0.84 \end{bmatrix} = [x \quad y]$$

$$[0.62x + 0.16y \quad 0.38x + 0.84y] = [x \quad y]$$

Yields: $0.62x + 0.16y = x$
$0.38x + 0.84y = y$

Combining like terms: $-0.38x + 0.16y = 0$
$0.38x - 0.16y = 0$

**Step 4:** Discard any redundant equations and include $x + y = 1$.

The Step 3 equations are the same so use only once.

The system to be solved is:

$x + y = 1$
$0.38x - 0.16y = 0$

**Step 5:** Solving the resulting system by elimination.

$x + \quad y = 1$     times 0.16     $0.16x + 0.16y = 0.16$
$0.38x - 0.16y = 0$                $\underline{0.38x - 0.16y = \quad 0}$

add together     $0.54x \qquad\quad = 0.16$

solve for $x$     $x = \dfrac{0.16}{0.54} = 0.2963$

Find $y$ by substitution:

$0.2963 + y = 1$
$y = 1 - 0.2963 = 0.7037$

Resulting matrix is $L = [0.2963 \quad 0.7037]$

**Step 6:** *Check the work by verifying that $LT = L$.*

$$LT = [0.2963 \quad 0.7037] \begin{bmatrix} 0.62 & 0.38 \\ 0.16 & 0.84 \end{bmatrix}$$

$$= [[0.296298 \quad 0.703702] \approx [0.2963 \quad 0.7037] = L$$

Toyonda Motors' long-range market share is predicted to be **29.6%**.

d)     We always assume that the current trends reflected in the transition matrix will be continued into the future and the survey accurately reflects future purchases.

# 8  Linear Programming

1.     **Solve the inequality for y:** $3x + y < 4 \rightarrow y < -3x + 4$

**Graph the line:** $y = -3x + 4$ is a dashed line (= is not part of the inequality) with slope $-3$ and $y$-intercept 4.  Start at (0, 4) on the $y$-axis and from that point rise $-3$ (move three units down) and run 1 (move one unit to the right).

$$\text{slope} = m = \frac{\text{rise}}{\text{run}} = -3 = \frac{-3}{1} = \frac{\text{three units down}}{\text{one unit right}}$$

**Shade in one side of the line:**  The graph of the inequality is the set of all points below the line (values of $y$ decrease if we move down).

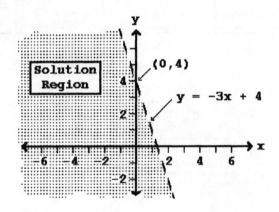

5.     **Solve the inequality for y:**  There is no $y$ in $x \geq 4$.

**Graph the line:**  $x = 4$ is a solid line (= is a part of the inequality) through any $x$-coordinate of 4 such as (4, 0), (4, 2) and (4, 3).  This makes a vertical line.

**Shade in one side of the line:**  The graph of the inequality is the set of all points to the right of the line (values of $x$ increase if we move to the right).

5. Continued.

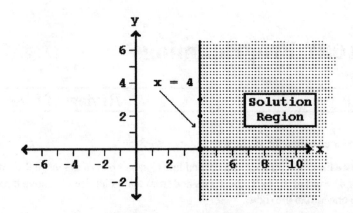

9.     a)

| Original Inequality | Slope-Intercept Form | Associated Equation | Graph of the Inequality |
|---|---|---|---|
| $y > 2x + 1$ | $y > 2x + 1$ | $y = 2x + 1$ | all points above the line with slope 2 and y-intercept 1 |
| $y \leq -x + 4$ | $y \leq -x + 4$ | $y = -x + 4$ | all points on or below the line with slope −1 and y-intercept 4 |

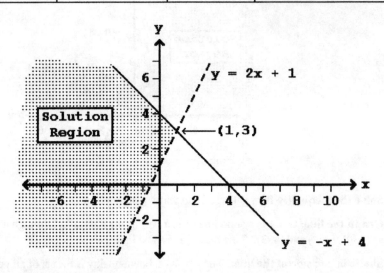

    b)     The region of solutions is unbounded.

9. Continued.

  c)    There is one corner point where $y = 2x + 1$ intersects $y = -x + 4$. Use the elimination
  method.

$$y = 2x + 1 \qquad \rightarrow \qquad y = 2x + 1$$
$$y = -x + 4 \qquad \rightarrow \qquad \underline{-y = \; x - 4}$$
$$0 = 3x - 3$$

$$-3x = -3 \qquad \rightarrow \qquad x = 1$$

$$y = 2x + 1 = 2(1) + 1 \; = 3$$

  Thus, the corner point is (1, 3).

13.  a)

| Original Inequality | Slope-Intercept Form | Associated Equation | Graph of the Inequality |
|---|---|---|---|
| $x + 2y \le 4$ | $2y \le -x + 4$ <br> $y \le -\dfrac{1}{2}x + 2$ | $y = -\dfrac{1}{2}x + 2$ | all points on or below the line with slope $-\frac{1}{2}$ and $y$-intercept 2 |
| $3x - 2y \le -12$ | $-2y \le -3x - 12$ <br> $y \ge \dfrac{3}{2}x + 6$ | $y = \dfrac{3}{2}x + 6$ | all points on or above the line with slope $\frac{3}{2}$ and $y$-intercept 6 |
| $x - y < -7$ | $-y < -x - 7$ <br> $y > x + 7$ | $y = x + 7$ | all points above the line with slope 1 and $y$-intercept 7 |

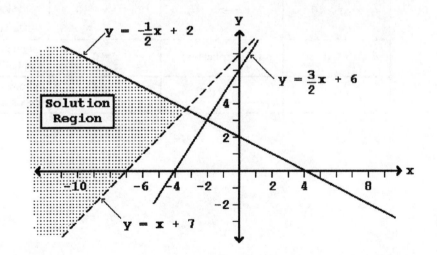

  b)    The region of solutions is unbounded.

13. Continued.

c)    There is one corner point where $x + 2y = 4$ intersects $x - y = -7$. Use the elimination method.

$$
\begin{array}{lcl}
x + 2y = 4 & \rightarrow & x + 2y = 4 \\
x - y = -7 & \rightarrow & \underline{-x + y = 7} \\
& & 3y = 11 \qquad \rightarrow \qquad y = \dfrac{11}{3} = 3\dfrac{2}{3}
\end{array}
$$

$$
x - y = -7 \quad \rightarrow \quad x - \dfrac{11}{3} = -7 \quad \rightarrow \quad x = -\dfrac{21}{3} + \dfrac{11}{3} = -\dfrac{10}{3} = -3\dfrac{1}{3}
$$

Thus, the corner point is $\left( -3\dfrac{1}{3},\ 3\dfrac{2}{3} \right)$.

17.   a)

| Original Inequality | Slope-Intercept Form | Associated Equation | Graph of the Inequality |
|---|---|---|---|
| $15x + 22y \le 510$ | $22y \le -15x + 510$ <br><br> $y \le -\dfrac{15}{22}x + \dfrac{510}{22}$ | $y = -\dfrac{15}{22}x + \dfrac{510}{22}$ | all points on or below the line with slope $-\frac{15}{22}$ and $y$-intercept $\frac{510}{22}$ |
| $35x + 12y \le 600$ | $12y \le -35x + 600$ <br><br> $y \le -\dfrac{35}{12}x + 50$ | $y = -\dfrac{35}{12}x + 50$ | all points on or below the line with slope $-\frac{35}{12}$ and $y$-intercept 50 |
| $x + y > 10$ | $y > -x + 10$ | $y = -x + 10$ | all points above the line with slope $-1$ and $y$-intercept 10 |
| $x \ge 0$ | (not applicable) | $x = 0$ | all points on or to the right of the $y$-axis |
| $y \ge 0$ | $y \ge 0$ | $y = 0$ | all points on or above the $x$-axis |

17. a)  Continued.

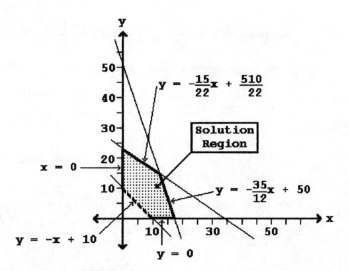

b)      The region of solutions is bounded.

c)      There are five corner points.  Starting at the origin $(0, 0)$ move clockwise around the region of solutions.

$P_1$:   Intersection of $x = 0$ and $y = -x + 10$.

Substituting $x = 0$ into $y = -x + 10$ gives $y = -(0) + 10 = 10$

$P_1 = (0, 10)$

$P_2$:    Intersection of $x = 0$ and $y = -\dfrac{15}{22}x + \dfrac{510}{22}$.

Substituting $x = 0$ into $y = -\dfrac{15}{22}(0) + \dfrac{510}{22} = \dfrac{510}{22} = 23\dfrac{2}{11}$

$P_2 = \left(0, 23\dfrac{2}{11}\right)$

17. c)  Continued.

$P_3$:  Intersection of $y = -\dfrac{15}{22}x + \dfrac{510}{22}$ and $y = -\dfrac{35}{12}x + 50$

Multiply both equations by $22(12) = 264$

$264y = -180x + 6{,}120$ → $264y = -180x + 6{,}120$
$264y = -770x + 13{,}200$ → $\underline{-264y = \phantom{-}770x - 13{,}200}$
$\phantom{264y = -770x +}0 = 590x - 7{,}080$

$-590x = -7{,}080$ → $x = \dfrac{7{,}080}{590} = 12$

$y = -\dfrac{35}{12}x + 50 = -\dfrac{35}{12}(12) + 50 = -35 + 50 = 15$

$P_3 = (12, 15)$

$P_4$:  Intersection of $y = 0$ and $y = -\dfrac{35}{12}x + 50.$

Substituting $y = 0$ into $y = -\dfrac{35}{12}(x) + 50$ gives $0 = -\dfrac{35}{12}x + 50$, or

$\dfrac{35}{12}x = 50$ → $35x = 50(12)$ → $x = \dfrac{600}{35} = 17\dfrac{1}{7}$

$P_4 = \left(17\dfrac{1}{7}, 0\right)$

$P_5$:  Intersection of $y = 0$ and $y = -x + 10.$

Substituting $y = 0$ into $y = -x + 10$ gives $0 = -x + 10$ → $x = 10$

$P_5 = (10, 0)$

21.  a)

| Original Inequality | Slope-Intercept Form | Associated Equation | Graph of the Inequality |
|---|---|---|---|
| $x - 2y + 16 \geq 0$ | $-2y \geq -x - 16$ <br> $y \leq \dfrac{1}{2}x + 8$ | $y = \dfrac{1}{2}x + 8$ | all points on or below the line with slope $\frac{1}{2}$ and $y$-intercept 8 |
| $3x + y \leq 30$ | $y \leq -3x + 30$ | $y = -3x + 30$ | all points on or below the line with slope $-3$ and $y$-intercept 30 |
| $x + y \leq 14$ | $y \leq -x + 14$ | $y = -x + 14$ | all points on or below the line with slope $-1$ and $y$-intercept 14 |
| $x \geq 0$ | (not applicable) | $x = 0$ | all points on or to the right of the $y$-axis |
| $y \geq 0$ | $y \geq 0$ | $y = 0$ | all points on or above the $x$-axis |

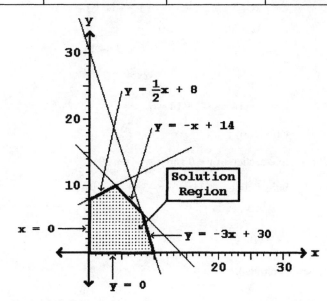

b)   The region of solutions is bounded.

c)   There are five corner points.  Starting at the origin $(0, 0)$ move clockwise around the region of solutions.

$P_1$:   Intersection of $x = 0$ and $y = 0$.        $P_1 = (0, 0)$

$P_2$:   The $y$-intercept of $y = \dfrac{1}{2}x + 8$.        $P_2 = (0, 8)$

21. c)  Continued.

$P_3$:      Intersection of $y = \dfrac{1}{2}x + 8$ and $y = -x + 14$.

Multiply both equations by the LCD of 2.

$$2y = 1x + 16 \qquad \rightarrow \qquad 2y = 1x + 16$$
$$2y = -2x + 28 \qquad \rightarrow \qquad \underline{-2y = 2x - 28}$$
$$0 = 3x - 12$$

$$-3x = -12 \qquad \rightarrow \qquad x = \dfrac{-12}{-3} = 4$$

$y = -x + 14 = -(4) + 14 = 10$

$P_3 = (4, 10)$

$P_4$:      Intersection of $y = -x + 14$ and $y = -3x + 30$.

$$y = -1x + 14 \qquad \rightarrow \qquad y = -1x + 14$$
$$y = -3x + 30 \qquad \rightarrow \qquad \underline{-y = 3x - 30}$$
$$0 = 2x - 16$$

$$-2x = -16 \qquad \rightarrow \qquad x = 8$$

$y = -x + 14 = -(8) + 14 = 6$

$P_4 = (8, 6)$

$P_5$:      Intersection of $y = 0$ and $y = -3x + 30$.

Substituting $y = 0$ into $y = -3x + 30$ gives
$$0 = -3x + 30 \qquad \rightarrow \qquad 3x = 30 \qquad \rightarrow \qquad x = 10$$

$P_5 = (10, 0)$

25.      Concept question omitted.

29. to 49.      Omitted. See exercises 1 to 21. Adjust the viewing window on your calculator if needed for a particular exercise.

| Section 8.1 | The Geometry of Linear Programming |
|---|---|

1.     Independent variables:

         $x$ = number of shrubs

         $y$ = number of trees

      Constraint:

$$\text{gallons of water used} \quad \leq 100$$
$$\text{(water for shrubs)} \quad + \quad \text{(water for trees)} \quad \leq 100$$
$$\text{(1 gallon per shrub)} \bullet \text{(number of shrubs)} + \text{(3 gallons per tree)} \bullet \text{(number of trees)} \leq 100$$
$$1x + 3y \leq 100$$

5.     Independent variables:

         $x$ = number of refrigerators

         $y$ = number of dishwashers

      Constraint:

$$\text{warehouse storage space} \quad \leq 1650$$
$$\text{(space for refrigerators)} \quad + \quad \text{(space for dishwashers)} \leq 1650$$
$$(63 \text{ ft}^3 \text{ per refrigerator}) \bullet \text{(number of refrigerators)}$$
$$+ (41 \text{ ft}^3 \text{ per dishwasher}) \bullet \text{(number of dishwashers)} \leq 1650$$
$$63x + 41y \leq 1650$$

9.     **Step 1:** *List the independent variables*

      Independent variables:

         $x$ = pounds of Morning Blend to be prepared each day

         $y$ = pounds of South American Blend prepared each day

      **Step 2:** *List the constraints as linear inequalities*

      Constraints:

$$\text{Mexican beans used} \quad \leq 100$$
$$\text{(Mexican in Morning)} + \text{(Mexican in South American)} \leq 100$$
$$\text{(one-third of Morning)} + \text{(two-thirds of South American)} \leq 100$$
$$\frac{1}{3}x + \frac{2}{3}y \leq 100$$

$$\text{Columbian beans used} \quad \leq 80$$
$$\text{(Columbian in Morning)} + \text{(Columbian in South American)} \leq 80$$
$$\text{(two-thirds of Morning)} + \text{(one-third of South American)} \leq 80$$
$$\frac{2}{3}x + \frac{1}{3}y \leq 80$$

      Pounds of coffee cannot be negative, so $x \geq 0$ and $y \geq 0$.

9. Continued.

**Step 3:** *Find the objective function (maximize profit)*

$z$ = (Morning profit) + (South American profit)
   = ($3 per pound)(pounds of Morning) + ($2.50 per pound)(pounds of South American)
$z = 3x + 2.5y$

**Step 4:** *Graph the region of possible solutions*

| Original Inequality | Slope-Intercept Form | Associated Equation | Graph of the Inequality |
|---|---|---|---|
| $\dfrac{1}{3}x + \dfrac{2}{3}y \le 100$ | $\dfrac{2}{3}y \le -\dfrac{1}{3}x + 100$ <br><br> $y \le -\dfrac{1}{2}x + 150$ | $y = -\dfrac{1}{2}x + 150$ | all points on or below the line with slope $-\dfrac{1}{2}$ and $y$-intercept 150 and $x$-intercept 300 |
| $\dfrac{2}{3}x + \dfrac{1}{3}y \le 80$ | $\dfrac{1}{3}y \le -\dfrac{2}{3}x + 80$ <br><br> $y \le -2x + 240$ | $y = -2x + 240$ | all points on or below the line with slope $-2$ and $y$-intercept 240 and $x$-intercept 120 |
| $x \ge 0$ | (not applicable) | $x = 0$ | all points on or to the right of the $y$-axis |
| $y \ge 0$ | $y \ge 0$ | $y = 0$ | all points on or above the $x$-axis |

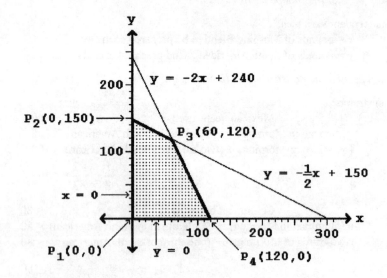

9. Continued.

**Step 5:** *Find all corner points and their z-values*

Use the elimination method to find $P_3$.

$$\frac{1}{3}x + \frac{2}{3}y = 100 \quad \rightarrow \quad x + 2y = 300 \quad \rightarrow \quad -2x - 4y = -600$$

$$\frac{2}{3}x + \frac{1}{3}y = 80 \quad\quad\quad \rightarrow \quad 2x + y = 240 \quad \rightarrow \quad \underline{2x + y = \ \ 240}$$

$$-3y = -360$$
$$y = 120$$

$$2x + (120) = 240$$
$$2x = 120$$
$$x = 60$$

The corner points are: $P_1(0, 0)$, $P_2(0, 150)$, $P_3(60, 120)$, $P_4(120, 0)$

| Point | Value of $z = 3x + 2.5y$ |
|---|---|
| $P_1(0, 0)$ | $z = 3(0) + 2.5(0) = 0$ |
| $P_2(0, 150)$ | $z = 3(0) + 2.5(150) = 375$ |
| $P_3(60, 120)$ | $z = 3(60) + 2.5(120) = 480$ |
| $P_4(120, 0)$ | $z = 3(120) + 2.5(0) = 360$ |

**Step 6:** *Find the maximum profit*

Pete's Coffees should produce 60 pounds of Morning Blend and 120 pounds of South American Blend each day to maximize their profit at $480.

13.

| | First Class Passengers | Economy Passengers | Cost | Maximum Flights |
|---|---|---|---|---|
| Orville 606 | 20 | 80 | $12,000 | 52 |
| Wilbur W-1112 | 80 | 120 | $18,000 | 30 |
| Minimum Needed | 1600 | 4800 | | |

**Step 1:** *List the independent variables*

Independent variables:

$x$ = flights using Orville 606
$y$ = flights using Wilbur W-1112

**13. Continued.**

**Step 2:** *List the constraints as linear inequalities*

Constraints:

$$
\begin{aligned}
\text{First class passengers} &\geq 1600 \\
\text{(First class on Orville)} + \text{(First class on Wilbur)} &\geq 1600 \\
\text{(20 on Orville)} \bullet \text{(flights of Orville)} + \text{(80 on Wilbur)} \bullet \text{(flights of Wilbur)} &\geq 1600 \\
20x + 80y &\geq 1600
\end{aligned}
$$

$$
\begin{aligned}
\text{Economy passengers} &\geq 4800 \\
\text{(Economy on Orville)} + \text{(Economy on Wilbur)} &\geq 4800 \\
\text{(80 on Orville)} \bullet \text{(flights of Orville)} + \text{(120 on Wilbur)} \bullet \text{(flights of Wilbur)} &\geq 4800 \\
80x + 120y &\geq 4800
\end{aligned}
$$

Number of flights on Orville $\leq 52$
$$x \leq 52$$

Number of flights on Wilbur $\leq 30$
$$y \leq 30$$

Number of flights cannot be negative, so $x \geq 0$ and $y \geq 0$.

**Step 3:** *Find the objective function (minimize cost)*

$z = $ (Orville cost) + (Wilbur cost)
$\quad = $ (\$12,000 per flight)$\bullet$(flights of Orville) + (\$18,000 per flight)$\bullet$(flights of Wilbur)

$z = 12{,}000x + 18{,}000y$

**Step 4:** *Graph the region of possible solutions*

| Original Inequality | Slope–Intercept Form | Associated Equation | Graph of the Inequality |
|---|---|---|---|
| $20x + 80y \geq 1600$ | $80y \geq -20x + 1600$ <br> $y \geq -\dfrac{1}{4}x + 20$ | $y = -\dfrac{1}{4}x + 20$ | all points on or below the line with slope $-\frac{1}{4}$ and $y$-intercept 20 and $x$-intercept 80 |
| $80x + 120y \geq 4800$ | $120y \geq -80x + 4800$ <br> $y \geq -\dfrac{2}{3}x + 40$ | $y = -\dfrac{2}{3}x + 40$ | all points on or below the line with slope $-\frac{2}{3}$ and $y$-intercept 40 and $x$-intercept 60 |
| $x \leq 52$ | (not applicable) | $x = 52$ | all points on or to the left of the vertical line through $x = 52$ |
| $y \leq 30$ | $y \leq 30$ | $y = 30$ | all points on or below the horizontal line through $y = 30$ |

13. Continued.

| $x \geq 0$ | (not applicable) | $x = 0$ | all points on or to the right of the $y$-axis |
|---|---|---|---|
| $y \geq 0$ | $y \geq 0$ | $y = 0$ | all points on or above the $x$-axis |

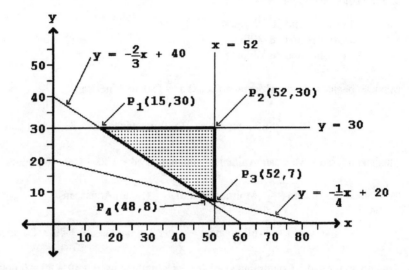

**Step 5:** *Find all corner points and their z-values*

Use the elimination method to find $P_4$.

$$20x + 80y = 1600 \quad \rightarrow \quad -80x - 320y = -6400$$
$$80x + 120y = 4800 \quad \rightarrow \quad \underline{80x + 120y = \ \ 4800}$$
$$-200y = -1600$$
$$y = 8$$

$$20x + 80(8) = 1600$$
$$20x = 1600 - 640$$
$$20x = 960$$
$$x = 48$$

The corner points are: $P_1(15, 30)$, $P_2(52, 30)$, $P_3(52, 7)$, $P_4(48, 8)$

| Point | Value of $z = 12{,}000x + 18{,}000y$ |
|---|---|
| $P_1(15, 30)$ | $z = 12{,}000(15) + 18{,}000(30) = 720{,}000$ |
| $P_2(52, 30)$ | $z = 12{,}000(52) + 18{,}000(30) = 1{,}164{,}000$ |
| $P_3(52, 7)$ | $z = 12{,}000(52) + 18{,}000(7) = 750{,}000$ |
| $P_4(48, 8)$ | $z = 12{,}000(48) + 18{,}000(8) = 720{,}000$ |

**13. Continued.**

Step 6: *Find the minimum cost*

To minimize its costs Global Air Lines should schedule between 15 and 48 flights on Orville 606 and between 8 and 30 flights on Wilbur W-1112 such that they satisfy the equation $L_2: y = -\frac{2}{3}x + 40$. The operating costs for using that type of scheduling would be $720,000. Some specific solutions might be:

> 15 Orvilles and 30 Wilburs
> 18 Orvilles and 28 Wilburs
> 21 Orvilles and 26 Wilburs

**17.** Mexican beans used $= \frac{1}{3}$ (Morning Blend) $+ \frac{2}{3}$ (South American)

$$= \frac{1}{3}(60 \text{ pounds}) + \frac{2}{3}(120 \text{ pounds}) = 20 + 80 = 100 \text{ pounds}$$

Mexican unused = Mexican available – Mexican used = $100 - 100 = 0$ pounds

Columbian beans used $= \frac{2}{3}$ (Morning Blend) $+ \frac{1}{3}$ (South American)

$$= \frac{2}{3}(60 \text{ pounds}) + \frac{1}{3}(120 \text{ pounds}) = 40 + 40 = 80 \text{ pounds}$$

Columbian unused = Columbian available – Columbian used = $80 - 80 = 0$ pounds

**21.** Constraints:
$$3x + y \geq 12$$
$$x + y \geq 6$$
$$x \geq 0 \text{ and } y \geq 0.$$

Objective function:
$$z = 2x + 3y$$

*Graph the region of possible solutions:*

| Original Inequality | Slope-Intercept Form | Associated Equation | Graph of the Inequality |
|---|---|---|---|
| $3x + y \geq 12$ | $y \geq -3x + 12$ | $y = -3x + 12$ | all points on or above the line with slope $-3$ and $y$-intercept 12 and $x$-intercept 4 |
| $x + y \geq 6$ | $y \geq -x + 6$ | $y = -x + 6$ | all points on or above the line with slope $-1$ and $y$-intercept 6 and $x$-intercept 6 |
| $x \geq 0$ | (not applicable) | $x = 0$ | all points on or to the right of the $y$-axis |
| $y \geq 0$ | $y \geq 0$ | $y = 0$ | all points on or above the $x$-axis |

21.  Continued.

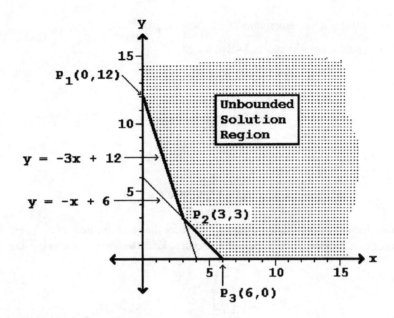

*Find all corner points and their z-values*

$P_1 = (0, 12)$

Use the elimination method to find $P_2$.

| | | |
|---|---|---|
| $3x + y = 12$ | | $3x + y = 12$ |
| $x + y = 6$ | (multiply by $-1$) | $\underline{-x - y = -6}$ |
| | | $2x\quad = 6 \quad \rightarrow \quad x = 3$ |

$x + y = 6 \quad \rightarrow \quad (3) + y = 6 \quad \rightarrow \quad y = 3$

$P_2 = (3, 3)$

$P_3 = (6, 0)$

The corner points are:  $P_1(0, 12), P_2(3, 3), P_3(6, 0)$

| Point | Value of $z = 2x + 3y$ |
|---|---|
| $P_1(0, 12)$ | $z = 2(0) + 3(12) = 36$ |
| $P_2(3, 3)$ | $z = 2(3) + 3(3) = 15$ |
| $P_3(6, 0)$ | $z = 2(6) + 3(0) = 12$ |

21. Continued.

*Determine if a maximum and/or minimum exists*

Let $z$ equal various values and solve for $y$.

$z = 12$ $\qquad$ $2x + 3y = 12$ $\qquad \rightarrow \qquad$ $3y = -2x + 12$ $\qquad \rightarrow \qquad$ $y = -\dfrac{2}{3}x + 4$

$\qquad$ slope $= -\dfrac{2}{3}$, $y$-intercept $= 4$, $x$-intercept $= 6$

$z = 36$ $\qquad$ $2x + 3y = 36$ $\qquad \rightarrow \qquad$ $3y = -2x + 36$ $\qquad \rightarrow \qquad$ $y = -\dfrac{2}{3}x + 12$

$\qquad$ slope $= -\dfrac{2}{3}$, $y$-intercept $= 12$, $x$-intercept $= 18$

Graph these lines (which are parallel) to show the relative values of $z$. The closer the line is to the upper right the larger the value of $z$, and the closer the line is to the origin the smaller the value of $z$.

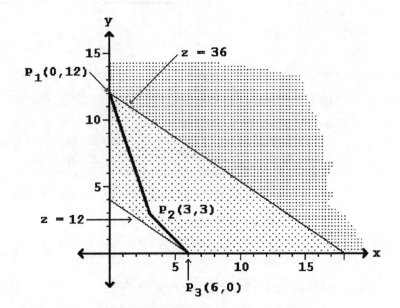

a)  $\quad$ There is no point at which a maximum will occur.

b)  $\quad$ There is no maximum value.

c)  $\quad$ The minimum occurs at $P_3(6, 0)$.

d)  $\quad$ The minimum value is 12.

| Section 8.2 | Introduction to the Simplex Method |
|---|---|

1.  The inequality is: $3x_1 + 2x_2 \leq 5$

    The equation is: $3x_1 + 2x_2 + s_1 = 5$

5.  a)  The inequality is: $6.99x_1 + 3.15x_2 + 1.98x_3 \leq \$42.15$

    b)  The equation is: $6.99x_1 + 3.15x_2 + 1.98x_3 + s_1 = \$42.15$

    c)  $x_1 = $ pounds of meat

    $x_2 = $ pounds of cheese

    $x_3 = $ loaves of bread

    $s_1 = $ unused money after the purchases have been made

9.  a)  The inequality is: $24x_1 + 36x_2 + 56x_3 + 72x_4 \leq 50,000$

    b)  The equation is: $24x_1 + 36x_2 + 56x_3 + 72x_4 + s_1 = 50,000$

    c)  $x_1 = $ number of twin beds

    $x_2 = $ number of double beds

    $x_3 = $ number of queen-size beds

    $x_4 = $ number of king-size beds

    $s_1 = $ cubic feet of unused space in the mattress warehouse after the 4 types of beds are stored

13. Non-identity matrix columns (value is 0): $x_1 = 0, x_2 = 0$

    Identity matrix columns (turn at 1 and read last column for answer):

    $s_1 = 7.8, s_2 = 9.3, s_3 = 0.5, z = 9.6$

    So $(x_1, x_2, s_1, s_2, s_3) = (0, 0, 7.8, 9.3, 0.5)$ with $z = 9.6$

17. Constraint equations:

    $C_1$: $3x_1 + 4x_2 + s_1 = 40$

    $C_2$: $4x_1 + 7x_2 + s_2 = 50$

    $(x_1, x_2, s_1, s_2 \geq 0)$

    Objective function equation:

    $z = 2x_1 + 4x_2$ becomes $-2x_1 - 4x_2 + 0s_1 + 0s_2 + 1z = 0$

17. Continued.

First simplex matrix:

$$
\begin{array}{ccccc}
x_1 & x_2 & s_1 & s_2 & z \\
\end{array}
$$
$$
\left[
\begin{array}{ccccc|c}
3 & 4 & 1 & 0 & 0 & 40 \\
4 & 7 & 0 & 1 & 0 & 50 \\
-2 & -4 & 0 & 0 & 1 & 0 \\
\end{array}
\right]
$$

Possible solution:
$$(x_1, x_2, s_1, s_2) = (0, 0, 40, 50) \text{ with } z = 0$$

21. Constraint equations:
$$C_1: \ 5x_1 + 3x_2 + 9x_3 + s_1 = 10$$
$$C_2: \ 12x_1 + 34x_2 + 100x_3 + s_2 = 10$$
$$C_3: \ 52x_1 + 7x_2 + 12x_3 + s_3 = 10$$
$$(x_1, x_2, x_3, s_1, s_2, s_3 \geq 0)$$

Objective function equation:
$$z = 4x_1 + 7x_2 + 9x_3 \text{ becomes } -4x_1 - 7x_2 - 9x_3 + 0s_1 + 0s_2 + 0s_3 + 1z = 0$$

First simplex matrix:

$$
\begin{array}{ccccccc}
x_1 & x_2 & x_3 & s_1 & s_2 & s_3 & z \\
\end{array}
$$
$$
\left[
\begin{array}{ccccccc|c}
5 & 3 & 9 & 1 & 0 & 0 & 0 & 10 \\
12 & 34 & 100 & 0 & 1 & 0 & 0 & 10 \\
52 & 7 & 12 & 0 & 0 & 1 & 0 & 10 \\
-4 & -7 & -9 & 0 & 0 & 0 & 1 & 0 \\
\end{array}
\right]
$$

Possible solution:
$$(x_1, x_2, x_3, s_1, s_2, s_3) = (0, 0, 0, 10, 10, 10) \text{ with } z = 0$$

25. **Step 1:** *Model the problem*

*List the independent variables*

Independent variables:

$x_1$ = pounds of Yusip Blend to be prepared each day

$x_2$ = pounds of Exotic Blend prepared each day

25. Continued.

*List the constraints as linear inequalities*

Constraints:

Costa Rican beans used $\leq 200$
(Costa Rican in Yusip) + (Costa Rican in Exotic) $\leq 200$
  (one-half of Yusip)   + (one-quarter of Exotic)  $\leq 200$

$C_1$: $\dfrac{1}{2}x_1 + \dfrac{1}{4}x_2 \leq 200$

Ethiopian beans used $\leq 330$
(Ethiopian in Yusip) +    (Ethiopian in Exotic)   $\leq 330$
 (one-half of Yusip)   + (three-quarters of Exotic) $\leq 330$

$C_2$: $\dfrac{1}{2}x_1 + \dfrac{3}{4}x_2 \leq 330$

Pounds of coffee cannot be negative, so $x_1 \geq 0$ and $x_2 \geq 0$.

*Find the objective function (maximize profit)*

$z = $ (Yusip profit) + (Exotic profit)
  $= (\$3.50$ per pound)(pounds of Yusip) + ($\$4$ per pound)(pounds of Exotic)
$z = 3.5x_1 + 4x_2$

**Step 2:** *Convert each constraint to an equation*

$C_1$: $\dfrac{1}{2}x_1 + \dfrac{1}{4}x_2 + s_1 = 200$, where $s_1 = $ unused Costa Rican beans

$C_2$: $\dfrac{1}{2}x_1 + \dfrac{3}{4}x_2 + s_2 = 330$, where $s_2 = $ unused Ethiopian beans

$(x_1, x_2, s_1, s_2 \geq 0)$

**Step 3:** *Rewrite the objective function*

$z = 3.5x_1 + 4x_2$ becomes $-3.5x_1 - 4x_2 + 0s_1 + 0s_2 + 1z = 0$

**Step 4:** *First simplex matrix*

$$
\begin{array}{ccccc}
x_1 & x_2 & s_1 & s_2 & z \\
\end{array}
$$

$$
\begin{bmatrix}
\frac{1}{2} & \frac{1}{4} & 1 & 0 & 0 & 200 \\
\frac{1}{2} & \frac{3}{4} & 0 & 1 & 0 & 330 \\
-3.5 & -4 & 0 & 0 & 1 & 0
\end{bmatrix}
$$

**Step 5:** *Find the possible solution*

$(x_1, x_2, s_1, s_2) = (0, 0, 200, 330)$ with $z = 0$

## Section 8.3 — The Simplex Method: Complete Problems

1.

$$
\begin{array}{ccccc}
x_1 & x_2 & s_1 & s_2 & z \\
\end{array}
$$

$$
\left[
\begin{array}{ccccc|c}
5 & 0 & 3 & 1 & 0 & 12 \\
3 & 0 & 19 & 0 & 10 & 22 \\
-21 & 1 & -48 & 0 & 0 & 19 \\
\end{array}
\right]
\begin{array}{l}
\leftarrow 12/3 = 4 \\
\leftarrow 22/19 \approx 1.2 \quad \leftarrow \text{pivot row} \\
\leftarrow \text{not a constraint row}
\end{array}
$$

$$\uparrow$$

pivot column

The most negative entry in the last row is $-48$ in column 3. Divide the last entry in each constraint row by the corresponding entry in column 3. The pivot row is row 2 which yielded the smallest positive quotient. The pivot entry is 19 in row 2, column 3.

5.   a)

$$
\begin{array}{ccccc}
x_1 & x_2 & s_1 & s_2 & z \\
\end{array}
$$

$$
\left[
\begin{array}{ccccc|c}
1 & 2 & 1 & 0 & 0 & 3 \\
4 & 1 & 0 & 1 & 0 & 2 \\
-6 & -4 & 0 & 0 & 1 & 0 \\
\end{array}
\right]
\begin{array}{l}
\leftarrow 3/1 = 3 \\
\leftarrow 2/4 = 0.5 \quad \leftarrow \text{pivot row} \\
\leftarrow \text{not a constraint row}
\end{array}
$$

$$\uparrow$$

pivot column

The most negative entry in the last row is $-6$ in column 1. Divide the last entry in each constraint row by the corresponding entry in column 1. The pivot row is row 2 which yielded the smallest positive quotient. The pivot entry is 4 in row 2, column 1.

b)

$$
\begin{array}{l}
\\
R2 \div 4 : R2 \\
\\
\end{array}
\left[
\begin{array}{cccccc}
1 & 2 & 1 & 0 & 0 & 3 \\
1 & \frac{1}{4} & 0 & \frac{1}{4} & 0 & \frac{1}{2} \\
-6 & -4 & 0 & 0 & 1 & 0 \\
\end{array}
\right]
$$

$$
\begin{array}{l}
-R2 + R1 : R1 \\
\\
6R2 + R3 : R3 \\
\end{array}
\left[
\begin{array}{cccccc}
0 & \frac{7}{4} & 1 & -\frac{1}{4} & 0 & \frac{5}{2} \\
1 & \frac{1}{4} & 0 & \frac{1}{4} & 0 & \frac{1}{2} \\
0 & -\frac{5}{2} & 0 & \frac{3}{2} & 1 & 3 \\
\end{array}
\right]
$$

c)   $(x_1, x_2, s_1, s_2) = \left( \frac{1}{2}, 0, 2\frac{1}{2}, 0 \right)$ with $z = 3$

9.    **Step 1:** *Model the problem*

Independent variables:

$x_1$ = loaves of bread produced each day

$x_2$ = number of cakes produced each day

Constraints:

$C_1$:  Time

5 friends(8 hr)(60 min/hr) = 2400 minutes per day

| | | |
|---|---|---|
| Working time per day | | $\leq 2400$ |
| (bread making time) + (cake making time) | | $\leq 2400$ |
| (time per loaf)(loaves) + (time per cake)(cakes) | | $\leq 2400$ |
| $50x_1$ + $30x_2$ | | $\leq 2400$ |

$C_2$:  Cost

| | |
|---|---|
| Daily costs | $\leq 190$ |
| (bread cost) + (cake cost) | $\leq 190$ |
| (cost per loaf)(loaves) + (cost per cake)(cakes) | $\leq 190$ |
| $0.90x_1$ + $1.50x_2$ | $\leq 190$ |

Objective Function:

$z$ = (bread profit) + (cake profit)

= (\$1.20 per loaf)(loaves of bread) + (\$4.00 per cake)(number of cakes)

$z = 1.2x_1 + 4x_2$

**Step 2:** *Find the constraint equations*

$C_1$:  $50x_1 + 30x_2 + s_1 = 2,400$  where $s_1$ = unused time

$C_2$:  $0.9x_1 + 1.5x_2 + s_2 = 190$  where $s_2$ = unused money

**Step 3:** *Rewrite the objective function equation*

$z = 1.2x_1 + 4x_2$ becomes  $-1.2x_1 - 4x_2 + 0s_1 + 0s_2 + 1z = 0$

**Step 4:** *Find the first simplex matrix*

| $x_1$ | $x_2$ | $s_1$ | $s_2$ | $z$ | |
|---|---|---|---|---|---|
| 50 | 30 | 1 | 0 | 0 | 2400 |
| 0.9 | 1.5 | 0 | 1 | 0 | 190 |
| −1.2 | −4 | 0 | 0 | 1 | 0 |

**Step 5:** *Find the possible solution*

$(x_1, x_2, s_1, s_2) = (0, 0, 2400, 190)$ with $z = 0$

9. Continued.

Step 6: *Pivot to find a better possible solution*

$$\begin{array}{ccccc} x_1 & x_2 & s_1 & s_2 & z \end{array}$$

$$\begin{bmatrix} 50 & 30 & 1 & 0 & 0 & 2400 \\ 0.9 & 1.5 & 0 & 1 & 0 & 190 \\ -1.2 & -4 & 0 & 0 & 1 & 0 \end{bmatrix}$$
$\leftarrow 2400/30 = 80 \qquad \leftarrow$ pivot row
$\leftarrow 190/1.5 \approx 126.7$
$\leftarrow$ not a constraint row

$\uparrow$
pivot column

$$R1 \div 30 : R1 \begin{bmatrix} 1.6667 & 1 & 0.0333 & 0 & 0 & 80 \\ 0.9 & 1.5 & 0 & 1 & 0 & 190 \\ -1.2 & -4 & 0 & 0 & 1 & 0 \end{bmatrix}$$

$$\begin{array}{ccccc} x_1 & x_2 & s_1 & s_2 & z \end{array}$$

$$\begin{array}{r} \\ -1.5R1 + R2 : R2 \\ 4R1 + R3 : R3 \end{array} \begin{bmatrix} 1.6667 & 1 & 0.0333 & 0 & 0 & 80 \\ -1.6 & 0 & -0.05 & 1 & 0 & 70 \\ 5.4667 & 0 & 0.1333 & 0 & 1 & 320 \end{bmatrix}$$

Step 7: *Determine the possible solution*

$(x_1, x_2, s_1, s_2) = (0, 80, 0, 70)$ with $z = 320$

*Check*:

$C_1$:  $50x_1 + 30x_2 + s_1 = 50(0) + 30(80) + 0 = 2,400$
$C_2$:  $0.9x_1 + 1.5x_2 + s_2 = 0.9(0) + 1.5(80) + 70 = 190$

$z = 1.2x_1 + 4x_2 = 1.2(0) + 4(80) = 320$

Step 8: *Determine if this maximizes the objective function*

**The last row contains no negative entries, so we do not pivot again.**

Step 9: *Interpret the final solution*

The five friends should make 80 cakes each day and no bread. This results in no unused hours and $70 in unused money. Their maximum profit will be $320.

13. **Step 1:** *Model the problem*

Independent variables:

$x_1$ = liters of House White wine produced each day

$x_2$ = liters of Premium White wine produced each day

$x_3$ = liters of Sauvignon Blanc wine produced each day

Constraints:

2 pounds of grapes = 1 liter of wine

$C_1$: French Colombard grapes

30,000 pounds of grapes = 15,000 liters of wine

| | French Colombard (FC) available | | | $\leq 15,000$ |
|---|---|---|---|---|
| (FC in House) + | (FC in Premium) | + | (FC in Sauvignon) | $\leq 15,000$ |
| (75% of House) + | (25% of Premium) | + | (0% of Sauvignon) | $\leq 15,000$ |
| $0.75x_1$ + | $0.25x_2$ | + | $0x_3$ | $\leq 15,000$ |

$C_2$: Sauvignon Blanc grapes

20,000 pounds of grapes = 10,000 liters of wine

| | Sauvignon Blanc (SB) available | | | $\leq 10,000$ |
|---|---|---|---|---|
| (SB in House) + | (SB in Premium) | + | (SB in Sauvignon) | $\leq 10,000$ |
| (25% of House) + | (75% of Premium) | + | (100% of Sauvignon) | $\leq 10,000$ |
| $0.25x_1$ + | $0.75x_2$ | + | $1x_3$ | $\leq 10,000$ |

Objective Function:

$z$ = (House profit) + (Premium profit) + (Sauvignon profit)

= ($1 per liter)(liters of House)

+ ($1.50 per liter)(liters of Premium) + ($2 per liter)(liters of Sauvignon Blanc)

$z = 1x_1 + 1.5x_2 + 2x_3$

**Step 2:** *Find the constraint equations*

$C_1$: $0.75x_1 + 0.25x_2 + 0x_3 + s_1 = 15,000$ where $s_1$ = unused FC grapes

$C_2$: $0.25x_1 + 0.75x_2 + 1x_3 + s_2 = 10,000$ where $s_2$ = unused SB grapes

**Step 3:** *Rewrite the objective function equation*

$z = 1x_1 + 1.5x_2 + 2x_3$ becomes $-1x_1 - 1.5x_2 - 2x_3 + 0s_1 + 0s_2 + 1z = 0$

**Step 4:** *Find the first simplex matrix*

| $x_1$ | $x_2$ | $x_3$ | $s_1$ | $s_2$ | $z$ | |
|---|---|---|---|---|---|---|
| 0.75 | 0.25 | 0 | 1 | 0 | 0 | 15,000 |
| 0.25 | 0.75 | 1 | 0 | 1 | 0 | 10,000 |
| −1.0 | −1.5 | −2 | 0 | 0 | 1 | 0 |

13. Continued.

**Step 5:** *Find the possible solution*

$(x_1, x_2, x_3, s_1, s_2) = (0, 0, 0, 15{,}000, 10{,}000)$ with $z = 0$

**Step 6:** *Pivot to find a better possible solution*

$$
\begin{array}{cccccc}
x_1 & x_2 & x_3 & s_1 & s_2 & z \\
\end{array}
$$

$$
\begin{bmatrix}
0.75 & 0.25 & 0 & 1 & 0 & 0 & 15{,}000 \\
0.25 & 0.75 & 1 & 0 & 1 & 0 & 10{,}000 \\
-1.0 & -1.5 & -2 & 0 & 0 & 1 & 0
\end{bmatrix}
\begin{array}{l}
\leftarrow 15{,}000/0 \text{ is undefined} \\
\leftarrow 10{,}000/1 = 10{,}000 \qquad \leftarrow \text{pivot row}
\end{array}
$$

$$\uparrow$$

pivot column

No division necessary

$$
\begin{array}{cccccc}
x_1 & x_2 & x_3 & s_1 & s_2 & z \\
\end{array}
$$

$$
2R2 + R3 : R3 \quad
\begin{bmatrix}
0.75 & 0.25 & 0 & 1 & 0 & 0 & 15{,}000 \\
0.25 & 0.75 & 1 & 0 & 1 & 0 & 10{,}000 \\
-0.5 & 0 & 0 & 0 & 2 & 1 & 20{,}000
\end{bmatrix}
$$

**Last row is not all positive, so pivot again.**

$$
\begin{array}{cccccc}
x_1 & x_2 & x_3 & s_1 & s_2 & z \\
\end{array}
$$

$$
\begin{bmatrix}
\mathbf{0.75} & 0.25 & 0 & 1 & 0 & 0 & 15{,}000 \\
0.25 & 0.75 & 1 & 0 & 1 & 0 & 10{,}000 \\
-0.5 & 0 & 0 & 0 & 2 & 1 & 20{,}000
\end{bmatrix}
\begin{array}{l}
\leftarrow 15{,}000/0.75 = 20{,}000 \qquad \leftarrow \text{pivot row} \\
\leftarrow 10{,}000/0.25 = 40{,}000
\end{array}
$$

$$\uparrow$$

pivot column

$$
R1 \div 0.75 : R1 \quad
\begin{bmatrix}
1 & 0.3333 & 0 & 1.3333 & 0 & 0 & 20{,}000 \\
0.25 & 0.75 & 1 & 0 & 1 & 0 & 10{,}000 \\
-0.5 & 0 & 0 & 0 & 2 & 1 & 20{,}000
\end{bmatrix}
$$

$$
\begin{array}{cccccc}
x_1 & x_2 & x_3 & s_1 & s_2 & z \\
\end{array}
$$

$$
\begin{array}{l}
-0.25R1 + R2 : R2 \\
0.5R1 + R3 : R3
\end{array}
\begin{bmatrix}
1 & 0.3333 & 0 & 1.3333 & 0 & 0 & 20{,}000 \\
0 & 0.6667 & 1 & -0.3333 & 1 & 0 & 5{,}000 \\
0 & 0.1667 & 0 & 0.6667 & 2 & 1 & 30{,}000
\end{bmatrix}
$$

**Last row is all positive, so we do not need to pivot again.**

13. Continued.

> **Step 7:** *Determine the possible solution*
>
> $(x_1, x_2, x_3, s_1, s_2) = (20,000, 0, 5000, 0, 0)$ with $z = 30,000$
>
> *Check:*
>
> $C_1$: $0.75x_1 + 0.25x_2 + 0x_3 + s_1 = 0.75(20,000) + 0.25(0) + 0(5000) + 0 = 15,000$
>
> $C_2$: $0.25x_1 + 0.75x_2 + 1x_3 + s_2 = 0.25(20,000) + 0.75(0) + 1(5000) + 0 = 10,000$
>
> $z = 1x_1 + 1.5x_2 + 2x_3 = 1(20,000) + 1.5(0) + 2(5000) = 20,000 + 10,000 = 30,000$
>
> **Step 8:** *Determine if this maximizes the objective function*
>
> The last row contains no negative entries, so we do not pivot again.
>
> **Step 9:** *Interpret the final solution*
>
> J & M Winery should make 20,000 liters of House White wine, no Premium White and 5000 liters of Sauvignon Blanc wine to maximize its profit. The maximum profit would be $30,000. There would be no unused grapes of either the French Colombard variety or the Sauvignon Blanc variety.

17. Omitted.

21. Omitted

25. *Constraint equations*

$C_1$: $3x_1 + 2x_2 + 5x_3 + 12x_4 + s_1 = 28$

$C_2$: $4x_1 + 5x_2 + 1x_3 + 7x_4 + s_2 = 32$

$C_3$: $1x_1 + 7x_2 + 9x_3 + 10x_4 + s_3 = 25$

*Rewrite the objective function equation*

$z = 25x_1 + 53x_2 + 18x_3 + 7x_4$ becomes $-25x_1 - 53x_2 - 18x_3 - 7x_4 + 0s_1 + 0s_2 + 0s_3 + 1z =$

0

*The first simplex matrix*

$$
\begin{array}{cccccccc}
x_1 & x_2 & x_3 & x_4 & s_1 & s_2 & s_3 & z \\
\end{array}
$$

$$
\begin{bmatrix}
3 & 2 & 5 & 12 & 1 & 0 & 0 & 0 & 28 \\
4 & 5 & 1 & 7 & 0 & 1 & 0 & 0 & 32 \\
1 & 7 & 9 & 10 & 0 & 0 & 1 & 0 & 25 \\
-25 & -53 & -18 & -7 & 0 & 0 & 0 & 1 & 0
\end{bmatrix}
\begin{array}{l}
\leftarrow 28/2 = 14 \\
\leftarrow 32/5 = 6.4 \\
\leftarrow 25/7 \approx 3.6 \quad \leftarrow \text{pivot row} \\
\\
\end{array}
$$

↑

pivot column

25. Continued.

*The first possible solution*

$(x_1, x_2, x_3, x_4, s_1, s_2, s_3) = (0, 0, 0, 0, 28, 32, 25)$ with $z = 0$

*Pivot to find the best possible solution*

Matrix dimensions are $4 \times 9$

Pivot on 7 in row 3, column 2

Divide Row3 by 7 or Multiply by 1/7
Add  $-2 \bullet$ Row3 + Row1 : Row1
Add  $-5 \bullet$ Row3 + Row2 : Row2
Add  $+53.0 \bullet$ Row3 + Row4 : Row4

**Column 1 ($x_1$) still has a negative in the last row, so pivot again.**

$$
\begin{array}{ccc}
\text{column 1} & \ldots & \text{last column}
\end{array}
$$

$$
\begin{bmatrix}
\frac{19}{7} & \ldots & \frac{146}{7} \\
\frac{23}{7} & \ldots & \frac{99}{7} \\
\frac{1}{7} & \ldots & \frac{25}{7} \\
-\frac{122}{7} & \ldots & \frac{1325}{7}
\end{bmatrix}
\begin{array}{l}
\leftarrow \frac{146}{7} \div \frac{19}{7} = \frac{146}{19} \approx 7.7 \\
\leftarrow \frac{99}{7} \div \frac{23}{7} = \frac{99}{23} \approx 4.3 \quad \leftarrow \text{pivot row} \\
\leftarrow \frac{25}{7} \div \frac{1}{7} = \frac{25}{1} = 25 \\
\\
\end{array}
$$

$\uparrow$
pivot column

Pivot on $\frac{23}{7}$ in Row 2, Column 1

Divide Row2 by $\frac{23}{7}$ or Multiply by $\frac{7}{23}$
Add  $-\frac{19}{7} \bullet$ Row2 + Row1 : Row1
Add  $-\frac{1}{7} \bullet$ Row2 + Row3 : Row3
Add  $+\frac{122}{7} \bullet$ Row2 + Row4 : Row4

**There are no more negative numbers in the last row**

*Determine and check the best possible solution*

$(x_1, x_2, x_3, x_4, s_1, s_2, s_3) = (4.3043, 2.9565, 0, 0, 9.1739, 0, 0)$ with $z = 264.3043$

*Check*:

$C_1$:  $3x_1 + 2x_2 + 5x_3 + 12x_4 + s_1 = 3(4.3043) + 2(2.9565) + 5(0) + 12(0) + (9.1739)$
$$= 27.9998 \approx 28$$

$C_2$:  $4x_1 + 5x_2 + 1x_3 + 7x_4 + s_2 = 4(4.3043) + 5(2.9565) + 1(0) + 7(0) + (0)$
$$= 31.9997 \approx 32$$

25. Continued.

$$C_3: \; 1x_1 + 7x_2 + 9x_3 + 10x_4 + s_3 = 1(4.3043) + 7(2.9565) + 9(0) + 10(0) + (0)$$
$$= 24.9998 \approx 25$$

$$z = 25x_1 + 53x_2 + 18x_3 + 7x_4 = 25(4.3043) + 53(2.9565) + 18(0) + 7(0)$$
$$= 264.302 \approx 264.3043$$

29.    From Example 2, *the first simplex matrix is*

$$
\begin{array}{ccccccccc}
x_1 & x_2 & x_3 & x_4 & x_5 & x_6 & s_1 & s_1 & z \\
\end{array}
$$

$$
\left[
\begin{array}{ccccccccc|c}
6.00 & 6.25 & 5.00 & 5.25 & 5.75 & 4.50 & 1 & 0 & 0 & 130 \\
230 & \mathbf{220} & 190 & 125 & 120 & 105 & 0 & 1 & 0 & 4000 \\
-495 & -500 & -430 & -245 & -250 & -250 & 0 & 0 & 1 & 0
\end{array}
\right]
\begin{array}{l}
\leftarrow \frac{130}{6.25} = 20.8 \\
\leftarrow \frac{4000}{220} = 18.2 \;\; \leftarrow \text{pivot row} \\
\;
\end{array}
$$

$$\underset{\text{pivot column}}{\uparrow}$$

The first possible solution is $(x_1, x_2, x_3, x_4, x_5, x_6, s_1, s_2) = (0, 0, 0, 0, 0, 0, 130, 4000)$ with $z = 0$.

*Pivot to find the best possible solution using the computer or calculator.*

Matrix dimensions are $3 \times 10$

Pivot on 220 in row 2, column 2

Divide Row2 by 220 or Multiply by 1/220
Add $-6.25 \bullet$ Row2 + Row1 : Row1
Add $+500 \bullet$ Row2 + Row3 : Row3

**Column 6 ($x_6$) still has a negative in the last row**

column 6    ...    last column

$$
\left[
\begin{array}{ccc}
\cdots & \frac{267}{176} & \cdots & \frac{180}{11} \\
\cdots & \frac{21}{44} & \cdots & \frac{200}{11} \\
\cdots & -\frac{125}{11} & \cdots & \frac{100{,}000}{11}
\end{array}
\right]
\begin{array}{l}
\leftarrow \frac{180}{11} \div \frac{267}{176} = \frac{960}{89} \approx 10.8 \;\; \leftarrow \text{pivot row} \\
\leftarrow \frac{200}{11} \div \frac{21}{44} = \frac{800}{21} \approx 38.1 \\
\;
\end{array}
$$

$$\underset{\text{pivot column}}{\uparrow}$$

Pivot on $\frac{267}{176}$ in Row 1, Column 6

Divide Row1 by $\frac{267}{176}$ or Multiply by $\frac{176}{267}$
Add $-\frac{21}{44} \bullet$ Row1 + Row2 : Row2
Add $+\frac{125}{11} \bullet$ Row1 + Row3 : Row3

29. Continued.

**Column 3 ($x_3$) still has a negative in the last row**

column 3   ...   last column

$$\begin{bmatrix} \cdots & -70/267 & \cdots & 960/89 \\ \cdots & 88/89 & \cdots & 1160/89 \\ \cdots & -310/267 & \cdots & 820{,}000/89 \end{bmatrix} \begin{array}{l} \leftarrow \text{Negative} \\ \leftarrow 1160/89 \div 88/89 = 1160/88 \approx 13.2 \leftarrow \text{pivot row} \\ {} \end{array}$$

           ↑

pivot column

Pivot on $88/89$ in Row 2, Column 3

Divide Row2 by $88/89$ or Multiply by $89/88$

Add $+70/267 \bullet$ Row2 + Row1 : Row1

Add $+310/267 \bullet$ Row2 + Row3 : Row3

**There are no more negative numbers in the last row**

*Determine and check the best possible solution*

$(x_1, x_2, x_3, x_4, x_5, x_6, s_1, s_2) = (0, 0, 13.1818, 0, 0, 14.2424, 0, 0)$ with $z = 9{,}228.7878$

*Check*:

$C_1$: $6.00x_1 + 6.25x_2 + 5.00x_3 + 5.25x_4 + 5.75x_5 + 4.50x_6 + s_1$

     $= 0 + 0 + 5(13.1818) + 0 + 0 + 4.50(14.2424) + 0 = 129.9998 \approx 130$

$C_2$: $230x_1 + 220x_2 + 190x_3 + 125x_4 + 120x_5 + 105x_6 + s_2$

     $= 0 + 0 + 190(13.1818) + 0 + 0 + 105(14.2424) + 0 = 3999.994 \approx 4000$

$z = 495x_1 + 500x_2 + 430x_3 + 245x_4 + 250x_5 + 250x_6$

     $= 0 + 0 + 430(13.1818) + 0 + 0 + 250(14.2424) = 9228.774 \approx 9228.79$

*Interpret the final solution*

Each week the craftsman should make 13.2 modern coffee tables, and 14.2 modern end tables, and nothing else. If he does that, he will use all the hours available and all the money available and make a weekly profit of approximately $9,228.79.

33.    **Step 1:** *Model the problem*

Independent variables:

      $x_1$ = number of sofas made each week

      $x_2$ = number of love seats made each week

      $x_3$ = number of easy chairs made each week

      $x_4$ = number of recliners made each week

| | **Carpentry** | **Upholstery** | **Coating** | **Profit** |
|---|---|---|---|---|
| Sofas | 4 hours | 8 hours | 20 min. = $\frac{1}{3}$ hr | $450 |
| Love Seats | 4 hours | 7.5 hours | 20 min. = $\frac{1}{3}$ hr | $350 |
| Easy Chairs | 3 hours | 6 hours | 15 min. = $\frac{1}{4}$ hr | $300 |
| Recliners | 6 hours | 6.5 hours | 15 min. = $\frac{1}{4}$ hr | $475 |
| Number of People | 9 | 15 | 1 | |
| Number of Hours Available (40 hours/week) | 360 hours | 600 hours | 40 hours | |

Constraints:

    $C_1$:  Carpentry Hours $\leq 360$

        Sofa hours + Love Seat hours + Easy Chair hours + Recliner hours $\leq 360$

        $4x_1 + 4x_2 + 3x_3 + 6x_4 \leq 360$

    $C_2$:  Upholstery hours $\leq 600$

        Sofa hours + Love Seat hours + Easy Chair hours + Recliner hours $\leq 600$

        $8x_1 + 7.5x_2 + 6x_3 + 6.5x_4 \leq 600$

    $C_3$:  Coating hours $\leq 40$

        Sofa hours + Love Seat hours + Easy Chair hours + Recliner hours $\leq 40$

        $\frac{1}{3}x_1 + \frac{1}{3}x_2 + \frac{1}{4}x_3 + \frac{1}{4}x_4 \leq 40$

Objective Function:

    $z$ = Sofa profit + Love Seat profit + Easy Chair profit + Recliner profit

    $z = 450x_1 + 350x_2 + 300x_3 + 475x_4$

**33. Continued.**

**Step 2:** *Find the constraint equations*

$C_1$: $4x_1 + 4x_2 + 3x_3 + 6x_4 + s_1 = 360$, where $s_1 =$ unused carpentry hours

$C_2$: $8x_1 + 7.5x_2 + 6x_3 + 6.5x_4 + s_2 = 600$, where $s_2 =$ unused upholstery hours

$C_3$: $\dfrac{1}{3}x_1 + \dfrac{1}{3}x_2 + \dfrac{1}{4}x_3 + \dfrac{1}{4}x_4 + s_3 = 40$, where $s_3 =$ unused coating hours

**Step 3:** *Rewrite the objective function equation*

$z = 450x_1 + 350x_2 + 300x_3 + 475x_4$ becomes

$$-450x_1 - 350x_2 - 300x_3 - 475x_4 + 0s_1 + 0s_2 + 0s_3 + 1z = 0$$

**Step 4:** *Find the first simplex matrix*

$$
\begin{array}{ccccccccc}
x_1 & x_2 & x_3 & x_4 & s_1 & s_2 & s_3 & z & \\
\end{array}
$$

$$
\left[
\begin{array}{cccccccc}
4 & 4 & 3 & 6 & 1 & 0 & 0 & 0 & 360 \\
8 & 7.5 & 6 & 6.5 & 0 & 1 & 0 & 0 & 600 \\
\tfrac{1}{3} & \tfrac{1}{3} & \tfrac{1}{4} & \tfrac{1}{4} & 0 & 0 & 1 & 0 & 40 \\
-450 & -350 & -300 & -475 & 0 & 0 & 0 & 1 & 0 \\
\end{array}
\right]
\begin{array}{l}
\leftarrow {}^{360}\!/_6 = 60 \;\; \leftarrow \text{pivot row} \\
\leftarrow {}^{600}\!/_{6.5} \approx 92.3 \\
\leftarrow {}^{40}\!/_{0.25} = 160 \\
\\
\end{array}
$$

$$\uparrow$$
$$\text{pivot column}$$

**Step 5:** *Find the possible solution*

$(x_1, x_2, x_3, x_4, s_1, s_2, s_3) = (0, 0, 0, 0, 360, 600, 40)$ with $z = 0$

**Step 6:** *Pivot to find the best possible solution*

Use the computer or calculator to find the best possible solution

Enter the matrix from above

Matrix dimensions are $4 \times 9$

Pivot on 6 in row 1, column 4

Divide Row1 by 6 or Multiply by $\frac{1}{6}$

Add $-6.5 \bullet$ Row1 + Row2 : Row2

Add $-0.25 \bullet$ Row1 + Row3 : Row3

Add $+475 \bullet$ Row1 + Row4 : Row4

33. Continued.

The most negative number in the last row is $-\frac{400}{3}$ in Column 1

column 1　...　last column

$$
\begin{bmatrix}
\frac{2}{3} & ... & 60 \\
\frac{11}{3} & ... & 210 \\
\frac{1}{6} & ... & 25 \\
-\frac{400}{3} & ... & 28,500
\end{bmatrix}
\begin{array}{l}
\leftarrow 60 \div \frac{2}{3} = 90 \\
\leftarrow 210 \div \frac{11}{3} = \frac{630}{11} \approx 57.3 \quad \leftarrow \text{pivot row} \\
\leftarrow 25 \div \frac{1}{6} = 150 \\
\phantom{x}
\end{array}
$$

　　　↑

pivot column

Pivot on $\frac{11}{3}$ in Row 2, Column 1

Divide Row2 by $\frac{11}{3}$ or Multiply by $\frac{3}{11}$

Add $-\frac{2}{3} \bullet$ Row2 + Row1:Row1

Add $-\frac{1}{6} \bullet$ Row2 + Row3:Row3

Add $+\frac{400}{3} \bullet$ Row2 + Row4 : Row4

**Last row is all positive, so we do not need to pivot again.**

**Step 7:** *Determine and check the best possible solution*

$(x_1, x_2, x_3, x_4, s_1, s_2, s_3) = (57.2727, 0, 0, 21.8181, 0, 0, 15.4545)$ with $z = \$36,136.3636$

*Check:*

$C_1$:　$4x_1 + 4x_2 + 3x_3 + 6x_4 + s_1 = 4(57.2727) + 4(0) + 3(0) + 6(21.8181) + (0)$
$$= 359.9994 \approx 360$$

$C_2$:　$8x_1 + 7.5x_2 + 6x_3 + 6.5x_4 + s_2 = 8(57.2727) + 7.5(0) + 6(0) + 6.5(21.8181) + (0)$
$$= 599.99925 \approx 600$$

$C_3$:　$\dfrac{1}{3}x_1 + \dfrac{1}{3}x_2 + \dfrac{1}{4}x_3 + \dfrac{1}{4}x_4 + s_3$

$$= \frac{1}{3}(57.2727) + \frac{1}{3}(0) + \frac{1}{4}(0) + \frac{1}{4}(21.8181) + (15.4545) = 39.99925 \approx 40$$

$z = 450x_1 + 350x_2 + 300x_3 + 475x_4 = 450(57.2727) + 350(0) + 300(0) + 475(21.8181)$
$$= 36,136.3125 \approx 36,136.36$$

**Step 8:** *Interpret the final solution*

The furniture firm should make 57.3 sofas and 21.8 recliners and nothing else. The estimated profit would be $36,136. There would be 15.5 unused protective coating hours.

| Chapter 8 | Review |

**1.    Solve the inequality for y:**

$$4x - 5y > 7 \quad \rightarrow \quad -5y > -4x + 7 \quad \rightarrow \quad y < \frac{4}{5}x - \frac{7}{5}$$

**Graph the line:** $y = \frac{4}{5}x - \frac{7}{5}$ is a dashed line (= is not part of the inequality) with slope $\frac{4}{5}$ and $y$-intercept $-\frac{7}{5}$. Start at $\left(0, -\frac{7}{5}\right)$ on the $y$-axis and from that point rise 4 (move four units up) and run 5 (move five units to the right).

$$\text{slope} = m = \frac{\text{rise}}{\text{run}} = \frac{4}{5} = \frac{\text{four units up}}{\text{five units right}}$$

**Shade in one side of the line:** The graph of the inequality is the set of all points below the line (values of $y$ decrease if we move down).

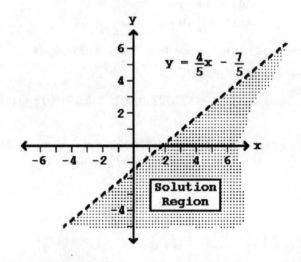

**5.**

| Original Inequality | Slope-Intercept Form | Associated Equation | Graph of the Inequality |
|---|---|---|---|
| $8x - 4y < 10$ | $-4y < -8x + 10$ $y > 2x - \frac{5}{2}$ | $y = 2x - \frac{5}{2}$ | all points above the line with slope 2 and $y$-intercept $-\frac{5}{2}$ |
| $3x + 5y \geq 7$ | $5y \geq -3x + 7$ $y \geq -\frac{3}{5}x + \frac{7}{5}$ | $y = -\frac{3}{5}x + \frac{7}{5}$ | all points on or above the line with slope $-\frac{3}{5}$ and $y$-intercept $\frac{7}{5}$ |

5. Continued.

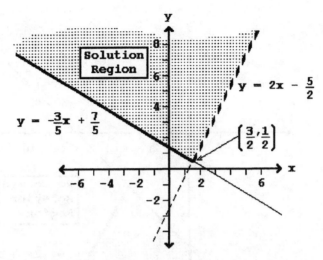

There is one corner point where the lines intersect. Use the elimination method.

$$y = 2x - \frac{5}{2} \qquad \rightarrow \qquad 10y = 20x - 25$$

$$y = -\frac{3}{5}x + \frac{7}{5} \qquad \rightarrow \qquad \underline{-10y = 6x - 14}$$

$$0 = 26x - 39$$

$$-26x = -39 \qquad \rightarrow \qquad x = \frac{-39}{-26} = \frac{3}{2}$$

$$y = 2x - \frac{5}{2} = 2\left(\frac{3}{2}\right) - \frac{5}{2} = \frac{6}{2} - \frac{5}{2} = \frac{1}{2}$$

Thus, the corner point is $\left(\frac{3}{2}, \frac{1}{2}\right)$.

9.

| Original Inequality | Slope-Intercept Form | Associated Equation | Graph of the Inequality |
|---|---|---|---|
| $x - y \geq 7$ | $-y \geq -x + 7$ $y \leq x - 7$ | $y = x - 7$ | all points on or below the line with slope 1 and $y$-intercept $-7$ |
| $5x + 3y \leq 9$ | $3y \leq -5x + 9$ $y \leq -\frac{5}{3}x + 3$ | $y = -\frac{5}{3}x + 3$ | all points on or below the line with slope $-\frac{5}{3}$ and $y$-intercept 3 |
| $x \geq 0$ | (not applicable) | $x = 0$ | all points on or to the right of the $y$-axis |
| $y \geq 0$ | $y \geq 0$ | $y = 0$ | all points on or above the $x$-axis |

9. Continued.

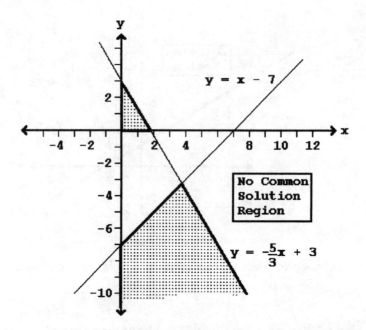

There are no corner points because there is not a common solution region for the four inequalities.

13. **Step 1:** *Model the problem*

Independent variables:

$x_1$ = number of Model 110 produced

$x_2$ = number of Model 330 produced

Constraints:

$C_1$: tweeters available $\leq 90$

| (tweeters in Model 110) | + | (tweeters in Model 330) | $\leq 90$ |
|---|---|---|---|
| (1 tweeter)(Model 110) | + | (2 tweeters)(Model 330) | $\leq 90$ |
| $1x_1$ | + | $2x_2$ | $\leq 90$ |

$C_2$: mid-range speakers available $\leq 60$

| (mid-range in Model 110) | + | (mid-range in Model 330) | $\leq 60$ |
|---|---|---|---|
| (1 mid-range)(Model 110) | + | (1 mid-range)(Model 330) | $\leq 60$ |
| $1x_1$ | + | $1x_2$ | $\leq 60$ |

$C_3$: woofers available $\leq 44$

| (woofers in Model 110) | + | (woofers in Model 330) | $\leq 44$ |
|---|---|---|---|
| (0 woofers)(Model 110) | + | (1 woofer)(Model 330) | $\leq 44$ |
| $0x_1$ | + | $1x_2$ | $\leq 44$ |

13. Continued.

Objective Function:
$$z = \text{(Model 110 income)} + \text{(Model 330 income)}$$
$$= (\$200)(\text{Model 110}) + (\$350)(\text{Model 330})$$
$$z = 200x_1 + 350x_2$$

**Step 2:** *Find the constraint equations*

$C_1$: $1x_1 + 2x_2 + s_1 = 90$  where $s_1 =$ unused tweeters
$C_2$: $1x_1 + 1x_2 + s_2 = 60$  where $s_2 =$ unused mid-range
$C_3$: $0x_1 + 1x_2 + s_3 = 44$  where $s_3 =$ unused woofers

**Step 3:** *Rewrite the objective function equation*

$$z = 200x_1 + 350x_2 \text{ becomes } -200x_1 - 350x_2 + 0s_1 + 0s_2 + 0s_3 + 1z = 0$$

**Step 4:** *Find the first simplex matrix*

$$
\begin{array}{cccccc}
x_1 & x_2 & s_1 & s_2 & s_3 & z \\
\end{array}
$$

$$
\left[
\begin{array}{cccccc|c}
1 & 2 & 1 & 0 & 0 & 0 & 90 \\
1 & 1 & 0 & 1 & 0 & 0 & 60 \\
0 & 1 & 0 & 0 & 1 & 0 & 44 \\
-200 & -350 & 0 & 0 & 0 & 1 & 0 \\
\end{array}
\right]
$$

**Step 5:** *Find the possible solution*

$$(x_1, x_2, s_1, s_2, s_3) = (0, 0, 90, 60, 44) \text{ with } z = 0$$

**Step 6:** *Pivot to find a better possible solution*

$$
\begin{array}{cccccc}
x_1 & x_2 & s_1 & s_2 & s_3 & z \\
\end{array}
$$

$$
\left[
\begin{array}{cccccc|c}
1 & 2 & 1 & 0 & 0 & 0 & 90 \\
1 & 1 & 0 & 1 & 0 & 0 & 60 \\
0 & 1 & 0 & 0 & 1 & 0 & 44 \\
-200 & -350 & 0 & 0 & 0 & 1 & 0 \\
\end{array}
\right]
\begin{array}{l}
\leftarrow 90/2 = 45 \\
\leftarrow 60/1 = 60 \\
\leftarrow 44/1 = 44 \quad \leftarrow \text{pivot row} \\
\phantom{x}
\end{array}
$$

$$\uparrow$$
pivot column

Pivot on 1 in row 3, column 2

13. Continued.

$$
\begin{array}{cc}
\begin{array}{r}
\\
-2R3 + R1 : R1 \\
-R3 + R2 : R2 \\
\\
350R3 + R4 : R4
\end{array}
&
\begin{array}{ccccccc}
x_1 & x_2 & s_1 & s_2 & s_3 & z & \\
\left[\begin{array}{cccccc|c}
1 & 0 & 1 & 0 & -2 & 0 & 2 \\
1 & 0 & 0 & 1 & -1 & 0 & 16 \\
0 & 1 & 0 & 0 & 1 & 0 & 44 \\
-200 & 0 & 0 & 0 & 350 & 1 & 15,400
\end{array}\right]
\end{array}
\end{array}
$$

$\leftarrow 2/1 = 2 \leftarrow$ pivot row
$\leftarrow 16/1 = 16$
$\leftarrow 44/0$ is undefined

↑
pivot column

**Last row is not all positive, so pivot again.**

Pivot on 1 in row 1, column 1

$$
\begin{array}{cc}
\begin{array}{r}
\\
-R1 + R2 : R2 \\
\\
200R1 + R4 : R4
\end{array}
&
\begin{array}{ccccccc}
x_1 & x_2 & s_1 & s_2 & s_3 & z & \\
\left[\begin{array}{cccccc|c}
1 & 0 & 1 & 0 & -2 & 0 & 2 \\
0 & 0 & -1 & 1 & 1 & 0 & 14 \\
0 & 1 & 0 & 0 & 1 & 0 & 44 \\
0 & 0 & 200 & 0 & -50 & 1 & 15,800
\end{array}\right]
\end{array}
\end{array}
$$

$\leftarrow 2/-2 = -1$
$\leftarrow 14/1 = 14 \leftarrow$ pivot row
$\leftarrow 44/1 = 44$

↑
pivot column

**Last row is not all positive, so pivot again.**

Pivot on 1 in row 2, column 5

$$
\begin{array}{cc}
\begin{array}{r}
2R2 + R1 : R1 \\
\\
-R2 + R3 : R3 \\
50R2 + R4 : R4
\end{array}
&
\begin{array}{ccccccc}
x_1 & x_2 & s_1 & s_2 & s_3 & z & \\
\left[\begin{array}{cccccc|c}
1 & 0 & -1 & 2 & 0 & 0 & 30 \\
0 & 0 & -1 & 1 & 1 & 0 & 14 \\
0 & 1 & 1 & -1 & 0 & 0 & 30 \\
0 & 0 & 150 & 50 & 0 & 1 & 16,500
\end{array}\right]
\end{array}
\end{array}
$$

**Last row is all positive, so we do not need to pivot again.**

**Step 7:** *Determine the possible solution*

$(x_1, x_2, s_1, s_2, s_3) = (30, 30, 0, 0, 14)$ with $z = 16,500$

*Check*:

$C_1$:  $1x_1 + 2x_2 + s_1 = 1(30) + 2(30) + 0 = 30 + 60 = 90$

$C_2$:  $1x_1 + 1x_2 + s_2 = 1(30) + 1(30) + 0 = 30 + 30 = 60$

$C_3$:  $0x_1 + 1x_2 + s_3 = 0(30) + 1(30) + 14 = 0 + 30 + 14 = 44$

$z = 200x_1 + 350x_2 = 200(30) + 350(30) = 6,000 + 10,500 = 16,500$

13. Continued.

**Step 8:** *Determine if this maximizes the objective function*

The last row contains no negative entries, so we have maximized the income.

**Step 9:** *Interpret the final solution*

Mowson Audio Co. should make 30 model 110 speaker assemblies and 30 model 330 assemblies to maximize its income. The maximum income will be $16,500. They would use all the tweeters and all the mid-range speakers, but would have 14 woofers left over.

# 9 Exponential and Logarithmic Functions

| Section 9.0A | Review of Exponentials and Logarithms |
|---|---|

1.    $v = \log_2 8$ can be rewritten as $2^v = 8$. Then since $2^v = 8 = 2^3$, $v = 3$.

5.    $\log_5 u = -2$ can be rewritten as $5^{-2} = u$. Then $u = 5^{-2} = \dfrac{1}{5^2} = \dfrac{1}{25}$.

9.    $\log_b 9 = 2$ can be rewritten as $b^2 = 9$. Then since $b^2 = 9 = 3^2$, $b = 3$ (base can only be positive).

13.    $K = \log_b H$ can be rewritten as $b^K = H$.

17.    $b^T = S$ can be rewritten as $T = \log_b S$

21.    a)    $e^{1.2} = 3.320116923$

        b)    $10^{1.2} = 15.84893192$

25.    a)    $\dfrac{1}{e^{1.5}} = e^{-1.5} = 0.2231301601$

        b)    $\dfrac{1}{10^{1.5}} = 10^{-1.5} = 0.0316227766$

29.    a)    $\dfrac{e^{4.7}}{2e^{5.1}} = \dfrac{e^{4.7} \cdot e^{-5.1}}{2} = \dfrac{e^{4.7-5.1}}{2} = \dfrac{e^{-0.4}}{2} = \dfrac{0.67032005}{2} = 0.335160023$

        b)    $\dfrac{10^{4.7}}{2(10)^{5.1}} = \dfrac{10^{4.7} \cdot 10^{-5.1}}{2} = \dfrac{10^{4.7-5.1}}{2} = \dfrac{10^{-0.4}}{2} = \dfrac{0.39810717}{2} = 0.1990535853$

33.    a)    $\ln(0.58) = -0.5447271754$

        b)    $\log(0.58) = -0.2365720064$

37.    a)    $\ln(4e^{0.02}) = \ln[(4)(1.02020134)] = \ln(4.08080536) = 1.406294361$

      b)    $\log[4(10)^{0.02}] = \log[(4)(1.04712855)] = \log(4.18851419) = 0.622059913$

41.    a)    $e^{2\ln 5} = e^{2(1.60943791)} = e^{3.21887583} = 25$

      b)    $10^{2\log 5} = 10^{2(0.69897000)} = 10^{1.39794001} = 25$

45.    a)    $\ln(10^{2.47}) = \ln(295.120923) = 5.68738518$

      b)    $\log(10^{2.47}) = \log(295.120923) = 2.47$

---

## Section 9.0B                  Review of Properties of Logarithms

1.    $\log(10^{7.5x}) = 7.5x$

5.    $10^{\log(3x+1)} = 3x + 1$

9.    $\log\left(\dfrac{x}{5}\right) = \log x - \log 5$           Division-Becomes-Subtraction

13.    $\log(1.0625^{x}) = x\log(1.0625)$         Exponent-Becomes-Multiplier

17.    $\ln\left(\dfrac{2x}{3}\right) = \ln(2x) - \ln 3$         Division-Becomes-Subtraction

             $= \ln 2 + \ln x - \ln 3$        Multiplication-Becomes-Addition

21.    $\ln(5x) + \ln 2 = \ln[(5x)(2)]$       Reverse Multiplication-Becomes-Addition

             $= \ln(10x)$              Simplify

25.    $2\ln(3x) - \ln 9 = \ln(3x)^2 - \ln 9$        Exponent-Becomes-Multiplier

             $= \ln\dfrac{(3x)^2}{9}$          Division-Becomes-Subtraction

             $= \ln\left(\dfrac{9x^2}{9}\right) = \ln(x^2)$        Simplify

29.    a)        $e^x = 0.25$

         $\ln(e^x) = \ln(0.25)$              Taking ln of each side

              $x = \ln(0.25)$                Inverse Property

              $x = -1.386294361$        Calculator

       *Check*: $e^x = e^{-1.386294361} = 0.25$

    b)         $10^x = 0.25$

       $\log(10^x) = \log(0.25)$         Taking log of each side

            $x = \log(0.25)$            Inverse Property

            $x = -0.602059991$       Calculator

       *Check*: $10^x = 10^{-0.602059991} = 0.25$

33.    a)      $1000\, e^{0.009x} = 1500$

           $e^{0.009x} = \dfrac{1500}{1000} = 1.5$          Isolate the exponential

       $\ln\left(e^{0.009x}\right) = \ln(1.5)$         Taking ln of each side

          $0.009x = \ln(1.5)$           Inverse Property

             $x = \dfrac{\ln(1.5)}{0.009}$

             $x = \dfrac{0.40546511}{0.009} = 45.05167868$      Calculator

      *Check*: $1000\, e^{0.009x} = 1000\, e^{0.009\,(45.05167868)}$

                   $= 1000\, e^{0.40546511} = 1000\,(1.5) = 1500$

    b)     $1000\,(10)^{0.009x} = 1500$

           $10^{0.009x} = \dfrac{1500}{1000} = 1.5$          Isolate the exponential

     $\log\left(10^{0.009x}\right) = \log(1.5)$         Taking log of each side

         $0.009x = \log(1.5)$           Inverse Property

            $x = \dfrac{\log(1.5)}{0.009}$

            $x = \dfrac{0.17609126}{0.009} = 19.56569545$      Calculator

      *Check*: $1000\,(10)^{0.009x} = 1000\,(10)^{0.009\,(19.56569545)}$

                   $= 1000\,(10)^{0.17609126} = 1000\,(1.5) = 1500$

37.  a)    $50 \, e^{-0.0016x} = 40$

$$e^{-0.0016x} = \frac{40}{50} = 0.8 \qquad\qquad\qquad \text{Isolate the exponential}$$

$$\ln\left(e^{-0.0016x}\right) = \ln(0.8) \qquad\qquad\qquad \text{Taking ln of each side}$$

$$-0.0016x = \ln(0.8) \qquad\qquad\qquad \text{Inverse Property}$$

$$x = \frac{\ln(0.8)}{-0.0016}$$

$$x = \frac{-0.22314355}{-0.0016} = 139.4647196 \qquad\qquad \text{Calculator}$$

*Check*:  $50 \, e^{-0.0016x} = 50 \, e^{-0.0016 \, (139.4647196)}$

$$= 50 \, e^{-0.22314355} = 50 \, (0.8) = 40$$

b)    $50 \, (10)^{-0.0016x} = 40$

$$10^{-0.0016x} = \frac{40}{50} = 0.8 \qquad\qquad\qquad \text{Isolate the exponential}$$

$$\log(10^{-0.0016x}) = \log(0.8) \qquad\qquad\qquad \text{Taking log of each side}$$

$$-0.0016x = \log(0.8) \qquad\qquad\qquad \text{Inverse Property}$$

$$x = \frac{\log(0.8)}{-0.0016}$$

$$x = \frac{-0.09691001}{-0.0016} = 60.56875813 \qquad\qquad \text{Calculator}$$

*Check*:  $50 \, (10)^{-0.0016x} = 50 \, (10)^{-0.0016 \, (60.56875813)}$

$$= 50 \, (10)^{-0.09691001} = 50 \, (0.8) = 40$$

41.  a)    $\log x = 1.85$

     $10^{\log x} = 10^{1.85}$             Exponentiate each side

        $x = 10^{1.85}$               Inverse Property

        $x = 70.79457844$       Calculator

*Check*:  $\log x = \log(70.79457844) = 1.85$

b)    $\ln x = 1.85$

     $e^{\ln x} = e^{1.85}$             Exponentiate each side

        $x = e^{1.85}$              Inverse Property

        $x = 6.359819523$       Calculator

*Check*:  $\ln x = \ln(6.359819523) = 1.85$

45.   a)                              $\log x = 1.8 + \log (3.6)$

$\log x - \log (3.6) = 1.8$                              Collect log terms

$\log \left( \dfrac{x}{3.6} \right) = 1.8$                              Division-Becomes-Subtraction

$10^{\log(x/3.6)} = 10^{1.8}$                              Exponentiate each side

$\dfrac{x}{3.6} = 10^{1.8}$                              Inverse Property

$x = 3.6 \, (10)^{1.8}$

$x = 3.6 \, (63.09573445)$              Calculator

$x = 227.144644$

*Check*:                              $\log x = 1.8 + \log (3.6)$
$\log (227.144644) = 1.8 + \log (3.6)$
$2.3563025 = 1.8 + 0.5563025$
$2.3563025 = 2.3563025$

b)                              $\ln x = 1.8 - \ln (3.6)$

$\ln x + \ln (3.6) = 1.8$                              Collect ln terms

$\ln [(x)(3.6)] = 1.8$                              Multiplication-Becomes-Addition

$e^{\ln (3.6x)} = e^{1.8}$                              Exponentiate each side

$3.6x = e^{1.8}$                              Inverse Property

$x = \dfrac{e^{1.8}}{3.6}$

$x = \dfrac{6.04964746}{3.6} = 1.680457629$              Calculator

*Check*:                              $\ln x = 1.8 - \ln (3.6)$
$\ln (1.680457629) = 1.8 - \ln (3.6)$
$0.51906615 = 1.8 - 1.28093385$
$0.51906615 = 0.51906615$

49.    Show that $\ln\left(\frac{A}{B}\right) = \ln A - \ln B$

Let $u = \ln A$ which can be rewritten as $e^u = A$
Let $v = \ln B$ which can be rewritten as $e^v = B$

$$\ln\left(\frac{A}{B}\right) = \ln\left(\frac{e^u}{e^v}\right) \qquad \text{Substitution}$$

$$= \ln\left(e^{u-v}\right) \qquad \text{Exponent law}$$

$$= u - v \qquad \text{Inverse property}$$

$$= \ln A - \ln B \qquad \text{Substitution}$$

53.    $[H^+] = 3.0 \times 10^{-4}$ moles per liter

$\text{pH} = -\log[H^+]$       pH formula
$\quad = -\log(3.0 \times 10^{-4})$       substituting $3.0 \times 10^{-4}$ for $[H^+]$
$\quad = 3.522878745$       using a calculator
$\quad = 3.5$       rounding to one decimal place

The pH is less than 7, so the substance is classified as an acid.

57.    $[H^+] = 1.3 \times 10^{-5}$ moles per liter

$\text{pH} = -\log[H^+]$       pH formula
$\quad = -\log(1.3 \times 10^{-5})$       substituting $1.3 \times 10^{-5}$ for $[H^+]$
$\quad = 4.886056648$       using a calculator
$\quad = 4.9$       rounding to one decimal place

61.    a)    $[H^+] = 3.5 \times 10^{-7}$ moles per liter

$\text{pH} = -\log[H^+]$       pH formula
$\quad = -\log(3.5 \times 10^{-7})$       substituting $3.0 \times 10^{-7}$ for $[H^+]$
$\quad = 6.455931955$       using a calculator
$\quad = 6.5$       rounding to one decimal place

The pH is between 5.5 and 7.5, so you could plant tomatoes.

b)    $[H^+] = 3.5 \times 10^{-4}$ moles per liter

$\text{pH} = -\log[H^+]$       pH formula
$\quad = -\log(3.5 \times 10^{-4})$       substituting $3.5 \times 10^{-4}$ for $[H^+]$
$\quad = 3.455931956$       using a calculator
$\quad = 3.5$       rounding to one decimal place

The pH is less than 5.5, so you should probably not plant tomatoes.

65.    From Section 6.8, Exercise 9,  $s = 2, n = 3$

$$s^d = n$$

$$2^d = 3$$

$$\log 2^d = \log 3$$

$$d \log 2 = \log 3$$

$$d = \frac{\log 3}{\log 2} = 1.5849625$$

*Check*:  $2^{1.5849625} = 3$

69.    From Section 6.8, Exercise 13,  $s = 4, n = 8$

$$s^d = n$$

$$4^d = 8$$

$$\log 4^d = \log 8$$

$$d \log 4 = \log 8$$

$$d = \frac{\log 8}{\log 4} = 1.5$$

*Check*:  $4^{1.5} = 8$

---

| **Section 9.1** | **Exponential Growth** |
| --- | --- |

1.     $p = 30\, e^{0.0198026273t}, t = 2007 - 2003 = 4$ years
       $p = 30\, e^{0.0198026273\,(4)} = 32.47296480 \approx 32.473$ thousand or 32,473
       The population of Anytown is predicted to be 32,473 in 2007.

5.     a)    In 1990, $t = 0$ and $p = 8240$ thousand so ordered pair is (0, 8240).
              In 2000, $t = 10$ and $p = 9258$ thousand so ordered pair is (10, 9258).

       b)    $\Delta t = 10 - 0 = 10$ years

       c)    $\Delta p = 9258 - 8240 = 1018$ thousand

       d)    $\dfrac{\Delta p}{\Delta t} = \dfrac{1018}{10} = 101.8$ thousand people per year

       e)    $\dfrac{\Delta p / \Delta t}{p} = \dfrac{101.8}{8240} = 0.012354368 \approx 1.23\%$ per year

9.    a)    In 1990, time 0, $(t, p) = (0, 8240)$
In 2000, $(t, p) = (10, 9258)$

Substitute the first ordered pair $(0, 8240)$

$$p = a\, e^{\,bt}$$
$$8240 = a\, e^{\,b(0)}$$
$$8240 = a\, e^{\,0}$$
$$8240 = a\,(1)$$
$$a = 8240 \quad \text{(the initial value of } p\text{)}$$

Rewrite model to $p = 8240\, e^{\,bt}$

Substitute the second ordered pair $(10, 9258)$

$$9258 = 8240\, e^{\,b(10)}$$

$$9258 = 8240\, e^{\,10b}$$

$$\frac{9258}{8240} = e^{\,10b}$$

$$1.1235437 = e^{\,10b}$$

$$\ln(1.1235437) = \ln\left(e^{\,10b}\right)$$

$$10\, b = \ln(1.1235437)$$

$$b = \frac{\ln(1.1235437)}{10} = \frac{0.1164877}{10} = 0.01164877$$

The model is $p = 8240\, e^{\,0.01164877t}$, $t$ in years.

b)    In 2005, $t = 2005 - 1990 = 15$ years

$$p = 8240\, e^{\,0.01164877\,(15)} = 9813.2341 \approx 9813 \text{ thousand}$$

The predicted population for 2005 is 9813 thousand people.

c)    In 2010, $t = 2010 - 1990 = 20$ years

$$p = 8240\, e^{\,0.01164877\,(20)} = 10{,}401.767 \approx 10{,}402 \text{ thousand}$$

The predicted population for 2010 is 10,402 thousand people.

9. Continued.

     d)     Double the 1990 population = 2(8240)

$$2(8240) = 8240 \, e^{\,0.01164877t}$$

$$\frac{2(8240)}{8240} = e^{\,0.01164877t}$$

$$2 = e^{\,0.01164877t}$$

$$\ln 2 = \ln(e^{\,0.01164877t})$$

$$0.01164877t = \ln 2$$

$$t = \frac{\ln 2}{0.01164877} = 59.5039 \approx 60 \text{ years}$$

In 2050 AD (1990 + 60 years) the population will have doubled.

13.     a)     Time 0, $(t, p) = (0, 2510)$
               Three days later, $(t, p) = (3, 5380)$

          Substitute the first ordered pair (0, 2510)

$$p = a \, e^{\,bt}$$

$$2510 = a \, e^{\,b(0)}$$

$$2510 = a \, e^{\,0}$$

$$2510 = a\,(1)$$

$$a = 2510 \qquad \text{(the initial value of } p)$$

Rewrite model to $p = 2510 \, a^{\,bt}$

Substitute the second ordered pair (3, 5380)

$$5380 = 2510 \, e^{\,b(3)}$$

$$5380 = 2510 \, e^{\,3b}$$

$$\frac{5380}{2510} = e^{\,3b}$$

$$\ln\left(\frac{5380}{2510}\right) = \ln(e^{\,3b})$$

$$3b = \ln\left(\frac{5380}{2510}\right)$$

$$b = 0.2541352$$

The model is $p = 2510 \, e^{\,0.2541352t}$, $t$ in days.

13. Continued.

    b)    In one week, $t = 7$ days

$$p = 2510\ e^{\,0.2541352(7)} = 14{,}868.27 \approx 14{,}870$$

In one week the predicted population is 14,870 fruit flies (rounded to the nearest 10).

    c)    Double the population = $2(2510)$

$$2(2510) = 2510\ e^{\,0.2541352t}$$

$$\frac{2(2510)}{2510} = e^{\,0.2541352t}$$

$$2 = e^{\,0.2541352t}$$

$$\ln 2 = \ln\left(e^{\,0.2541352t}\right)$$

$$0.2541352t = \ln 2$$

$$t = \frac{\ln 2}{0.2541352} = 2.72747 \approx 2.7 \text{ days}$$

The population will double in approximately 2.7 days.

17.    a)    In 1995, time 0, $(t, A) = (0, 90.9)$, $A$ in quadrillion Btu's
In 1999, $(t, A) = (4, 96.9)$, $A$ in quadrillion Btu's

Substitute the first ordered pair $(0, 90.9)$

$$A = a\,e^{\,bt}$$

Since $a$ is the initial value of $A$, $a = 90.9$.

Rewrite model to $A = 90.9\ e^{\,bt}$

Substitute the second ordered pair $(4, 96.9)$

$$96.9 = 90.9\ e^{\,b(4)}$$

$$96.9 = 90.9\ e^{\,4b}$$

$$\frac{96.9}{90.9} = e^{\,4b}$$

$$\ln\left(\frac{96.9}{90.9}\right) = \ln\left(e^{\,4b}\right)$$

$$4b = \ln\left(\frac{96.9}{90.9}\right) = \ln 1.0660066 = 0.0639195$$

$$b = \frac{0.0639195}{4} = 0.0159798$$

The model is $A = 90.9\ e^{\,0.0159798t}$, $t$ in years.

17. Continued.

    b)    In 2000, $t = 2000 - 1995 = 5$ years

$$A = 90.9\, e^{\,0.0159798\,(5)} = 98.46089 \approx 98.5$$

In 2000, the predicted energy consumed was 98.5 quadrillion Btu's.

    c)    Omitted.

21.    a)    In 1790, time 0, $(t, p) = (0, 3{,}929{,}214)$
                In 1800, $(t, p) = (10, 5{,}308{,}483)$
                Substitute the first ordered pair $(0, 3{,}929{,}214)$

$$p = a\, e^{\,bt}$$

Since $a$ is the initial value of $p$, $a = 3{,}929{,}214$.

Rewrite model to $p = 3{,}929{,}214\, e^{\,bt}$

Substitute the second ordered pair $(10, 5{,}308{,}483)$

$$5{,}308{,}483 = 3{,}929{,}214\, e^{\,b(10)}$$

$$5{,}308{,}483 = 3{,}929{,}214\, e^{\,10b}$$

$$\frac{5{,}308{,}483}{3{,}929{,}214} = e^{\,10b}$$

$$\ln\left(\frac{5{,}308{,}483}{3{,}929{,}214}\right) = e^{\,10b}$$

$$10b = \ln\left(\frac{5{,}308{,}483}{3{,}929{,}214}\right) = \ln 1.3510292 = 0.3008667$$

$$b = 0.3008667$$

The model is $p = 3{,}929{,}214\, e^{\,0.3008667t}$, $t$ in years.

    b)    The 1810 prediction would be more accurate.

    c)    In 1810, $t = 1810 - 1790 = 20$ years

$$p = 3{,}929{,}214\, e^{\,0.3008667\,(20)} = 7{,}171{,}915.747 \approx 7{,}171{,}916$$

In 1810, the predicted population was 7,171,916 people.

In 2000, $t = 2000 - 1790 = 210$ years

$$p = 3{,}929{,}214\, e^{\,0.3008667\,(210)} = 2.179041 \times 10^{\,9}$$

In 2000, the predicted population was approximately 2,179,041,000 people.

25.    In 1980, time 0, $(t, p) = (0, 4.478 \text{ billion})$
       In 1991, $(t, p) = (11, 5.423 \text{ billion})$

The model is $p = 4.478 \, e^{\, 0.0174066t}$

Double the population $= 2(4.478)$

$$2(4.478) = 4.478 \, e^{\, 0.0174066t}$$

$$2 = e^{\, 0.0174066t}$$

$$\ln 2 = \ln (e^{\, 0.0174066t})$$

$$0.0174066t = \ln 2$$

$$t = \frac{\ln 2}{0.0174066} = 39.8209 \approx 40 \text{ years}$$

The article claims the world population will double in 39 years. We calculate it will double in approximately 40 years which is a difference of one year or an error of $\frac{1}{39} = 0.025641 \approx 2.6\%$.

29.    From Exercise 7:  In 1990, time 0, $(t, p) = (0, 19{,}342)$
                         In 2000, $(t, p) = (10, 21{,}200)$

Substitute $FV = 21{,}200$, $P = 19{,}342$, $n = 10$ in

$$FV = P(1 + i)^n \qquad\qquad\qquad \text{Solve for } (1 + i)$$

$$21{,}200 = 19{,}342(1 + i)^{10}$$

$$\frac{21{,}200}{19{,}342} = (1 + i)^{10} \qquad\qquad \text{Dividing by 19,342}$$

$$\ln \left( \frac{21{,}200}{19{,}342} \right) = \ln (1 + i)^{10} \qquad\qquad \text{Taking ln of each side}$$

$$\ln \left( \frac{21{,}200}{19{,}342} \right) = 10 \ln (1 + i) \qquad\qquad \text{Exponent-Becomes-Multiplier}$$

$$\frac{1}{10} \left[ \ln \left( \frac{21{,}200}{19{,}342} \right) \right] = \ln (1 + i) \qquad\qquad \text{Dividing by 10}$$

$$0.00917222 = \ln (1 + i)$$

$$e^{\, 0.00917222} = 1 + i \qquad\qquad \text{Apply the definition}$$

$$1.0092144 = 1 + i$$

$$FV = 19{,}342 \, (1.0092144)^n$$

33.   From Exercise 9, $p = 8240 \, e^{\,0.0116487t} = 8240 \, (e^{\,0.0116487})^{\,t}$
      From Exercise 31, $FV = 8240 \, (1.0117169)^{\,n}$

      Since $a = P$, $e^{\,0.0116487} = 1.0117168$, and $n = t$, the two models are the same.

37.   From Section 5.3, Exercise 47a

$$FV = \$500{,}000, \; \text{pymt} = \$200 \text{ monthly}, \; i = \frac{5\%}{12} = \frac{0.05}{12}$$

$$FV = \text{pymt} \, \frac{(1+i)^{\,n} - 1}{i} \qquad \text{Calculate } n.$$

$$500{,}000 = 200 \, \frac{\left(1 + \dfrac{0.05}{12}\right)^{\,n} - 1}{\dfrac{0.05}{12}}$$

$$2500 \left(\frac{0.05}{12}\right) = \left(1 + \frac{0.05}{12}\right)^{\,n} - 1 \qquad \text{divide by 200 and multiply by denominator}$$

$$2500 \left(\frac{0.05}{12}\right) + 1 = \left(1 + \frac{0.05}{12}\right)^{\,n}$$

$$\ln\left[2500 \left(\frac{0.05}{12}\right) + 1\right] = n \ln\left(1 + \frac{0.05}{12}\right)$$

$$n = \frac{\ln\left[2500 \left(\dfrac{0.05}{12}\right) + 1\right]}{\ln\left(1 + \dfrac{0.05}{12}\right)}$$

$$= \frac{\ln(11.416667)}{\ln(1.0041667)} = \frac{2.4350743}{0.004158} = 586 \text{ months or 48 years, 10 months.}$$

---

## Section 9.2                                   Exponential Decay

1.   $Q = 20 \, e^{\,-0.086643397t}$, $t = 3$ weeks $= 21$ days

     $Q = 20 \, e^{\,-0.086643397 \,(21)} = 3.24209893 \approx 3.2$

     After 3 weeks approximately 3.2 grams of iodine-131 are left.

5.   Half-life of silicon-31 is 2.6 hours

a)   Ordered pairs $(t, Q)$ are $(0, a)$ and $(2.6, \frac{a}{2})$

Omit substituting the first pair since we are not using a specific value.

Model is $Q = a\,e^{bt}$

Substitute the second ordered pair $(2.6, \frac{a}{2})$

$$\frac{a}{2} = a\,e^{b(2.6)}$$

$$\frac{1}{2} = e^{2.6b} \qquad\qquad \text{Dividing by } a$$

$$0.5 = e^{2.6b}$$

$$\ln(0.5) = \ln(e^{2.6b})$$

$$2.6b = \ln(0.5)$$

$$b = \frac{\ln(0.5)}{2.6} = -0.266595069$$

The model is $Q = 50\,e^{-0.266595069t}$, $t$ in hours.

b)   In one hour, $t = 1$, $a = 50$ milligrams

$$Q = 50\,e^{-0.266595069\,(1)} = 38.299159$$

In one hour the predicted amount of silicon-31 is 38.3 milligrams.

c)   In one day, $t = 24$ hours, $a = 50$ milligrams

$$Q = 50\,e^{-0.266595069\,(24)} = 0.0832207$$

In one day the predicted amount of silicon-31 is 0.08 milligrams.

d)   Ordered pairs at one hour:  $(0, 50)$ and $(1, 38.299159)$

$$\frac{\Delta Q}{\Delta t} = \frac{38.299159 - 50}{1 - 0} = \frac{-11.70084}{1} = -11.70084 \approx -11.7 \text{ milligrams per hour}$$

e)   $$\frac{\Delta Q/\Delta t}{Q} = \frac{-11.70084}{50} = -0.2340168 \approx -23.4\% \text{ per hour}$$

f)   Ordered pairs at one day:  $(0, 50)$ and $(24, 0.0832207)$

$$\frac{\Delta Q}{\Delta t} = \frac{0.0832207 - 50}{24 - 0} = \frac{-49.916779}{24} = -2.079866 \approx -2.1 \text{ milligrams per hour}$$

5. Continued.

g) $\dfrac{\Delta Q/\Delta t}{Q} = \dfrac{-2.079866}{50} = -0.0415973 \approx -4.2\%$ per hour

h) Radioactive substances decay faster when there is more substance present. A larger quantity is lost during the first part of the day than during the last part of the day and when the relative changes are computed the relative decay rate will be greater for shorter periods of time.

9. Half-life of plutonium-241 is 13 years

Ordered pairs $(t, Q)$ are $(0, a)$ and $(13, \frac{a}{2})$

Omit substituting the first pair since we are not using a specific value.

Model is $Q = a\,e^{\,bt}$

Substitute second ordered pair $(13, \frac{a}{2})$

$$\frac{a}{2} = a\,e^{\,b(13)}$$

$$\frac{1}{2} = e^{\,13b} \qquad\qquad \text{dividing by } a$$

$$0.5 = e^{\,13b}$$

$$\ln(0.5) = \ln(e^{\,13b})$$

$$13b = \ln(0.5)$$

$$b = \frac{\ln(0.5)}{13} = -0.053319014$$

The model is $Q = a\,e^{-0.053319014t}$, $t$ in years.

$Q = 100$ grams, $a = 500$ grams, find $t$

$$100 = 500\,e^{-0.053319014t}$$

$$\frac{100}{500} = e^{-0.053319014t}$$

$$0.2 = e^{-0.053319014t}$$

$$\ln(0.2) = \ln(e^{-0.053319014t})$$

$$-0.053319014t = \ln(0.2)$$

$$t = \frac{\ln(0.2)}{-0.053319014} = 30.185065 \approx 30.2 \text{ years}$$

It would take 30.2 years for plutonium-241 to decay from 500 grams to 100 grams.

13.   Half-life of plutonium-239 is 24,400 years

Ordered pairs $(t, Q)$ are $(0, a)$ and $(24,400, \frac{a}{2})$

Omit substituting the first pair since we are not using a specific value.

   Model is $Q = a e^{bt}$

Substitute second ordered pair $(24,400, \frac{a}{2})$

$$\frac{a}{2} = a e^{b(24,400)}$$

$$\frac{1}{2} = e^{24,400b} \qquad \text{dividing by } a$$

$$0.5 = e^{24,400b}$$

$$\ln(0.5) = \ln(e^{24,400b})$$

$$24,400b = \ln(0.5)$$

$$b = \frac{\ln(0.5)}{24,400} = -0.00002840767$$

The model is $Q = a e^{-0.00002840767t}$, $t$ in years.

If 90% of its radioactivity is lost, 10% remains.

$Q = 0.10$, $a = 1.00$, find $t$

$$0.1 = 1.0\, e^{-0.00002840767t}$$

$$0.1 = e^{-0.00002840767t}$$

$$\ln(0.1) = \ln(e^{-0.00002840767t})$$

$$-0.00002840767t = \ln(0.1)$$

$$t = \frac{\ln(0.1)}{-0.00002840767} = 81,055.0455 \approx 81,055 \text{ years}$$

It would take 81,055 years for plutonium-239 to lose 90% of its radioactivity.

17.   The radiocarbon dating model is $Q = a e^{-0.000120968t}$, $t$ in years. $t = 5250$

$$Q = a e^{-0.000120968\,(5250)} = a\,(0.529892) \approx 0.53a$$

You would expect to find 53% of the original amount of carbon-14.

21.   The radiocarbon dating model is $Q = a\,e^{-0.000120968t}$, t in years.

There is 84% of the expected carbon-14 remaining, $Q = 0.84a$.

$$0.84a = a\,e^{-0.000120968t}$$

$$0.84 = e^{-0.000120968t}$$

$$\ln(0.84) = \ln(e^{-0.000120968t})$$

$$-0.000120968t = \ln(0.84)$$

$$t = \frac{\ln(0.84)}{-0.000120968} = 1441.3183 \approx 1441 \text{ years}$$

The age of the roof material and therefore the age of the Mayan codex would be approximately 1441 years.

25.   The radiocarbon dating model is $Q = a\,e^{-0.000120968t}$, $t$ in years.

There is 70% of the expected carbon-14 remaining, $Q = 0.70a$.

$$0.70a = a\,e^{-0.000120968t}$$

$$0.70 = e^{-0.000120968t}$$

$$\ln(0.70) = \ln(e^{-0.000120968t})$$

$$-0.000120968t = \ln(0.70)$$

$$t = \frac{\ln(0.70)}{-0.000120968} = 2948.5066 \approx 2949 \text{ years}$$

The age of the parchment would be approximately 2949 years.

29.   The radiocarbon dating model is $Q = a\,e^{-0.000120968t}$, $t$ in years.   $t = 5730$

$$Q = a\,e^{-0.000120968\,(5730)} = a(0.5000) = 0.5a$$

You would expect to find 50% of the original amount of carbon-14.  (5730 is the half-life of carbon-14.)

33.  The radiocarbon dating model is $Q = a\,e^{-0.000120968t}$, $t$ in years.

If 63.5% of the carbon-14 is lost, then $100\% - 63.5\% = 36.5\%$ of the expected carbon-14 is remaining, $Q = 0.365a$.

$$0.365a = a\,e^{-0.000120968t}$$

$$0.365 = e^{-0.000120968t}$$

$$\ln(0.365) = \ln(e^{-0.000120968t})$$

$$-0.000120968t = \ln(0.365)$$

$$t = \frac{\ln(0.365)}{-0.000120968} = 8331.6077 \approx 8332 \text{ years}$$

In the 1980s, the age of the flute would have been approximately 8332 years.

| Section 9.3 | Logarithmic Scales |
|---|---|

1.  $A = 3.9 \times 10^4 \ \mu\text{m}$ at 100 km from the epicenter, find $M$.

$$M = \log A - \log A_0$$

$$= \log(3.9 \times 10^4) - (-3.0) \qquad \text{From Figure 9.25}$$

$$= 4.591064607 + 3.0$$

$$= 7.59106 \approx 7.6$$

The magnitude of the earthquake was 7.6 on the Richter scale.

5. $M_1 = 8.3 =$ San Francisco (1906), $M_2 = 7.1 =$ San Francisco (1989)

a) Use the magnitude comparison formula

$$M_1 - M_2 = \log\left(\frac{A_1}{A_2}\right)$$

$$8.3 - 7.1 = \log\left(\frac{A_1}{A_2}\right)$$

$$1.2 = \log\left(\frac{A_1}{A_2}\right)$$

$$10^{1.2} = 10^{\log(A_1/A_2)}$$

$$10^{1.2} = \frac{A_1}{A_2}$$

$$A_1 = 10^{1.2} A_2 = 15.84893\, A_2 \approx 16\, A_2$$

The 1906 earthquake caused about 16 times as much earth movement as the 1989 earthquake.

b) Use the energy formula

$$\log E \approx 11.8 + 1.45M$$

**For 1906**

$$\log E_1 \approx 11.8 + 1.45\,(8.3) = 23.835$$
$$10^{\log E_1} \approx 10^{23.835}$$
$$E_1 \approx 10^{23.835}$$

**For 1989**

$$\log E_2 \approx 11.8 + 1.45\,(7.1) = 22.095$$
$$10^{\log E_2} \approx 10^{22.095}$$
$$E_2 \approx 10^{22.095}$$

Comparing energies: $\dfrac{E_1}{E_2} \approx \dfrac{10^{23.835}}{10^{22.095}} = 10^{1.74} = 54.954087$

$$E_1 \approx 55\, E_2$$

The 1906 earthquake released 55 times as much energy as the 1989 earthquake.

9.    $M_1 = 7.8 =$ Turkey,  $M_2 = 7.7 =$ Iran

   a)    Use the magnitude comparison formula

$$M_1 - M_2 = \log\left(\frac{A_1}{A_2}\right)$$

$$7.8 - 7.7 = \log\left(\frac{A_1}{A_2}\right)$$

$$0.1 = \log\left(\frac{A_1}{A_2}\right)$$

$$10^{0.1} = 10^{\log(A_1/A_2)} = \frac{A_1}{A_2}$$

$$A_1 = 10^{0.1} A_2 = 1.258925\, A_2 \approx 1.3\, A_2$$

The earthquake in Turkey caused about 1.3 times as much earth movement as the earthquake in Iran.

   b)    Use the energy formula

$$\log E \approx 11.8 + 1.45M$$

**For Turkey**

$$\log E_1 \approx 11.8 + 1.45\,(7.8) = 23.110$$

$$10^{\log E_1} \approx 10^{23.110}$$

$$E_1 \approx 10^{23.110}$$

**For Iran**

$$\log E_2 \approx 11.8 + 1.45\,(7.7) = 22.965$$

$$10^{\log E_2} \approx 10^{22.965}$$

$$E_2 \approx 10^{22.965}$$

**Comparing energies:**  $\dfrac{E_1}{E_2} \approx \dfrac{10^{23.110}}{10^{22.965}} = 10^{0.145} = 1.396368$

$$E_1 \approx 1.4\, E_2$$

The earthquake in Turkey released 1.4 times as much energy as the earthquake in Iran.

13.   $M_1 = 7.1$ = final estimate,  $M_2 = 7.0$ = original estimate

a)   Use the magnitude comparison formula

$$M_1 - M_2 = \log\left(\frac{A_1}{A_2}\right)$$

$$7.1 - 7.0 = \log\left(\frac{A_1}{A_2}\right)$$

$$0.1 = \log\left(\frac{A_1}{A_2}\right)$$

$$10^{0.1} = 10^{\log(A_1/A_2)} = \frac{A_1}{A_2}$$

$$A_1 = 10^{0.1} A_2 = 1.258925\, A_2 \approx 1.26\, A_2$$

The change in magnitude readings from 7.0 to 7.1 corresponds to a 26% increase in earth movement.

b)   Use the energy formula

$$\log E \approx 11.8 + 1.45M$$

**For final estimate**

$$\log E_2 \approx 11.8 + 1.45\,(7.1) = 22.095$$
$$10^{\log E_2} \approx 10^{22.095}$$
$$E_2 \approx 10^{22.095}$$

**For original estimate**

$$\log E_2 \approx 11.8 + 1.45\,(7.0) = 21.950$$
$$10^{\log E_2} \approx 10^{21.950}$$
$$E_2 \approx 10^{21.950}$$

**Comparing energies:**   $\dfrac{E_1}{E_2} \approx \dfrac{10^{22.095}}{10^{21.950}} = 10^{0.145} = 1.396368$

$$E_1 \approx 1.396368\, E_2 \approx 1.40\, E_2$$

The change in magnitude from 7.0 to 7.1 increased the energy release by 40%.

17.   Use the decibel rating definition with $I_0 \approx 10^{-16}$ watts/cm$^2$.
$I = 10^{-9}$ watts/cm$^2$

$$D = 10 \log \left( \frac{I}{I_0} \right)$$

$$= 10 \log \left( \frac{10^{-9}}{10^{-16}} \right) = 10 \log \left( 10^{-9-(-16)} \right) = 10 \log \left( 10^7 \right) = 10 \, (7) = 70$$

The decibel rating of the television is 70 dB.

21.   Use the decibel gain formula
$I_1 = 10^{-13}$ watts/cm$^2$, $I_2 = 10^{-14}$ watts/cm$^2$

$$D_1 - D_2 = 10 \log \left( \frac{I_1}{I_2} \right)$$

$$= 10 \log \left( \frac{10^{-13}}{10^{-14}} \right) = 10 \log \left( 10^{-13-(-14)} \right) = 10 \log \left( 10^1 \right) = 10 \, (1) = 10$$

The decibel gain is 10 dB.

25.   $I_2$ = sound intensity of single singer
$I_1$ = sound intensity with additional singers
$D_1 = 81$ dB, $D_2 = 74$ dB

$$D_1 - D_2 = 10 \log \left( \frac{I_1}{I_2} \right)$$

$$81 - 74 = 10 \log \left( \frac{I_1}{I_2} \right)$$

$$7 = 10 \log \left( \frac{I_1}{I_2} \right)$$

$$0.7 = \log \left( \frac{I_1}{I_2} \right)$$

$$10^{0.7} = 10^{\log(I_1/I_2)}$$

$$\frac{I_1}{I_2} = 10^{0.7} = 5.0118723 \approx 5 \text{ total singers}$$

Four singers have joined the original singer for a total of 5 singers.

29.    $I_2$ = sound intensity of single trumpet

       $I_1$ = sound intensity with additional trumpets

       $D_1 = 85.8$ dB, $D_2 = 78$ dB

$$D_1 - D_2 = 10 \log \left( \frac{I_1}{I_2} \right)$$

$$85.8 - 78 = 10 \log \left( \frac{I_1}{I_2} \right)$$

$$7.8 = 10 \log \left( \frac{I_1}{I_2} \right)$$

$$0.78 = \log \left( \frac{I_1}{I_2} \right)$$

$$10^{0.78} = 10^{\log (I_1 / I_2)}$$

$$\frac{I_1}{I_2} = 10^{0.78} = 6.02559586 \approx 6 \text{ trumpets}$$

There are five additional trumpets for a total of 6 trumpets.

33.    $D_1 = 105$ dB = original volume

       $D_2 = 60$ dB = reduced volume

$$D_1 - D_2 = 10 \log \left( \frac{I_1}{I_2} \right)$$

$$105 - 60 = 10 \log \left( \frac{I_1}{I_2} \right)$$

$$45 = 10 \log \left( \frac{I_1}{I_2} \right)$$

$$4.5 = \log \left( \frac{I_1}{I_2} \right)$$

$$10^{4.5} = 10^{\log (I_1 / I_2)}$$

33. Continued.

$$\frac{I_1}{I_2} = 10^{4.5} = 31{,}622.78 \approx 31{,}623$$

$$I_1 \approx 31{,}623\, I_2 \text{ or } I_2 \approx \frac{1}{31{,}623}\, I_1 \approx 0.00003162\, I_1$$

The original volume was about 32,000 times the sound intensity of the reduced volume.  Or, the reduced volume would be 0.003% as intense as the original volume.

---

| **Chapter 9** | **Review** |

1.    $x = \log_3 81$ can be rewritten as $3^x = 81$.
$$3^x = 81 = 3^4 \text{, so } x = 4$$

5.    $\ln(e^x) = x$ and $e^{\ln x} = x$ are the inverse properties of the natural logarithm.

9.    $\ln(A^n) = n(\ln A)$ is the Exponent-Becomes-Multiplier property of the natural logarithm.

13.    $\log(x + 2)$ cannot be rewritten using the properties of logarithms.

17.    $\log(5x) + \log x^2 - \log x = 12$

$\log[5x\,(x^2)] - \log x = 12$           Multiplication-Becomes-Addition

$\log\left(\dfrac{5x^3}{x}\right) = 12$           Division-Becomes-Subtraction

$\log(5x^2) = 12$           Simplify

$10^{\log(5x^2)} = 10^{12}$           Exponentiate each side

$5x^2 = 10^{12}$           Inverse Property

$x^2 = \dfrac{10^{12}}{5} = 2 \times 10^{11}$

$x = 447{,}213.6$           Taking square root

21.    **Closest:** $A = 25 \ \mu$m at 20 km from the epicenter, $A_0 = -1.7$

$$
\begin{aligned}
M &= \log A - \log A_0 \\
&= \log (25) - (-1.7) \qquad \text{From Figure 9.24} \\
&= 1.397940 + 1.7 \\
&= 3.097940 \approx 3.1
\end{aligned}
$$

**Second:** $A = 2 \ \mu$m at 60 km from the epicenter, $A_0 = -2.8$

$$
\begin{aligned}
M &= \log A - \log A_0 \\
&= \log (2) - (-2.8) \qquad \text{From Figure 9.24} \\
&= 0.30103 + 2.8 \\
&= 3.10103 \approx 3.1
\end{aligned}
$$

The magnitude of the earthquake was 3.1 on the Richter scale.

25.    $I_2$ = sound intensity of single trumpet
$I_1$ = sound intensity with additional trumpets
$D_1 = 84$ dB, $D_2 = 78$ dB

$$
D_1 - D_2 = 10 \log \left( \frac{I_1}{I_2} \right)
$$

$$
84 - 78 = 10 \log \left( \frac{I_1}{I_2} \right)
$$

$$
6 = 10 \log \left( \frac{I_1}{I_2} \right)
$$

$$
0.6 = \log \left( \frac{I_1}{I_2} \right)
$$

$$
10^{0.6} = 10^{\log (I_1 / I_2)}
$$

$$
\frac{I_1}{I_2} = 10^{0.6} = 3.98107 \approx 4 \text{ trumpets}
$$

There are three additional trumpets making a total of four trumpets.

# 10 Calculus

| Section 10.0 | Review of Ratios, Parabolas, and Functions |
|---|---|

1. Similar triangles: 4 corresponds to 2, 5 corresponds to $a$, and 7 corresponds to $b$

   <u>Solving for $a$</u>

   $$\frac{4}{2} = \frac{5}{a}$$
   $$4a = 5(2)$$
   $$4a = 10$$
   $$a = \frac{10}{4} = \frac{5}{2}$$

   <u>Solving for $b$</u>

   $$\frac{4}{2} = \frac{7}{b}$$
   $$4b = 7(2)$$
   $$4b = 14$$
   $$b = \frac{14}{4} = \frac{7}{2}$$

5. Similar triangles: 6 corresponds to 9, 8 corresponds to $x$, and 10 corresponds to $(10 + y)$

   <u>Solving for $x$</u>

   $$\frac{6}{9} = \frac{8}{x}$$
   $$6x = 8(9)$$
   $$6x = 72$$
   $$x = \frac{72}{6} = 12$$

   <u>Solving for $y$</u>

   $$\frac{6}{9} = \frac{10}{10 + y}$$
   $$6(10 + y) = 9(10)$$
   $$60 + 6y = 90$$
   $$6y = 90 - 60 = 30$$
   $$y = \frac{30}{6} = 5$$

9. Student-teacher ratio of 15 to 1

   a) $\dfrac{15 \text{ students}}{1 \text{ teacher}}$ or $\dfrac{1 \text{ teacher}}{15 \text{ students}}$

   b) $\dfrac{15 \text{ students}}{1 \text{ teacher}} = \dfrac{5430 \text{ students}}{x \text{ teachers}}$

   $$15x = 1(5430)$$
   $$x = \frac{5430}{15} = 362 \text{ teachers or faculty members}$$

13. Odometer: beginning = 101,569.3 at 10 A.M., ending = 101,633.5 at 12 noon.

 a) Change in Distance = (ending − beginning) odometer = 101,633.5 − 101,569.3 = 64.2 miles

 Change in Time = 12 noon − 10 A.M. = 2 hours

 $$\text{Average rate} = \frac{\text{change in distance}}{\text{change in time}} = \frac{\Delta d}{\Delta t} = \frac{64.2 \text{ mi}}{2 \text{ hr}} = \frac{32.1 \text{ mi}}{1 \text{ hr}} = 32.1 \text{ mph (miles per hour)}$$

 b) $$\frac{32.1 \text{ mi}}{1 \text{ hr}} = \frac{x}{4 \text{ hr}}$$

 $$x = (32.1 \text{ mi/hr})(4 \text{ hr}) = 128.4 \text{ miles}$$

 c) distance = rate • time = (32.1 mi/hr) (4 hr) = 128.4 mi

17. Odometer: beginning = 5.4, ending = 332.5, gallons = 13.3

 a) Miles traveled = (ending − beginning) odometer = 332.5 − 5.4 = 327.1 miles

 $$\frac{\text{miles traveled}}{\text{gallons consumed}} = \frac{327.1 \text{ mi}}{13.3 \text{ gal}} = 24.59398 \approx 24.6 \text{ mpg (miles per gallon)}$$

 b) $$\frac{24.59398 \text{ miles}}{1 \text{ gallon}} = \frac{x}{15 \text{ gallons}}$$

 $$x = 24.59398 \,(15) = 368.9097 \approx 369 \text{ miles}$$

 c) miles = (mpg) • (gallon) = 24.59398 (15) = 368.9097 ≈ 369 miles

21. See Figure 10.2. $b = 756$ feet, $S = 342$ feet, $p = 6$ feet, $s = 9$ feet

 Similar triangles: $h$ corresponds to $p$, $\left(\dfrac{1}{2}b + S\right)$ corresponds to $s$

 $$\frac{h}{p} = \frac{\frac{1}{2}b + S}{s}$$

 $$\frac{h}{6} = \frac{\frac{1}{2}(756) + 342}{9} = \frac{378 + 342}{9} = \frac{720}{9}$$

 $$9h = 6(720) = 4320$$

 $$h = \frac{4320}{9} = 480 \text{ ft}$$

25.  $y = -x^2 - 4x + 3$

    a)     Substituting 0, 1, and 2 for $x$ yields the following points:

| $x$ | $y = -x^2 - 4x + 3$ | $y$ | $(x, y)$ |
|---|---|---|---|
| 0 | $-(0)^2 - 4(0) + 3 = -0 - 0 + 3$ | 3 | $(0, 3)$ |
| 1 | $-(1)^2 - 4(1) + 3 = -1 - 4 + 3$ | $-2$ | $(1, -2)$ |
| 2 | $-(2)^2 - 4(2) + 3 = -4 - 8 + 3$ | $-9$ | $(2, -9)$ |

Look at the slope of the secant lines to determine which direction the vertex lies:

| Points | Slope of Secant Line | |
|---|---|---|
| $(0, 3)$ and $(1, -2)$ | $m = \dfrac{\Delta y}{\Delta x} = \dfrac{-2-3}{1-0} = -5$ | slope is closer to 0 and line is less steep |
| $(1, -2)$ and $(2, -9)$ | $m = \dfrac{\Delta y}{\Delta x} = \dfrac{-9-(-2)}{2-1} = -7$ | |

The vertex lies to the left of $x = 0$ since $-5$ is closer to zero than $-7$.

| $x$ | $y = -x^2 - 4x + 3$ | $y$ | $(x, y)$ |
|---|---|---|---|
| $-1$ | $-(-1)^2 - 4(-1) + 3 = -1 + 4 + 3$ | 6 | $(-1, 6)$ |
| $-2$ | $-(-2)^2 - 4(-2) + 3 = -4 + 8 + 3$ | 7 | $(-2, 7)$ |
| $-3$ | $-(-3)^2 - 4(-3) + 3 = -9 + 12 + 3$ | 6 | $(-3, 6)$ |
| $-4$ | $-(-4)^2 - 4(-4) + 3 = -16 + 16 + 3$ | 3 | $(-4, 3)$ |
| $-5$ | $-(-5)^2 - 4(-5) + 3 = -25 + 20 + 3$ | $-2$ | $(-5, -2)$ |

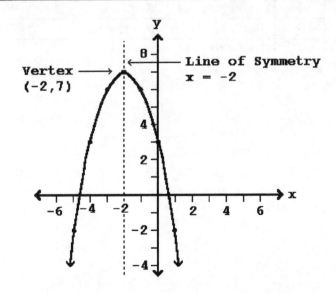

25. Continued.

   b)  The $y$-values are the same at $x = -1$ and $x = -3$, so the equation of the line of symmetry is $x = -2$.

   c)  The vertex is the point $(-2, 7)$.

29.  $y = 8x - x^2 - 14$ or, in standard form, $y = -x^2 + 8x - 14$

   a)  Substituting 0, 1, and 2 for $x$ yields the following points:

| $x$ | $y = -x^2 + 8x - 14$ | $y$ | $(x, y)$ |
|---|---|---|---|
| 0 | $-(0)^2 + 8(0) - 14 = 0 + 0 - 14$ | $-14$ | $(0, -14)$ |
| 1 | $-(1)^2 + 8(1) - 14 = -1 + 8 - 14$ | $-7$ | $(1, -7)$ |
| 2 | $-(2)^2 + 8(2) - 14 = -4 + 16 - 14$ | $-2$ | $(2, -2)$ |

Look at the slope of the secant lines to determine which direction the vertex lies:

| **Points** | **Slope of Secant Line** | |
|---|---|---|
| $(0, -14)$ and $(1, -7)$ | $m = \dfrac{\Delta y}{\Delta x} = \dfrac{-7 - (-14)}{1 - 0} = 7$ | |
| $(1, -7)$ and $(2, -2)$ | $m = \dfrac{\Delta y}{\Delta x} = \dfrac{-2 - (-7)}{2 - 1} = 5$ | slope is closer to 0 and line is less steep |

The vertex lies to the right of $x = 2$ since 5 is closer to zero than 7.

| $x$ | $y = -x^2 + 8x - 14$ | $y$ | $(x, y)$ |
|---|---|---|---|
| 3 | $-(3)^2 + 8(3) - 14 = -9 + 24 - 14$ | 1 | $(3, 1)$ |
| 4 | $-(4)^2 + 8(4) - 14 = -16 + 32 - 14$ | 2 | $(4, 2)$ |
| 5 | $-(5)^2 + 8(5) - 14 = -25 + 40 - 14$ | 1 | $(5, 1)$ |

29. a)  Continued.

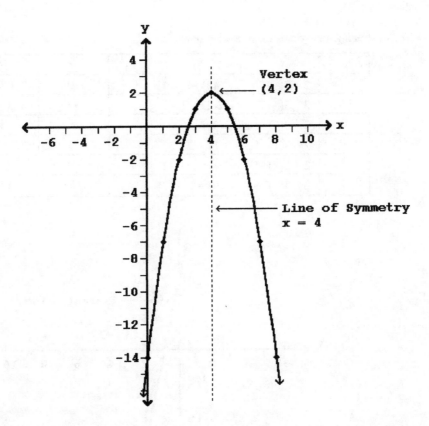

b)      The y-values are the same at $x = 3$ and $x = 5$, so the equation of the line of symmetry is $x = 4$.

c)      The vertex is the point $(4, 2)$.

33.     $y = 2 - 3x^2$ or, in standard form, $y = -3x^2 + 2; x = 1, x = 3$

a)      $x = 1: \ y = -3(1)^2 + 2 = -3 + 2 = -1$
        $x = 3: \ y = -3(3)^2 + 2 = -3(9) + 2 = -27 + 2 = -25$

Using points $(1, -1)$ and $(3, -25)$, the slope is

$$m = \frac{\Delta y}{\Delta x} = \frac{-25 - (-1)}{3 - 1} = \frac{-24}{2} = -12$$

33. Continued.

b) Substituting for $x$ yields the following points:

| $x$ | $y = 2 - 3x^2$ | $y$ | $(x, y)$ |
|---|---|---|---|
| $-2$ | $2 - 3(-2)^2 = 2 - 3(4) = 2 - 12$ | $-10$ | $(-2, -10)$ |
| $-1$ | $2 - 3(-1)^2 = 2 - 3(1) = 2 - 3$ | $-1$ | $(-1, -1)$ |
| $0$ | $2 - 3(0)^2 = 2 - 3(0) = 2 - 0$ | $2$ | $(0, 2)$ |
| $1$ | $2 - 3(1)^2 = 2 - 3(1) = 2 - 3$ | $-1$ | $(1, -1)$ |
| $2$ | $2 - 3(2)^2 = 2 - 3(4) = 2 - 12$ | $-10$ | $(2, -10)$ |
| $3$ | $2 - 3(3)^2 = 2 - 3(9) = 2 - 27$ | $-25$ | $(3, -25)$ |

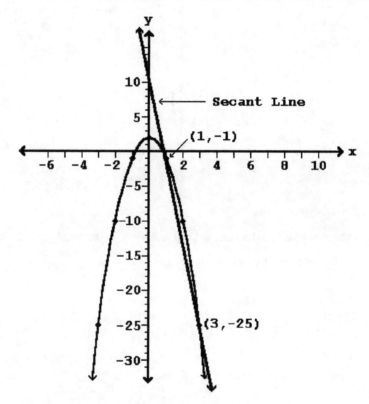

37.   $f(x) = 8x - 11$
$f(4)$ means "substitute 4 for $x$ in function $f$"
$f(4) = 8(4) - 11 = 32 - 11 = 21$

41.   $f(x) = 8x - 11$
$f(x + 3)$ means "substitute $(x + 3)$ for $x$ in function $f$"
$f(x + 3) = 8(x + 3) - 11 = 8x + 24 - 11 = 8x + 13$

45.   $f(x) = 8x - 11$

$f(x + \Delta x)$ means "substitute $(x + \Delta x)$ for $x$ in function $f$"

$$f(x + \Delta x) = 8(x + \Delta x) - 11 = 8x + 8\Delta x - 11$$

---

## Section 10.1                    The Antecedents of Calculus

1.   $y = \dfrac{1}{4}x^2 = (x^2)/4$

a)

| $x$ | $y = (x^2)/4$ | $y$ | $(x, y)$ |
|---|---|---|---|
| $-3$ | $(-3)^2/4 = (9)/4$ | $9/4$ | $(-3, 9/4)$ |
| $-2$ | $(-2)^2/4 = (4)/4$ | $1$ | $(-2, 1)$ |
| $-1$ | $(-1)^2/4 = (1)/4$ | $1/4$ | $(-1, 1/4)$ |
| $0$ | $(0)^2/4 = (0)/4$ | $0$ | $(0, 0)$ |
| $1$ | $(1)^2/4 = (1)/4$ | $1/4$ | $(1, 1/4)$ |
| $2$ | $(2)^2/4 = (4)/4$ | $1$ | $(2, 1)$ |
| $3$ | $(3)^2/4 = (9)/4$ | $9/4$ | $(3, 9/4)$ |
| $4$ | $(4)^2/4 = (16)/4$ | $4$ | $(4, 4)$ |

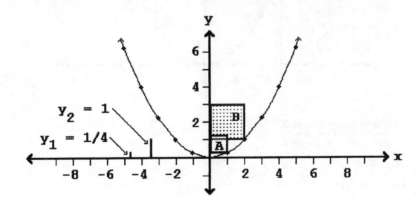

b)   At point $(1, 1/4)$ the line segment is of length $y_1 = 1/4$. The base of the square is of length $x_1 = 1$, so the area of the square is $1^2 = 1$. See square A above.

At point $(2, 1)$, the line segment is of length $y_2 = 1$. The base of the square is of length $x_2 = 2$, so the area of the square is $2^2 = 4$. See square B above.

1. Continued.

c)   $\dfrac{\text{length of line segment 1}}{\text{length of line segment 2}} = \dfrac{y_1}{y_2} = \dfrac{\frac{1}{4}}{1} = \dfrac{1}{4}$

$\dfrac{\text{area of square A}}{\text{area of square B}} = \dfrac{1}{4}$

d)   See computations in 1.(a) above for graph.

At point (3, 9/4) the line segment is of length $y_3 = 9/4$.  The base of the square is of length $x_3 = 3$, so the area of the square is $3^2 = 9$.  See square C below.

At point (4,4), the line segment is of length $y_4 = 4$.  The base of the square is of length $x_4 = 4$, so the area of the square is $4^2 = 16$.  See square D below.

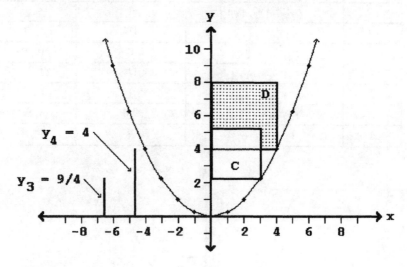

e)   $\dfrac{\text{length of line segment 3}}{\text{length of line segment 4}} = \dfrac{y_3}{y_4} = \dfrac{\frac{9}{4}}{4} = \left(\dfrac{9}{4}\right)\left(\dfrac{1}{4}\right) = \dfrac{9}{16}$

$\dfrac{\text{area of square C}}{\text{area of square D}} = \dfrac{9}{16}$

5.  $y = x^2 + 1$

a)

| $x$ | $y = x^2 + 1$ | $y$ | $(x, y)$ |
|---|---|---|---|
| $-3$ | $(-3)^2 + 1 = 9 + 1$ | 10 | $(-3, 10)$ |
| $-2$ | $(-2)^2 + 1 = 4 + 1$ | 5 | $(-2, 5)$ |
| $-1$ | $(-1)^2 + 1 = 1 + 1$ | 2 | $(-1, 2)$ |
| 0 | $(0)^2 + 1 = 0 + 1$ | 1 | $(0, 1)$ |
| 1 | $(1)^2 + 1 = 1 + 1$ | 2 | $(1, 2)$ |
| 2 | $(2)^2 + 1 = 4 + 1$ | 5 | $(2, 5)$ |
| 3 | $(3)^2 + 1 = 9 + 1$ | 10 | $(3, 10)$ |

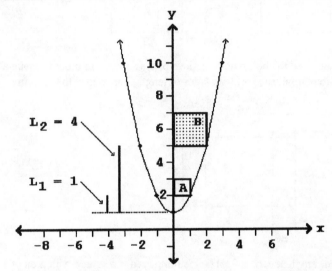

b)   At point (1, 2) the line segment is of length $L_1 = 1$. The base of the square is of length $x_1 = 1$, so the area of the square is $1^2 = 1$. See square A above.

At point (2, 5), the line segment is of length $L_2 = 4$. The base of the square is of length $x_2 = 2$, so the area of the square is $2^2 = 4$. See square B above.

c)   $\dfrac{\text{length of line segment 1}}{\text{length of line segment 2}} = \dfrac{L_1}{L_2} = \dfrac{1}{4}$

$\dfrac{\text{area of square A}}{\text{area of square B}} = \dfrac{1}{4}$

9.    a)    A square and 12 roots are equal to 45 units. To solve this problem take 1/2 the roots which would give 6. Add the square of this number (36) to 45 which gives 81. Take the square root of 81 which is 9 and subtract 1/2 of the roots (or 6). Hence the root is 3 and the square is 9.

     b)    **al-Khowarizmi's Geometric Justification**: Construct a square with unknown roots. Construct a rectangle with the same unknown root as a side and the other side of length 12. Divide the rectangle into 4 parts each with the unknown length and the other length equal to 12 divided by 4 which is 3.

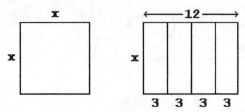

Each strip is attached to one of the four sides of the square. In order to make a complete large square, four small squares with sides of length 3 are added to the figure.

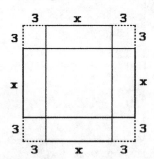

The area of one small square is 9. The resulting area is a square with area of 45 plus 4(9) = 36 which is 81. Hence each root of the large square is 9. But each side of the large square is the unknown root plus 6, therefore the unknown root is 3 and the area of the square is 9.

     c)    **Modern version**

Solve $x^2 + 12x = 45$ for $x^2$.

$$\frac{1}{2}(12) = 6$$      Take half of the roots

$$6^2 = 36$$      Square result

$$x^2 + 12x + 36 = 45 + 36$$      Add result to both sides

$$x^2 + 12x + 36 = 81$$      Simplify

$$(x + 6)^2 = 9^2$$      Factor perfect squares

$$x + 6 = 9$$      Take positive square root only

$$x = 3$$      Solve equation

$$x^2 = 9$$      Square result

13.  a)    A square and 2 roots are equal to 80. To solve this problem take 1/2 the roots which would be 1. Add the square of this number to 80 which gives 81. Take the square root of 81, which is 9, and subtract 1/2 of the roots or 1. Hence the root is 8 and the square is 64.

b)    **al-Khowarizmi's geometric justification**: Construct a square with unknown side x. Construct a rectangle with the unknown width x and length of 2. Divide the rectangle into 4 equal rectangles each x wide and 1/2 unit long.

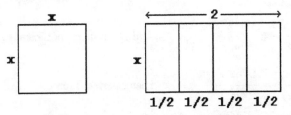

Each small rectangle is attached to a side of the initial square to make a figure with area of 80. Four small squares of area 1/4 square units are added to each corner of the figure making a complete larger square with area of 81.

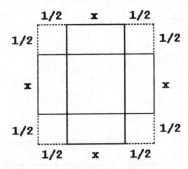

Hence, the larger square has a side of 9. But each side is the unknown plus 1, so the unknown root is 8 and the area of the square is 64.

c)    **Modern version**

Solve $x^2 + 2x = 80$ for $x^2$.

| | |
|---|---|
| $\dfrac{1}{2}(2) = 1$ | Take half of the roots |
| $1^2 = 1$ | Square result |
| $x^2 + 2x + 1 = 80 + 1$ | Add result to both sides |
| $x^2 + 2x + 1 = 81$ | Simplify |
| $(x + 1)^2 = 9^2$ | Factor perfect squares |
| $x + 1 = 9$ | Take positive square root only |
| $x = 8$ | Solve equation |
| $x^2 = 64$ | Square result |

17.    Solve $ax^2 + bx = c$ for $x$.

$$\frac{ax^2}{a} + \frac{bx}{a} = \frac{c}{a}$$    Divide by $a$.

$$x^2 + \frac{b}{a}x = \frac{c}{a}$$    Simplify.

$$x^2 + \frac{b}{a}x + \frac{b^2}{4a^2} = \frac{c}{a} + \frac{b^2}{4a^2}$$    Add 1/2 of the $x$ coefficient squared to both sides.

$$\left(x + \frac{b}{2a}\right)^2 = \frac{4ac + b^2}{4a^2}$$    Factor perfect square.

$$x + \frac{b}{2a} = \sqrt{\frac{4ac + b^2}{4a^2}}$$    Take positive square roots.

$$x + \frac{b}{2a} = \frac{\sqrt{4ac + b^2}}{2a}$$    Simplify.

$$x = \frac{-b + \sqrt{4ac + b^2}}{2a}$$    Solve equation.

---

## Section 10.2                                        Four Problems

1.    Time in motion until time B = AB = 10 seconds,
      Speed at time B = BC = 4 feet per second

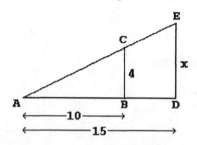

a)    distance = area of triangle $ABC = \frac{1}{2}bh$

$$d = \frac{1}{2}(10)(4) = 20 \text{ feet}$$

1.  Continued.

b)  time in motion until time $D = AD = 15$ seconds,
    speed at time $D = x$

$$\frac{\text{speed at time } B}{\text{speed at time } D} = \frac{\text{time in motion until time } B}{\text{time in motion until time } D}$$

$$\frac{4}{x} = \frac{10}{15}$$

$$10x = 4(15) = 60$$

$$x = \frac{60}{10} = 6 \text{ feet per second}$$

c)  distance $=$ area of triangle $ADE = \frac{1}{2}bh$

$$d = \frac{1}{2}(15)(6) = 45 \text{ feet}$$

d)  average speed $= \dfrac{\text{distance traveled in 15 seconds}}{15 \text{ seconds}}$

average speed $= \dfrac{45 \text{ feet}}{15 \text{ second}} = 3 \text{ feet per second}$

5.

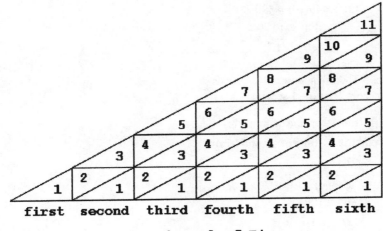

first    second    third    fourth    fifth    sixth

Interval of Time

5. Continued.

| Interval of Time | Distance Traveled during that Interval | Total Distance Traveled |
|---|---|---|
| first | 1 | 1 |
| second | 3 | $1 + 3 = 4$ |
| third | 5 | $1 + 3 + 5 = 9$ |
| fourth | 7 | $1 + 3 + 5 + 7 = 16$ |
| fifth | 9 | $1 + 3 + 5 + 7 + 9 = 25$ |
| sixth | 11 | $1 + 3 + 5 + 7 + 9 + 11 = 36$ |

9. First Trip:  distance = 40 feet, water = 7 ounces
   Second Trip:  distance = 10 feet, water = $x$

$$\frac{\text{distance from first trip}}{\text{distance from second trip}} = \frac{(\text{water from first trip})^2}{(\text{water from second trip})^2}$$

$$\frac{40}{10} = \frac{7^2}{x^2}$$

$$\frac{40}{10} = \frac{49}{x^2}$$

$$40x^2 = 49(10) = 490$$

$$x^2 = \frac{490}{40} = \frac{49}{4}, \text{ so } x = \frac{7}{2} = 3.5 \text{ ounces}$$

13. First Trip:  distance = 1600 feet, time = 10 seconds
    Second Trip:  distance = ?, time = 1 second

$$\frac{\text{distance from first trip}}{\text{distance from second trip}} = \frac{(\text{time of first trip})^2}{(\text{time of second trip})^2}$$

$$\frac{1600}{x} = \frac{10^2}{1^2}$$

$$\frac{1600}{x} = \frac{100}{1}$$

$$100x = 1600$$

$$x = \frac{1600}{100} = 16 \text{ feet}$$

17.  a)  $y = 2x^2$, $P$ is at (3, 18), so $Q$ is at $(3-e, 18-a)$     Solve for $\dfrac{a}{e}$

$$y = 2x^2$$
$$18 - a = 2(3-e)^2$$
$$18 - a = 2(9 - 6e + e^2)$$
$$18 - a = 18 - 12e + 2e^2$$
$$2e^2 - 12e + a = 0$$
$$-12e + a = 0 \qquad \text{eliminate powers of } e \text{ and } a$$
$$a = 12e$$

$$\frac{a}{e} = 12 = m = \text{slope of tangent line at (3, 18)}$$

b)  Equation of the tangent line:

$$y - y_1 = m(x - x_1) \qquad \text{Solve for } y \text{ where } (x_1, y_1) = (3, 18)$$

$$y - 18 = 12(x - 3)$$
$$y - 18 = 12x - 36$$
$$y = 12x - 18$$

c)

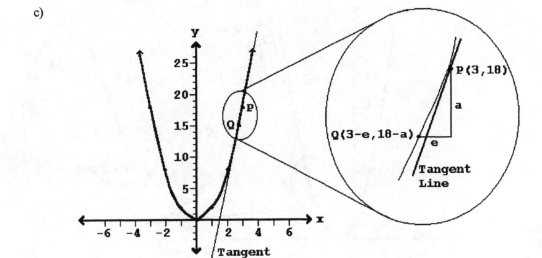

21.    a)    $y = x^2 - 2x + 1$, $P$ is at $(3, 4)$, so $Q$ is at $(3 - e, 4 - a)$.    Solve for $\dfrac{a}{e}$.

$$y = x^2 - 2x + 1$$
$$4 - a = (3 - e)^2 - 2(3 - e) + 1$$
$$4 - a = 9 - 6e + e^2 - 6 + 2e + 1$$
$$4 - a = 4 - 4e + e^2$$
$$e^2 - 4e + a = 0$$
$$-4e + a = 0 \qquad \text{eliminate powers of } e \text{ and } a$$
$$a = 4e$$
$$\frac{a}{e} = 4 = m = \text{slope of tangent line at } (3, 4)$$

b)    Solve for the tangent line:

$$y - y_1 = m(x - x_1) \qquad \text{Solve for } y \text{ where } (x_1, y_1) = (3, 4)$$

$$y - 4 = 4(x - 3)$$
$$y - 4 = 4x - 12$$
$$y = 4x - 8$$

c)

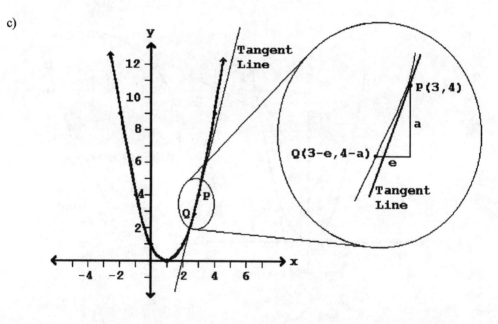

25.   a)   $y = x^2 + 4x + 6$, $P$ is at $(0, 6)$, so $Q$ is at $(0 - e, 6 - a)$     Solve for $\dfrac{a}{e}$.

$$y = x^2 + 4x + 6$$
$$6 - a = (0 - e)^2 + 4(0 - e) + 6$$
$$6 - a = (-e)^2 + 4(-e) + 6$$
$$6 - a = e^2 - 4e + 6$$
$$-a = e^2 - 4e$$
$$e^2 - 4e + a = 0$$
$$-4e + a = 0 \qquad \text{eliminate powers of } e \text{ and } a$$
$$a = 4e$$
$$\dfrac{a}{e} = 4 = m = \text{slope of tangent line at } (0, 6)$$

b)   Solve for the tangent line:

$$y - y_1 = m(x - x_1) \qquad \text{Solve for } y \text{ where } (x_1, y_1) = (0, 6)$$

$$y - 6 = 4(x - 0)$$
$$y - 6 = 4x - 0$$
$$y = 4x + 6$$

c)

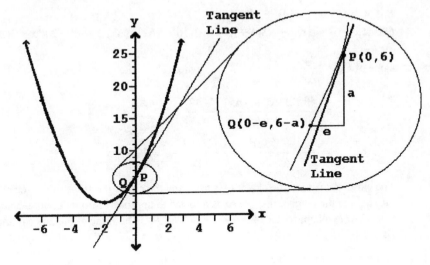

| Section 10.3 | Newton and Tangent Lines |
|---|---|

1.    $y = 3x^2$ at $(4, 48)$

    a)    **Newton's Method:**

        $P$ is at $x_1 = 4$, $y_1 = 3(4)^2 = 3(16) = 48$

        $Q$ is at $x_2 = 4 + o$, $y_2 = 3(4 + o)^2 = 3(16 + 8o + o^2) = 48 + 24o + 3o^2$

        slope of $PQ = \dfrac{\Delta y}{\Delta x} = \dfrac{y_2 - y_1}{x_2 - x_1}$    where $PQ$ is the secant line

$$= \frac{(48 + 24o + 3o^2) - 48}{(4 + o) - 4}$$

$$= \frac{24o + 3o^2}{o} = \frac{o(24 + 3o)}{o} = 24 + 3o$$

        Let $o = 0$, then slope of tangent line $= 24 + 0 = 24$

    b)    **Cauchy's method:**

        $P$ is at $x_1 = 4$, $y_1 = 3(4)^2 = 48$

        $Q$ is at $x_2 = 4 + \Delta x$,  $y_2 = 3(4 + \Delta x)^2 = 3(16 + 8\Delta x + (\Delta x)^2) = 48 + 24\Delta x + 3(\Delta x)^2$

        slope of $PQ = \dfrac{\Delta y}{\Delta x} = \dfrac{y_2 - y_1}{x_2 - x_1}$    where $PQ$ is the secant line

$$= \frac{(48 + 24\Delta x + 3(\Delta x)^2) - 48}{(4 + \Delta x) - 4}$$

$$= \frac{24\Delta x + 3(\Delta x)^2}{\Delta x} = \frac{\Delta x(24 + 3\Delta x)}{\Delta x} = 24 + 3\Delta x$$

        The closer $\Delta x$ is to 0, the closer $Q$ is to $P$, and the closer the slope of the secant line $PQ$ is to the slope of the tangent line. If $\Delta x$ is allowed to approach zero without reaching zero, then $(24 + 3\Delta x)$ will approach 24 without reaching 24. The slope of the tangent line at $(3, 48)$ is 24.

    c)    **Equation of tangent line:**

        $y - y_1 = m(x - x_1)$    Solve for $y$ where $(x_1, y_1) = (4, 48)$

$$y - 48 = 24(x - 4)$$
$$y - 48 = 24x - 96$$
$$y = 24x - 48$$

5.  $y = 2x^2 - 5x + 1$  at $x = 7$

    a)  **Newton's Method:**

        $P$ is at $x_1 = 7, y_1 = 2(7)^2 - 5(7) + 1 = 2(49) - 35 + 1 = 64$

        $Q$ is at $x_2 = 7 + o, y_2 = 2(7 + o)^2 - 5(7 + o) + 1$

$$= 2(49 + 14o + o^2) - 35 - 5o + 1$$
$$= 98 + 28o + 2o^2 - 34 - 5o$$
$$= 64 + 23o + 2o^2$$

        slope of $PQ = \dfrac{\Delta y}{\Delta x} = \dfrac{y_2 - y_1}{x_2 - x_1}$     where $PQ$ is the secant line

$$= \frac{(64 + 23o + 2o^2) - 64}{(7 + o) - 7}$$

$$= \frac{23o + 2o^2}{o} = \frac{o(23 + 2o)}{o} = 23 + 2o$$

        Let $o = 0$, then slope of tangent line $= 23 + 0 = 23$

    b)  **Cauchy's method:**

        $P$ is at $x_1 = 7, y_1 = 64$    (See Exercise 5.a)

        $Q$ is at $x_2 = 7 + \Delta x, y_2 = 2(7 + \Delta x)^2 - 5(7 + \Delta x) + 1$

$$= 2(49 + 14\Delta x + (\Delta x)^2) - 35 - 5\Delta x + 1$$
$$= 98 + 28\Delta x + 2(\Delta x)^2 - 34 - 5\Delta x$$
$$= 64 + 23\Delta x + 2(\Delta x)^2$$

        slope of $PQ = \dfrac{\Delta y}{\Delta x} = \dfrac{y_2 - y_1}{x_2 - x_1}$     where $PQ$ is the secant line

$$= \frac{(64 + 23\Delta x + 2(\Delta x)^2) - 64}{(7 + \Delta x) - 7}$$

$$= \frac{23\Delta x + 2(\Delta x)^2}{\Delta x} = \frac{\Delta x(23 + 2\Delta x)}{\Delta x} = 23 + 2\Delta x$$

        The closer $\Delta x$ is to 0, the closer $Q$ is to $P$, and the closer the slope of the secant line $PQ$ is to the slope of the tangent line. If $\Delta x$ is allowed to approach zero without reaching zero, then $(23 + 2\Delta x)$ will approach 23 without reaching 23. The slope of the tangent line at $(7, 64)$ is 23.

    c)  **Equation of tangent line:**

        $y - y_1 = m(x - x_1)$    Solve for $y$ where $(x_1, y_1) = (7, 64)$

$$y - 64 = 23(x - 7)$$
$$y - 64 = 23x - 161$$
$$y = 23x - 97$$

9. $y = x^3 - x^2$ at $x = 1$

a) **Newton's Method:**

$P$ is at $x_1 = 1, y_1 = (1)^3 - (1)^2 = (1) - (1) = 0$

$Q$ is at $x_2 = 1 + o, y_2 = (1 + o)^3 - (1 + o)^2$

$$= (1 + o)(1 + 2o + o^2) - (1 + 2o + o^2)$$
$$= (1 + 3o + 3o^2 + o^3) - 1 - 2o - o^2$$
$$= 0 + o + 2o^2 + o^3$$

slope of $PQ = \dfrac{\Delta y}{\Delta x} = \dfrac{y_2 - y_1}{x_2 - x_1}$     where $PQ$ is the secant line

$$= \frac{(0 + o + 2o^2 + o^3) - 0}{(1 + o) - 1}$$

$$= \frac{o + 2o^2 + o^3}{o} = \frac{o(1 + 2o + o^2)}{o} = 1 + 2o + o^2$$

Let $o = 0$, then slope of tangent line $= 1 + 0 + 0 = 1$

b) **Cauchy's Method:**

$P$ is at $x_1 = 1, y_1 = 0$    (See Exercise 9.a)

$Q$ is at $x_2 = 1 + \Delta x, y_2 = (1 + \Delta x)^3 - (1 + \Delta x)^2$

$$= (1 + 3\Delta x + 3(\Delta x)^2 + (\Delta x)^3) - (1 + 2\Delta x + (\Delta x)^2)$$
$$= 1 + 3\Delta x + 3(\Delta x)^2 + (\Delta x)^3 - 1 - 2\Delta x - (\Delta x)^2$$
$$= 0 + \Delta x + 2(\Delta x)^2 + (\Delta x)^3$$

slope of $PQ = \dfrac{\Delta y}{\Delta x} = \dfrac{y_2 - y_1}{x_2 - x_1}$    where $PQ$ is the secant line

$$= \frac{(0 + \Delta x + 2(\Delta x)^2 + (\Delta x)^3) - 0}{(1 + \Delta x) - 1}$$

$$= \frac{\Delta x + 2(\Delta x)^2 + (\Delta x)^3}{\Delta x} = \frac{\Delta x(1 + 2\Delta x + (\Delta x)^2)}{\Delta x} = 1 + 2\Delta x + (\Delta x)^2$$

The closer $\Delta x$ is to 0, the closer $Q$ is to $P$, and the closer the slope of the secant line $PQ$ is to the slope of the tangent line. If $\Delta x$ is allowed to approach zero without reaching zero, then $(1 + 2\Delta x + (\Delta x)^2)$ will approach 1 without reaching 1. The slope of the tangent line at $(1, 0)$ is 1.

c) **Equation of tangent line:**

$y - y_1 = m(x - x_1)$    Solve for $y$ where $(x_1, y_1) = (1, 0)$

$y - 0 = 1(x - 1)$

$y - 0 = x - 1$

$y = x - 1$

## Section 10.4                 Newton on Falling Objects and the Derivative

1.    $t = 1$ second

    a)    $d = 16t^2$        Calculate $d =$ distance traveled
         $= 16(1)^2 = 16(1) = 16$ feet

    b)    $s = 32t$        Calculate $s =$ instantaneous speed
         $= 32(1) = 32$ feet/second

    c)    Average speed $= \dfrac{\text{change in distance}}{\text{change in time}} = \dfrac{\Delta d}{\Delta t}$

$$= \frac{\text{distance at 1 second } - \text{ distance at 0 seconds}}{1 \text{ second } - 0 \text{ seconds}}$$

$$= \frac{16 - 0}{1 - 0} = \frac{16}{1} = 16 \text{ feet/second}$$

5.    $t = 8$ seconds

    a)    $d = 16t^2 = 16(8)^2 = 16(64) = 1024$ feet

    b)    $s = 32t = 32(8) = 256$ feet/second

    c)    Average speed $= \dfrac{\text{change in distance}}{\text{change in time}} = \dfrac{\Delta d}{\Delta t}$

$$= \frac{\text{distance at 8 second } - \text{ distance at 0 seconds}}{8 \text{ second } - 0 \text{ seconds}}$$

$$= \frac{1024 - 0}{8 - 0} = \frac{1024}{8} = 128 \text{ feet/second}$$

9.    $s = f(t) = 32t, \ f(t + \Delta t) = 32(t + \Delta t) = 32t + 32\Delta t$

$$f'(t) \text{ or } \frac{ds}{dt} = \lim_{\Delta t \to 0} \frac{f(t + \Delta t) - f(t)}{\Delta t} = \frac{\text{change in speed}}{\text{change in time}}$$

$$= \lim_{\Delta t \to 0} \frac{(32t + 32\Delta t) - 32t}{\Delta t} = \lim_{\Delta t \to 0} \frac{32\Delta t}{\Delta t} = \lim_{\Delta t \to 0} 32 = 32$$

The "second fluxion of gravity" is the rate at which instantaneous speed is changing with respect to time. (This rate is the acceleration due to gravity.)

13. $f(x) = -7x + 42$, $f(x + \Delta x) = -7(x + \Delta x) + 42 = -7x - 7\Delta x + 42$

$$\frac{df}{dx} = \lim_{\Delta x \to 0} \frac{f(x + \Delta x) - f(x)}{\Delta x}$$

$$= \lim_{\Delta x \to 0} \frac{(-7x - 7\Delta x + 42) - (-7x + 42)}{\Delta x}$$

$$= \lim_{\Delta x \to 0} \frac{-7x - 7\Delta x + 42 + 7x - 42}{\Delta x} = \lim_{\Delta x \to 0} \frac{-7\Delta x}{\Delta x} = \lim_{\Delta x \to 0} -7 = -7$$

17. $f(x) = -7x + 42$, where $m = -7$, $b = 42$

$$\frac{df}{dx} = m = -7$$

21. $f(x) = 3x^2 - 11$, $f(x + \Delta x) = 3(x + \Delta x)^2 - 11$
$$= 3(x^2 + 2x(\Delta x) + (\Delta x)^2) - 11 = 3x^2 + 6x(\Delta x) + 3(\Delta x)^2 - 11$$

$$\frac{df}{dx} = \lim_{\Delta x \to 0} \frac{f(x + \Delta x) - f(x)}{\Delta x}$$

$$= \lim_{\Delta x \to 0} \frac{(3x^2 + 6x(\Delta x) + 3(\Delta x)^2 - 11) - (3x^2 - 11)}{\Delta x}$$

$$= \lim_{\Delta x \to 0} \frac{3x^2 + 6x(\Delta x) + 3(\Delta x)^2 - 11 - 3x^2 + 11}{\Delta x}$$

$$= \lim_{\Delta x \to 0} \frac{6x(\Delta x) + 3(\Delta x)^2}{\Delta x} = \lim_{\Delta x \to 0} \frac{\Delta x(6x + 3\Delta x)}{\Delta x}$$

$$= \lim_{\Delta x \to 0} (6x + 3\Delta x)$$

$$= 6x + 3(0) = 6x$$

25. $f(x) = 11x^2$, where $a = 11$, $b = 0$, $c = 0$

$$\frac{df}{dx} = 2ax + b = 2(11)x + 0 = 22x$$

29. $f(x) = 3x^2 - 2x + 7$, where $a = 3$, $b = -2$, $c = 7$

$$\frac{df}{dx} = 2ax + b = 2(3)x + (-2) = 6x - 2$$

33.  a)  $f(x) = x^2 - 2x + 3$, where $a = 1$, $b = -2$, $c = 3$

$$\frac{df}{dx} = 2ax + b = 2(1)x + (-2) = 2x - 2$$

b)  Find the slope by substitution in $df/dx$.

At $x = 1$, $m = \dfrac{df}{dx} = 2x - 2 = 2(1) - 2 = 2 - 2 = 0$

At $x = 2$, $m = \dfrac{df}{dx} = 2x - 2 = 2(2) - 2 = 4 - 2 = 2$

c)  Find the tangent lines:

At $x = 1$, $y = (1)^2 - 2(1) + 3 = 1 - 2 + 3 = 2$, and $m = 0$

At $x = 2$, $y = (2)^2 - 2(2) + 3 = 4 - 4 + 3 = 3$, and $m = 2$

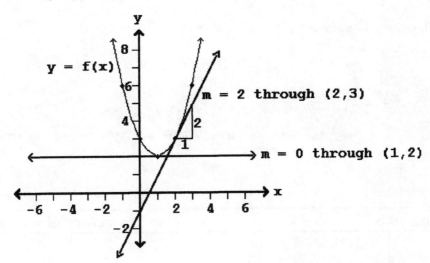

37.  $f(x) = 3x^3 + 4x - 1, \ f(x + \Delta x) = 3(x + \Delta x)^3 + 4(x + \Delta x) - 1$
$$= 3(x^3 + 3x^2(\Delta x) + 3x(\Delta x)^2 + (\Delta x)^3) + 4x + 4\Delta x - 1$$
$$= 3x^3 + 9x^2(\Delta x) + 9x(\Delta x)^2 + 3(\Delta x)^3 + 4x + 4\Delta x - 1$$

$$\frac{df}{dx} = \lim_{\Delta x \to 0} \frac{f(x + \Delta x) - f(x)}{\Delta x}$$

$$= \lim_{\Delta x \to 0} \frac{(3x^3 + 9x^2(\Delta x) + 9x(\Delta x)^2 + 3(\Delta x)^3 + 4x + 4\Delta x - 1) - (3x^3 + 4x - 1)}{\Delta x}$$

$$= \lim_{\Delta x \to 0} \frac{9x^2(\Delta x) + 9x(\Delta x)^2 + 3(\Delta x)^3 + 4\Delta x}{\Delta x}$$

$$= \lim_{\Delta x \to 0} \frac{\Delta x(9x^2 + 9x(\Delta x) + 3(\Delta x)^2 + 4)}{\Delta x}$$

$$= \lim_{\Delta x \to 0} (9x^2 + 9x(\Delta x) + 3(\Delta x)^2 + 4) = 9x^2 + 9x(0) + 3(0)^2 + 4 = 9x^2 + 4$$

41.  $f(x) = 2x^3 + x^2$, where $a = 2, \ b = 1, \ c = 0, \ d = 0$

$$\frac{df}{dx} = 3ax^2 + 2bx + c = 3(2)x^2 + 2(1)x + 0 = 6x^2 + 2x$$

45.  $f(x) = 5x^4 - 3x^2 + 2x - 7$, where $a = 5, \ b = 0, \ c = -3, \ d = 2, \ e = -7$
$$\frac{df}{dx} = 4ax^3 + 3bx^2 + 2cx + d = 4(5)x^3 + 3(0)x^2 + 2(-3)x + 2 = 20x^3 - 6x + 2$$

---

| **Section 10.5** | **The Trajectory of a Cannonball** |
|---|---|

1.  $s = 250$ ft/sec, $c = 4$ ft, angle of elevation $= 30°$

a)  $\quad y = -\dfrac{64x^2}{3s^2} + \dfrac{x}{\sqrt{3}} + c$  $\qquad\qquad$ 30° angle formula

$\quad = -\dfrac{64x^2}{3(250)^2} + \dfrac{x}{\sqrt{3}} + 4$  $\qquad\qquad$ substituting

$\quad = -\dfrac{64x^2}{187,500} + 0.5774x + 4$

$\quad = -0.0003x^2 + 0.5774x + 4$  $\qquad\qquad$ rounding

1. Continued.

b)   The distance that the ball travels is where $y = 0$.

$0 = -0.0003x^2 + 0.5774x + 4$, $a = -0.0003$, $b = 0.5774$, $c = 4$

$$x = \frac{-b \pm \sqrt{b^2 - 4ac}}{2a}$$

$$x = \frac{-0.5774 \pm \sqrt{(0.5774)^2 - 4(-0.0003)(4)}}{2(-0.0003)}$$

$$= \frac{-0.5774 \pm \sqrt{0.3381908}}{-0.0006} = \frac{-0.5774 \pm 0.58154171}{-0.0006}$$

$$x = \frac{-0.5774 + 0.58154}{-0.0006} = \frac{0.004142}{-0.0006} = -6.903 \approx -7 \text{ feet}$$

$$\text{or } x = \frac{-0.5774 - 0.58154}{-0.0006} = \frac{-1.15894}{-0.0006} = 1931.570 \approx 1932 \text{ feet}$$

The negative value does not fit this situation, so the cannonball will travel 1932 feet.

c)   Find slope of the tangent line and let it equal zero.

$$m = \frac{df}{dx} = 2ax + b \quad \text{where } a = -0.0003,\, b = 0.5774$$

$$= 2(-0.0003)x + 0.5774$$

$m = -0.0006x + 0.5774$   slope of the tangent line

$0 = -0.0006x + 0.5774$   (solve for $x$ with $m = 0$)

$0.0006x = 0.5774$

$$x = \frac{0.5774}{0.0006} = 962.3333 \approx 962.3 \text{ feet}$$

Substitute $x = 962.3$ in equation of path for maximum height.

$$y = -0.0003(962.3)^2 + 0.5774(962.3) + 4 = -277.81 + 555.63 + 4 = 281.82 \approx 282 \text{ feet}$$

d)   Substitute $x = 1900$ in equation of path to find the height of the cannonball.

$$y = -0.0003(1900)^2 + 0.5774(1900) + 4$$
$$= -1083.00 + 1{,}097.06 + 4 = 18.06 \approx 18 \text{ feet}$$

Since the height of the cannonball at 1900 feet from the cannon is 18 feet, the cannonball would not clear the 80-foot wall.

1. Continued.

e)

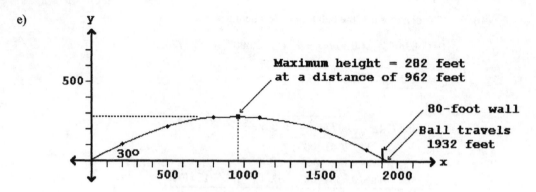

Maximum height = 282 feet
at a distance of 962 feet

80-foot wall

Ball travels
1932 feet

5.    $s = 325$ ft/sec, $c = 5$ ft, angle of elevation $= 60°$

a)   $y = -\dfrac{64x^2}{s^2} + \sqrt{3}\,x + c$         $60°$ angle formula

$\quad\; = -\dfrac{64x^2}{(325)^2} + \sqrt{3}\,x + 5$         substituting

$\quad\; = -\dfrac{64x^2}{105,625} + 1.7321x + 5$

$y = -0.0006x^2 + 1.7321x + 5$         rounding

b)   The distance that the ball travels is where $y = 0$.

$0 = -0.0006x^2 + 1.7321x + 5$, $a = -0.0006$, $b = 1.7321$, $c = 5$

$x = \dfrac{-b \pm \sqrt{b^2 - 4ac}}{2a}$

$x = \dfrac{-1.7321 \pm \sqrt{(1.7321)^2 - 4(-0.0006)(5)}}{2(-0.0006)}$

$\quad = \dfrac{-1.7321 \pm \sqrt{3.0121704}}{-0.0012} = \dfrac{-1.7321 \pm 1.735561}{-0.0012}$

$x = \dfrac{-1.7321 + 1.7356}{-0.0012} = \dfrac{0.0035}{-0.0012} = -2.9 \approx -3$ feet

or $x = \dfrac{-1.7321 - 1.7356}{-0.0012} = \dfrac{-3.4677}{-0.0012} = 2889.75 \approx 2890$ feet

The negative value does not fit this situation, so the cannonball will travel 2890 feet.

5. Continued.

   c)     Find slope of the tangent line and let it equal zero.

$$m = \frac{df}{dx} = 2ax + b \qquad \text{where } a = -0.0006, \, b = 1.7321$$

$$= 2(-0.0006)x + 1.7321$$

$$m = -0.0012x + 1.7321 \qquad \text{slope of the tangent line}$$
$$0 = -0.0012x + 1.7321 \qquad \text{(solve for } x \text{ with } m = 0)$$
$$0.0012x = 1.7321$$

$$x = \frac{1.7321}{0.0012} = 1443.4167 \approx 1443 \text{ feet}$$

Substitute $x = 1443$ in equation of path for maximum height.

$$y = -0.0006(1443)^2 + 1.7321(1443) + 5 = -1249.35 + 2499.42 + 5 = 1255.07 \approx 1255 \text{ feet}$$

   d)     Substitute $x = 1900$ in equation of path to find the height of the cannonball.

$$y = -0.0006(1900)^2 + 1.7321(1900) + 5 = -2166 + 3290.99 + 5 = 1129.99 \approx 1130 \text{ feet}$$

Since the height of the cannonball at 1900 feet from the cannon is 1130 feet, the cannonball would clear the 80-foot wall.

   e)

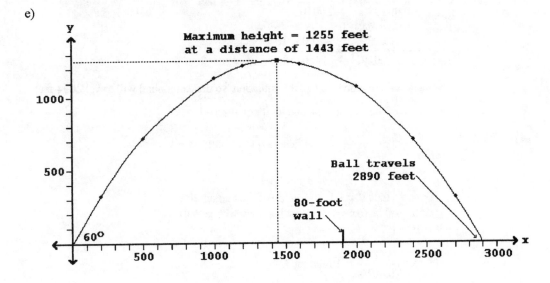

9. $s = 250$ ft/sec, $c = 4$ ft, angle of elevation $= 45°$

a) $y = -\dfrac{32x^2}{s^2} + x + c$

$= -\dfrac{32x^2}{(250)^2} + x + 4$

$= -\dfrac{32x^2}{62,500} + x + 4$

$y = -0.0005x^2 + x + 4$

b) The distance that the ball travels is where $y = 0$.

$0 = -0.0005x^2 + x + 4$, $a = -0.0005$, $b = 1$, $c = 4$

$x = \dfrac{-b \pm \sqrt{b^2 - 4ac}}{2a}$

$x = \dfrac{-1 \pm \sqrt{(1)^2 - 4(-0.0005)(4)}}{2(-0.0005)} = \dfrac{-1 \pm \sqrt{1.008}}{-0.0010} = \dfrac{-1 \pm 1.003992}{-0.0010}$

$x = \dfrac{-1 + 1.003992}{-0.0010} = \dfrac{0.003992}{-0.0010} = -3.99 \approx -4$ feet

or $x = \dfrac{-1 - 1.003992}{-0.0010} = \dfrac{-2.003992}{-0.0010} = 2003.99 \approx 2004$ feet

The negative value does not fit this situation, so the cannonball will travel 2004 feet.

c) Find slope of the tangent line and let it equal zero.

$m = \dfrac{df}{dx} = 2ax + b$        where $a = -0.0005$, $b = 1$

$= 2(-0.0005)x + 1$

$m = -0.0010x + 1$      slope of the tangent line
$0 = -0.0010x + 1$      (solve for $x$ with $m = 0$)
$0.0010x = 1$

$x = \dfrac{1}{0.0010} = 1000$ feet

Substitute $x = 1000$ in equation of path for maximum height.

$y = -0.0005(1000)^2 + 1000 + 4 = -500 + 1000 + 4 = 504$ feet

9. Continued.

    d)    Substitute $x = 1100$ in equation of path to find the height of the cannonball.

$$y = -0.0005(1100)^2 + 1100 + 4 = -605 + 1100 + 4 = 499 \text{ feet}$$

Since the height of the cannonball at 1100 feet from the cannon is 499 feet, the cannonball would clear the 70-foot wall.

    e)

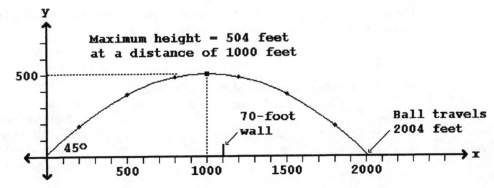

13.    $s = 240$ ft/sec, $c = 4$ ft, angle of elevation $= 45°$

    a)    $$y = -\frac{32x^2}{s^2} + x + c$$

$$= -\frac{32x^2}{(240)^2} + x + 4$$

$$= -\frac{32x^2}{57,600} + x + 4$$

$$y = -0.0006x^2 + x + 4$$

    b)    Find $x$ when $y = 20$

$$20 = -0.0006x^2 + x + 4$$

$$0 = -0.0006x^2 + x - 16, \ a = -0.0006, \ b = 1, \ c = -16$$

$$x = \frac{-b \pm \sqrt{b^2 - 4ac}}{2a}$$

$$x = \frac{-1 \pm \sqrt{(1)^2 - 4(-0.0006)(-16)}}{2(-0.0006)} = \frac{-1 \pm \sqrt{0.9616}}{-0.0012} = \frac{-1 \pm 0.9806}{-0.0012}$$

13. b) Continued.

$$x = \frac{-1 + 0.9806}{-0.0012} = \frac{-0.0194}{-0.0012} = 16.167 \approx 16 \text{ feet}$$

$$\text{or } x = \frac{-1 - 0.9806}{-0.0012} = \frac{-1.9806}{-0.0012} = 1650.5 \text{ feet}$$

The cannon should be placed 1651 feet from the target in order for the cannonball to hit it and keep the crew as far as possible from the enemy.

c)

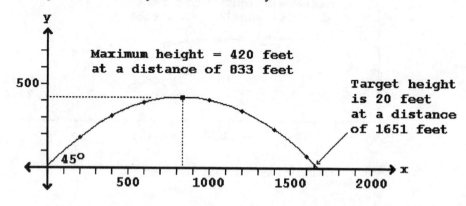

17.  a)  $c = 3$ ft, angle of elevation $= 30°$, $x = 400$, $y = 10$. Find $s$.

$$y = -\frac{64x^2}{3s^2} + \frac{x}{\sqrt{3}} + c \qquad 30° \text{ angle formula}$$

$$10 = -\frac{64(400)^2}{3s^2} + \frac{400}{\sqrt{3}} + 3 \qquad \text{substituting}$$

$$10 = -\frac{3,413,333}{s^2} + 230.9401 + 3$$

$$10 - 233.940 = -\frac{3,413,333}{s^2}$$

$$-223.940s^2 = -3,413,333$$

$$s^2 = \frac{-3,413,333}{-223.940} = 15,242.170$$

$$s = 123.459 \approx 123 \text{ ft/sec}$$

17. Continued.

b)    $y = -\dfrac{64x^2}{3s^2} + \dfrac{x}{\sqrt{3}} + c$          30° angle formula

$= -\dfrac{64x^2}{3(123)^2} + \dfrac{x}{\sqrt{3}} + 3$          substituting

$= -\dfrac{64x^2}{45,387} + 0.5774x + 3$

$y = -0.0014x^2 + 0.5774x + 3$          rounding

c)    Find slope of the tangent line and let it equal zero.

$m = \dfrac{df}{dx} = 2ax + b = 2(-0.0014)x + 0.5774$      when $a = -0.0014$, $b = 0.5774$

$m = -0.0028x + 0.5774$          slope of the tangent line
$0 = -0.0028x + 0.5774$          (solve for $x$ with $m = 0$)
$0.0028x = 0.5774$

$x = \dfrac{0.5774}{0.0028} = 206.2143 \approx 206$ feet

Substitute $x = 206$ in equation of path for maximum height.

$y = -0.0014(206)^2 + 0.5774(206) + 3 = -59.41 + 118.94 + 3 = 62.53 \approx 63$ feet

d)    $c = 3$ ft, angle of elevation $= 45°$, $x = 400$, $y = 10$.  Find $s$.

$y = -\dfrac{32x^2}{s^2} + x + c$      45° angle formula

$10 = -\dfrac{32(400)^2}{s^2} + 400 + 3$

$10 = -\dfrac{5,120,000}{s^2} + 403$

$10 - 403 = -\dfrac{5,120,000}{s^2}$

$-393s^2 = -5,120,000$

$s^2 = \dfrac{-5,120,000}{-393} = 13027.9898$

$s = 114.14 \approx 114$ ft/sec

21.   a)    Since there are 180° in a triangle, the
            remaining angle is:

            $x = 180° - (30° + 90°)$
            $\quad = 180° - 120°$
            $\quad = 60°$

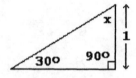

      b)    All three angles are 60° (see drawing).

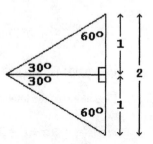

      c)    The length of the vertical side = $1 + 1 = 2$.

      d)    Since all three angles are 60°, the triangle is an equilateral triangle and all three sides are
            equal.  Each side of the larger triangle would equal 2.

      e)    $a = 1, b = ?, c = 2$

            $a^2 + b^2 = c^2$
            $1^2 + b^2 = 2^2$
            $1 + b^2 = 4$
            $b^2 = 3$
            $b = \sqrt{3}$

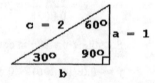

25.   From Exercise 17, find distance that the ball travels, $(y = 0)$.

            $0 = -0.0014x^2 + 0.5774x + 3, a = -0.0014, b = 0.5774, c = 3$

            $$x = \frac{-b \pm \sqrt{b^2 - 4ac}}{2a}$$

            $$x = \frac{-0.5774 \pm \sqrt{(0.5774)^2 - 4(-0.0014)(3)}}{2(-0.0014)}$$

            $$= \frac{-0.5774 \pm \sqrt{0.350190}}{-0.0028} = \frac{-0.5774 \pm 0.591769}{-0.0028}$$

25. Continued.

$$x = \frac{-0.5774 + 0.591769}{-0.0028} = \frac{-0.014369}{-0.0028} = -5.132 \approx -5 \text{ feet}$$

$$\text{or } x = \frac{-0.5774 - 0.591769}{-0.0028} = \frac{-1.169169}{-0.0028} = 417.5604 \approx 417 \text{ feet}$$

The negative value does not fit this situation, so the baseball will travel 417 feet.

Find $t$ when $x = 417$ feet, $s = 123$ ft/sec

$$x = \frac{\sqrt{3}\,s}{2} \cdot t \qquad \text{From Exercise 22}$$

$$417 = \frac{\sqrt{3}\,(123)}{2} \cdot t$$

$$t = \frac{2(417)}{\sqrt{3}\,(123)} = \frac{834}{213.0422} = 3.9147 \approx 3.9 \text{ sec}$$

29.    $s = 200$ ft/sec, $c = 5$ ft, angle of elevation $= 20°$

a)    $y = -16\dfrac{x^2}{(s\cos\theta)^2} + x\tan\theta + c$    general angle formula

$$y = -16\frac{x^2}{(200\cos 20°)^2} + x\tan 20° + 5 \qquad \text{substituting}$$

$$= -16\frac{x^2}{35320.88886} + 0.3639702343\,x + 5$$

$$y = -0.0004529897x^2 + 0.3639702343x + 5$$

$$y = -0.0005x^2 + 0.3640x + 5 \qquad \text{rounding}$$

b)    The distance that the ball travels is where $y = 0$.

$$0 = -0.0005x^2 + 0.3640x + 5,\; a = -0.0005,\, b = 0.3640,\, c = 5$$

$$x = \frac{-b \pm \sqrt{b^2 - 4ac}}{2a}$$

$$x = \frac{-0.3640 \pm \sqrt{(0.3640)^2 - 4(-0.0005)(5)}}{2(-0.0005)}$$

$$= \frac{-0.3640 \pm \sqrt{0.142496}}{-0.0010} = \frac{-0.3640 \pm 0.377486}{-0.0010}$$

29. b) Continued.

$$x = \frac{-0.3640 + 0.377486}{-0.0010} = \frac{-0.013486}{-0.0010} = -13.486 \approx -13 \text{ feet}$$

$$\text{or } x = \frac{-0.3640 - 0.377486}{-0.0010} = \frac{-0.741486}{-0.0010} = 741.486 \approx 741 \text{ feet}$$

The negative value does not fit this situation, so the cannonball will travel 741 feet.

c)    Find slope of the tangent line and let it equal zero.

$$m = \frac{df}{dx} = 2ax + b \quad \text{where } a = -0.0005, b = 0.3640$$

$$= 2(-0.0005)x + 0.3640$$

$$m = -0.0010x + 0.3640 \qquad \text{slope of the tangent line}$$
$$0 = -0.0010x + 0.3640 \qquad \text{(solve for } x \text{ with } m = 0)$$
$$0.0010x = 0.3640$$

$$x = \frac{0.3640}{0.0010} = 364 \text{ feet}$$

Substitute $x = 364$ in equation of path for maximum height.

$$y = -0.0005(364)^2 + 0.3640(364) + 5 = -66.248 + 132.496 + 5 = 71.248 \approx 71 \text{ feet}$$

d)

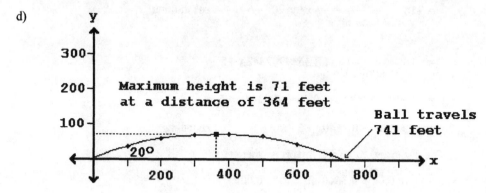

| **Section 10.6** | **Newton and Areas** |
|---|---|

1.   a)   Find the function $A(x)$ whose derivative is $f(x) = 2 = 2x^0$.

FIRST GUESS (power): Increase the power by 1. Need $x^1$ because the derivative of $x^1$ is 1. However, we want the derivative to be 2.

SECOND GUESS (coefficient): Place a 2 in front of the $x$ to make $2x$. The derivative of $2x$ is 2.

ANSWER: The antiderivative of $f(x) = 2$ is $2x + c$

$$\text{or} \qquad A(x) = \int 2\, dx = 2x + c$$

b)

c)   $A(x) = 2x + c$    so    $A(3) = 2(3) + c = 6 + c$

This is the area bounded below by the x-axis, above by the line $f(x) = 2x$, on the left by the y-axis, and on the right by a vertical line at $x = 3$.

d)   $A(x) = 2x + c$    so    $A(4) = 2(4) + c = 8 + c$

This is the area bounded below by the x-axis, above by the line $f(x) = 2x$, on the left by the y-axis, and on the right by a vertical line at $x = 4$.

e)   $A(4) - A(3) = (8 + c) - (6 - c) = 2$

This is the area bounded below by the x-axis, above by the line $f(x) = 2x$, on the left by a vertical line $x = 3$, and on the right by a vertical line at $x = 4$.

5.    a)    Find the function $A(x)$ whose derivative is $f(x) = x + 2 = x^1 + 2x^0$

FIRST GUESS (powers): Increase the powers by 1. Need $x^2 + x^1$ because the derivative of $x^2 + x^1$ is $2x + 1$. However, we want the derivative to be $x + 2$.

SECOND GUESS (coefficients): Place a $\frac{1}{2}$ in front of the $x^2$ and a 2 in front of the $x$ to make $\frac{1}{2}x^2 + 2x$. The derivative of $\frac{1}{2}x^2 + 2x$ is $x + 2$.

ANSWER: The antiderivative of $f(x) = x + 2$ is $\frac{1}{2}x^2 + 2x + c$

or    $A(x) = \int (x+2)dx = \frac{1}{2}x^2 + 2x + c$

b)

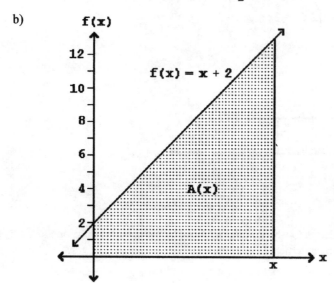

c)    $A(x) = \frac{1}{2}x^2 + 2x + c$    so    $A(3) = \frac{1}{2}(3)^2 + 2(3) + c = \frac{9}{2} + 6 + c = 10\frac{1}{2} + c$

This is the area bounded below by the $x$-axis, above by the line $f(x) = x + 2$, on the left by the $y$-axis, and on the right by a vertical line at $x = 3$.

d)    $A(x) = \frac{1}{2}x^2 + 2x + c$    so    $A(4) = \frac{1}{2}(4)^2 + 2(4) + c = 8 + 8 + c = 16 + c$

This is the area bounded below by the $x$-axis, above by the line $f(x) = x + 2$, on the left by the $y$-axis, and on the right by a vertical line at $x = 4$.

e)    $A(4) - A(3) = (16 + c) - \left(10\frac{1}{2} + c\right) = 5\frac{1}{2}$

This is the area bounded below by the $x$-axis, above by the line $f(x) = x + 2$, on the left by a vertical line $x = 3$, and on the right by a vertical line at $x = 4$.

9. Find the function whose derivative is $f(x) = 5 = 5x^0$.

FIRST GUESS (power): Increase the power by 1. Need $x^1$ because the derivative of $x^1$ is 1. However, we want the derivative to be 5.

SECOND GUESS (coefficient): Place a 5 in front of the $x$ to make $5x$. The derivative of $5x$ is 5.

ANSWER: The antiderivative of $f(x) = 5$ is $5x + c$    or    $\int 5\,dx = 5x + c$

13. The region is shown in the following graph

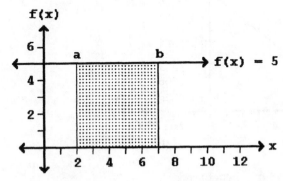

a)   $A(x) = \int 5\,dx = 5x + c$      from Exercise 1.

To find the area of a region find $A$(right boundary) $- A$(left boundary)
left boundary $= a = 2$, right boundary $= b = 7$

$$\begin{aligned}
A(b) - A(a) &= A(7) - A(2) \\
&= [5(7) + c] - [5(2) + c] \\
&= (35 + c) - (10 + c) \\
&= 35 + c - 10 - c \\
&= 25
\end{aligned}$$

b)   The region is a rectangle with base $= 5$ and height $= 5$.

$A = bh = 5(5) = 25$

17. Find the function whose derivative is $g(x) = 8x + 7 = 8x^1 + 7x^0$

FIRST GUESS (powers): Increase the powers by 1. Need $x^2 + x^1$ because the derivative of $x^2 + x^1$ is $2x + 1$. However, we want the derivative to be $8x + 7$.

SECOND GUESS (coefficients): Place a 4 in front of the $x^2$ and a 7 in front of the $x$ to make $4x^2 + 7x$. The derivative of $4x^2 + 7x$ is $8x + 7$.

ANSWER: The antiderivative of $g(x) = 8x + 7$ is $4x^2 + 7x + c$    or    $\int (8x+7)dx = 4x^2 + 7x + c$

21.   The region is shown in the following graph

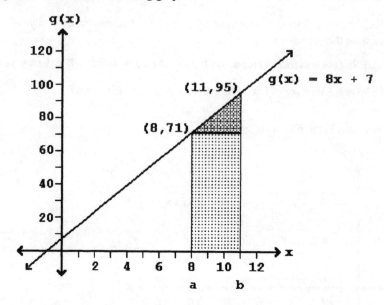

$A(x) = \int (8x+7)dx = 4x^2 + 7x + c$    from Exercise 9.

To find the area of a region calculate   $A$(right boundary) – $A$(left boundary)
left boundary = $a$ = 8,  right boundary = $b$ = 11

$$A(b) - A(a) = A(11) - A(8)$$
$$= [4(11)^2 + 7(11) + c] - [4(8)^2 + 7(8) + c]$$
$$= (484 + 77 + c) - (256 + 56 + c)$$
$$= 561 + c - 312 - c$$
$$= 249$$

*Check the solution:*
    The region is a rectangle with base = 3 and height = 71, plus a triangle of base = 3 and
    height = 95 – 71 = 24.

$$A = A(\text{rectangle}) + A(\text{triangle}) = bh + \frac{1}{2}bh = 3(71) + \frac{1}{2}(3)(24) = 213 + 36 = 249$$

25.  a)     $f(x) = x$, where $m = 1, b = 0$

$$\frac{df}{dx} = m = 1$$

b)     Find the function whose derivative is $f(x) = x = 1x^1$.

FIRST GUESS (power): Increase the power by 1. Need $x^2$ because the derivative of $x^2$ is $2x$. However, we want the derivative to be $x$.

SECOND GUESS (coefficient): Place a $\frac{1}{2}$ in front of the $x^2$ to make $\frac{1}{2}x^2$. The derivative of $\frac{1}{2}x^2$ is $x$.

ANSWER: The antiderivative of $f(x) = x$ is $\frac{1}{2}x^2 + c$     or     $\int x \, dx = \frac{1}{2}x^2 + c$

29.  a)     $m(x) = 8x$, where $a = 8, b = 0$

$$\frac{dm}{dx} = a = 8$$

b)     Find the function whose derivative is $m(x) = 8x = 8x^1$

FIRST GUESS (powers): Increase the powers by 1. Need $x^2$ because the derivative of $x^2$ is $2x$. However, we want the derivative to be $8x$.

SECOND GUESS (coefficients): Place a 4 in front of the $x^2$ to make $4x^2$. The derivative of $4x^2$ is $8x$.

ANSWER: The antiderivative of $m(x) = 8x$ is $4x^2 + c$     or     $\int 8x \, dx = 4x^2 + c$

33.  Find the function whose derivative is $f(x) = 3x^2$

FIRST GUESS (powers): Increase the powers by 1. Need $x^3$ because the derivative of $x^3$ is $3x^2$.

SECOND GUESS (coefficients): Not necessary.

ANSWER: The antiderivative of $f(x) = 3x^2$ is $x^3 + c$     or     $A(x) = \int 3x^2 \, dx = x^3 + c$

To find the area of a region calculate $A$(right boundary) $- A$(left boundary)

a)     left boundary $= a = 1$, right boundary $= b = 4$

$$\begin{aligned}
A(b) - A(a) &= A(4) - A(1) \\
&= [(4)^3 + c] - [(1)^3 + c] \\
&= (64 + c) - (1 + c) \\
&= 64 + c - 1 - c \\
&= 63
\end{aligned}$$

33. Continued.

    b)    left boundary = $a = 3$, right boundary = $b = 100$

$$A(b) - A(a) = A(100) - A(3)$$
$$= [(100)^3 + c] - [(3)^3 + c]$$
$$= (1,000,000 + c) - (27 + c)$$
$$= 1,000,000 + c - 27 - c$$
$$= 999,973$$

37.    Find the function whose derivative is $f(x) = 5x^6 + 3 = 5x^6 + 3x^0$

FIRST GUESS (powers): Increase the powers by 1. Need $x^7 + x^1$ because the derivative of $x^7 + x^1$ is $7x^6 + 1$. However, we want the derivative to be $5x^6 + 3$.

SECOND GUESS (coefficients): Place $\frac{5}{7}$ in front of the $x^7$ and a 3 in front of the $x$ to make $\frac{5}{7}x^7 + 3x$.

ANSWER:    $A(x) = \int (5x^6 + 3)dx = \frac{5}{7}x^7 + 3x + c$

To find the area of a region calculate $A$(right boundary) $- A$(left boundary)

    a)    left boundary = $a = 4$, right boundary = $b = 7$

$$A(b) - A(a) = A(7) - A(4)$$
$$= \left(\frac{5}{7}(7)^7 + 3(7) + c\right) - \left(\frac{5}{7}(4)^7 + 3(4) + c\right)$$
$$= (588,245 + 21 + c) - (11,702.86 + 12 + c)$$
$$= 588,266 + c - 11,714.86 - c$$
$$= 576,551.14$$

    b)    left boundary = $a = 3$, right boundary = $b = 10$

$$A(b) - A(a) = A(10) - A(3)$$
$$= \left(\frac{5}{7}(10)^7 + 3(10) + c\right) - \left(\frac{5}{7}(3)^7 + 3(3) + c\right)$$
$$= (7,142,857 + 30 + c) - (1,562 + 9 + c)$$
$$= 7,142,887 + c - 1,571 - c$$
$$= 7,141,316$$

41.  $g(x) = 8x + 7 = 8x + 7x^0$      Break into two separate problems added together

Use $\int ax^n \, dx = \dfrac{a}{n+1} x^{n+1} + c$      where $a_1 = 8$, $n_1 = 1$ and $a_2 = 7$, $n_2 = 0$

$\int (8x^1 + 7x^0) dx = \int 8x^1 \, dx + \int 7x^0 \, dx$

$= \dfrac{8}{1+1} x^{1+1} + \dfrac{7}{0+1} x^{0+1} + c = \dfrac{8}{2} x^2 + \dfrac{7}{1} x^1 + c = 4x^2 + 7x + c$

45.  $f(x) = 5x^6 - 3x^2 + 13$      Break into three separate problems added together

Use $\int ax^n \, dx = \dfrac{a}{n+1} x^{n+1} + c$      where $a_1 = 5$, $n_1 = 6$, $a_2 = -3$, $n_2 = 2$, and $a_3 = 13$, $n_3 = 0$

$\int (5x^6 - 3x^2 + 13) dx = \int 5x^6 \, dx - \int 3x^2 \, dx + \int 13x^0 \, dx$

$= \dfrac{5}{6+1} x^{6+1} - \dfrac{3}{2+1} x^{2+1} + \dfrac{13}{0+1} x^{0+1} + c$

$= \dfrac{5}{7} x^7 - \dfrac{3}{3} x^3 + \dfrac{13}{1} x^1 + c = \dfrac{5}{7} x^7 - x^3 + 13x + c$

---

## Section 10.7                                                              Conclusion

Projects Only

---

## Chapter 10                                                                   Review

1.    Omitted.

5.    Time in motion until time $B = AB = 30$ seconds,
      Speed at time $B = BC = 8$ feet per second

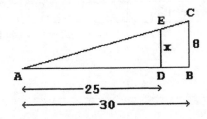

5. Continued.

a)  distance = area of triangle $ABC = \dfrac{1}{2}bh$

$d = \dfrac{1}{2}(30)(8) = 120$ feet

b)  time in motion until time $D = AD = 25$ seconds
Speed at time $D = DE = x$

$$\dfrac{\text{speed at time } B}{\text{speed at time } D} = \dfrac{\text{time in motion until time } B}{\text{time in motion until time } D}$$

$$\dfrac{8}{x} = \dfrac{30}{25}$$

$$30x = 8(25) = 200$$

$$x = \dfrac{200}{30} = 6.6667 \approx 6.7 \text{ feet per second}$$

c)  distance = area of triangle $ADE = \dfrac{1}{2}bh$

$d = \dfrac{1}{2}(25)\left(\dfrac{20}{3}\right) = \dfrac{500}{6} = 83.3333 \approx 83.3$ feet

d)  average speed $= \dfrac{\text{distance traveled in 25 seconds}}{25 \text{ seconds}}$

average speed $= \dfrac{83.3}{25} = 3.333 \approx 3.3$ feet per second

9.  $y = x^2 + 4x + 11$

a)  Substituting for $x$ yields the following points:

| $x$ | $y = x^2 + 4x + 11$ | $y$ | $(x, y)$ |
|---|---|---|---|
| 0 | $(0)^2 + 4(0) + 11 = 0 + 0 + 11$ | 11 | $(0, 11)$ |
| 1 | $(1)^2 + 4(1) + 11 = 1 + 4 + 11$ | 16 | $(1, 16)$ |
| 2 | $(2)^2 + 4(2) + 11 = 4 + 8 + 11$ | 23 | $(2, 23)$ |

Look at the slope of the secant lines to determine which direction the vertex lies:

| Points | Slope of Secant Line | |
|---|---|---|
| $(0, 11)$ and $(1, 16)$ | $m = \dfrac{\Delta y}{\Delta x} = \dfrac{16 - 11}{1 - 0} = 5$ | slope is closer to 0 and line is less steep |
| $(1, 16)$ and $(2, 23)$ | $m = \dfrac{\Delta y}{\Delta x} = \dfrac{23 - 16}{2 - 1} = 7$ | |

9. a) Continued.

The vertex lies to the left of x = 0 since 5 is closer to zero than 7.

| x | $y = x^2 + 4x + 11$ | y | (x, y) |
|---|---|---|---|
| −1 | $(-1)^2 + 4(-1) + 11 = 1 - 4 + 11$ | 8 | (−1, 8) |
| −2 | $(-2)^2 + 4(-2) + 11 = 4 - 8 + 11$ | 7 | (−2, 7) |
| −3 | $(-3)^2 + 4(-3) + 11 = 9 - 12 + 11$ | 8 | (−3, 8) |
| −4 | $(-4)^2 + 4(-4) + 11 = 16 - 16 + 11$ | 11 | (−4, 11) |

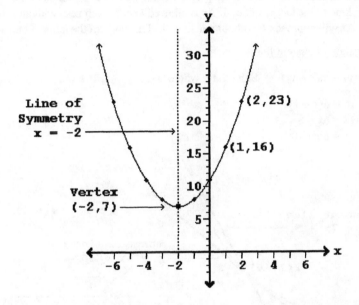

b) The y-values are the same at x = −3 and x = −1, so the equation of the line of symmetry is x = −2.

c) The vertex is the point (−2, 7).

d) Using points (1, 16) and (2, 23), the slope of the secant line is

$$m = \frac{\Delta y}{\Delta x} = \frac{23-16}{2-1} = 7$$

e) **Cauchy's method:**

$P$ is at $x_1 = 1, y_1 = 16$

$Q$ is at $x_2 = 1 + \Delta x, \ y_2 = (1 + \Delta x)^2 + 4(1 + \Delta x) + 11$

$$= (1 + 2\Delta x + (\Delta x)^2) + 4 + 4\Delta x + 11 = 16 + 6\Delta x + (\Delta x)^2$$

9. e) Continued.

$$\text{slope of } PQ = \frac{\Delta y}{\Delta x} = \frac{y_2 - y_1}{x_2 - x_1} \qquad \text{where } PQ \text{ is the secant line}$$

$$= \frac{(16 + 6\Delta x + (\Delta x)^2) - 16}{(1 + \Delta x) - 1}$$

$$= \frac{6\Delta x + (\Delta x)^2}{\Delta x} = \frac{\Delta x(6 + \Delta x)}{\Delta x} = 6 + \Delta x$$

The closer $\Delta x$ is to 0, the closer $Q$ is to $P$, and the closer the slope of the secant line $PQ$ is to the slope of the tangent line. If $\Delta x$ is allowed to approach zero without reaching zero, then $6 + \Delta x$ will approach 6 without reaching 6. The slope of the tangent line at $(1, 16)$ is 6.

f)    <u>Equation of tangent line:</u>

$$y - y_1 = m(x - x_1) \qquad \text{Solve for } y \text{ where } (x_1, y_1) = (1, 16)$$

$$y - 16 = 6(x - 1)$$
$$y - 16 = 6x - 6$$
$$y = 6x + 10$$

g)

13.    $f(x) = 2x - 15, \ f(x + \Delta x) = 2(x + \Delta x) - 15 = 2x + 2\Delta x - 15$

$$\frac{df}{dx} = \lim_{\Delta x \to 0} \frac{f(x + \Delta x) - f(x)}{\Delta x}$$

$$= \lim_{\Delta x \to 0} \frac{(2x + 2\Delta x - 15) - (2x - 15)}{\Delta x}$$

$$= \lim_{\Delta x \to 0} \frac{2x + 2\Delta x - 15 - 2x + 15}{\Delta x} = \lim_{\Delta x \to 0} \frac{2\Delta x}{\Delta x} = \lim_{\Delta x \to 0} 2 = 2$$

17.     $s = 200$ ft/sec, $c = 4.5$ ft, angle of elevation $= 30°$

a)      $y = -\dfrac{64x^2}{3s^2} + \dfrac{x}{\sqrt{3}} + c$          30° angle formula

      $= -\dfrac{64x^2}{3(200)^2} + \dfrac{x}{\sqrt{3}} + 4.5$          substituting

      $= -\dfrac{64x^2}{120,000} + 0.5774x + 4.5$

      $y = -0.0005x^2 + 0.5774x + 4.5$          rounding

b)      The distance that the ball travels is where $y = 0$.

      $0 = -0.0005x^2 + 0.5774x + 4.5,\ a = -0.0005,\ b = 0.5774,\ c = 4.5$

      $x = \dfrac{-b \pm \sqrt{b^2 - 4ac}}{2a}$

      $x = \dfrac{-0.5774 \pm \sqrt{(0.5774)^2 - 4(-0.0005)(4.5)}}{2(-0.0005)}$

      $= \dfrac{-0.5774 \pm \sqrt{0.3423908}}{-0.001} = \dfrac{-0.5774 \pm 0.585142}{-0.001}$

      $x = \dfrac{-0.5774 + 0.585142}{-0.001} = \dfrac{0.007742}{-0.001} = -7.7$ feet

      or $x = \dfrac{-0.5774 - 0.585142}{-0.001} = \dfrac{-1.1625}{-0.001} = 1162.5$ feet

The negative value does not fit this situation, so the cannonball will travel 1163 feet.

c)      Find slope of the tangent line and let it equal zero.

      $m = \dfrac{df}{dx} = 2ax + b$   where $a = -0.0005,\ b = 0.5774$

      $= 2(-0.0005)x + 0.5774$

      $m = -0.001x + 0.5774$     slope of the tangent line

      $0 = -0.001x + 0.5774$     (solve for $x$ with $m = 0$)

      $0.001x = 0.5774$

      $x = \dfrac{0.5774}{0.001} = 577.4$ feet

17. c) Continued.

Substitute $x = 577.4$ in equation of path for maximum height

$y = -0.0005(577.4) + 0.5774(577.4) + 4.5 = -166.70 + 333.39 + 4.5 = 171.19 \approx 171$ feet

d)    Substitute $x = 100$ in equation of path for height

$y = -0.0005(100) + 0.5774(100) + 4.5 = -5 + 57.74 + 4.5 = 57.24 \approx 57$ feet

Since the height at 100 feet from the cannon is 57 feet, the cannonball will clear the 45-foot wall.

e)

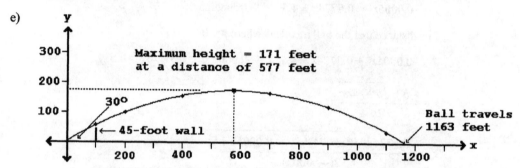

# Appendix E  Dimensional Analysis

| Appendix E | Dimensional Analysis |
|---|---|

1.  Standard Conversion:  1 yard = 3 feet

a)  $12 \text{ feet} = 12 \text{ feet}\left(\dfrac{1 \text{ yard}}{3 \text{ feet}}\right) = 4 \text{ yards}$

b)  $12 \text{ yards} = 12 \text{ yards}\left(\dfrac{3 \text{ feet}}{1 \text{ yard}}\right) = 12(3) \text{ feet} = 36 \text{ feet}$

5.  Standard Conversions:  1 mile = 5280 feet, 1 foot = 12 inches

$2 \text{ miles} = 2 \text{ miles}\left(\dfrac{5280 \text{ feet}}{1 \text{ mile}}\right) = 2(5280 \text{ feet})\left(\dfrac{12 \text{ inches}}{1 \text{ foot}}\right) = 2(5280)(12) \text{ inches} = 126{,}720 \text{ inches}$

9.  Standard Conversion:   1 centiliter = $\dfrac{1}{100}$ liter, or 100 centiliters = 1 liter

a)  $2 \text{ centiliters} = 2 \text{ centiliters}\left(\dfrac{1 \text{ liter}}{100 \text{ centiliters}}\right) = \left(\dfrac{2(1) \text{ liter}}{100}\right) = 0.02 \text{ liters}$

b)  $2 \text{ liters} = 2 \text{ liters}\left(\dfrac{100 \text{ centiliters}}{1 \text{ liter}}\right) = 2(100 \text{ centiliters}) = 200 \text{ centiliters}$

13.  Standard Conversions:  1 mile = 5280 feet, 1 hour = 60 minutes, 1 minute = 60 seconds

a)  $\dfrac{60 \text{ miles}}{\text{hour}} = \dfrac{60 \text{ miles}}{\text{hour}}\left(\dfrac{5280 \text{ feet}}{1 \text{ mile}}\right) = \dfrac{60(5280) \text{ feet}}{\text{hour}}$

$\dfrac{60(5280) \text{ feet}}{\text{hour}} = \dfrac{60(5280) \text{ feet}}{\text{hour}}\left(\dfrac{1 \text{ hour}}{60 \text{ minutes}}\right) = 5280 \text{ feet/minute}$

$\dfrac{5280 \text{ feet}}{\text{minute}} = \dfrac{5280 \text{ feet}}{\text{minute}}\left(\dfrac{1 \text{ minute}}{60 \text{ seconds}}\right) = \dfrac{5280 \text{ feet}}{60 \text{ seconds}} = 88 \text{ feet/second}$

13. Continued.

   b)   Distance = 80 feet, rate = 88 feet/second, find time.

   $$d = rt \text{ or } t = \frac{d}{r}$$

   $$t = \frac{80 \text{ feet}}{88 \text{ feet}/1 \text{ second}} = 80 \text{ feet}\left(\frac{1 \text{ second}}{88 \text{ feet}}\right) = \frac{80}{88} \text{ seconds } = 0.90909... \text{ seconds} \approx 0.9 \text{ seconds}$$

17.   Standard Conversion:  1 year = 365 days

   a)   $$\frac{5.75\%}{1 \text{ year}} = \frac{5.75\%}{1 \text{ year}}\left(\frac{1 \text{ year}}{365 \text{ days}}\right) = \frac{5.75\%}{365 \text{ days}} = 0.0157534247\%/\text{day}$$

   b)   Principal = $10,000,

   Interest = (rate/day)(principal)(days/month)

   $$= \left(\frac{0.0157534247\%}{\text{day}}\right)(\$10,000)\left(\frac{30 \text{ days}}{\text{Month of September}}\right)$$

   $$= \frac{(0.000157534247)(\$10,000)(30)}{\text{Month of September}} = 47.260274 \approx \$47.26 \text{ for September}$$

   c)   Principal = $10,000

   Interest = (rate/day)(principal)(days/month)

   $$= \left(\frac{0.0157534247\%}{\text{day}}\right)(\$10,000)\left(\frac{31 \text{ days}}{\text{Month of October}}\right)$$

   $$= \frac{(0.000157534247)(\$10,000)(31)}{\text{Month of October}} = 48.835616 \approx \$48.84 \text{ for October}$$

21.   $W = 4$ feet, $L = 6$ feet, $H = 11$ inches

   $V = L \bullet W \bullet H$          Formula for volume of a rectangular solid

   $$V = (6 \text{ feet})(4 \text{ feet})(11 \text{ inches})\left(\frac{1 \text{ foot}}{12 \text{ inches}}\right) = (6 \text{ feet})(4 \text{ feet})\left(\frac{11}{12} \text{ feet}\right) = \frac{264}{12} \text{feet}^3 = 22 \text{ feet}^3$$

   Convert cubic feet to gallons:

   $$22 \text{ feet}^3 = 22 \text{ feet}^3\left(\frac{7.48 \text{ gallons}}{1 \text{ foot}^3}\right) = 22(7.48 \text{ gallons}) = 164.56 \text{ gallons}$$